Lecture Notes on Multidisciplinary Industrial Engineering

Series Editor

J. Paulo Davim, Department of Mechanical Engineering, University of Aveiro, Aveiro, Portugal

"Lecture Notes on Multidisciplinary Industrial Engineering" publishes special volumes of conferences, workshops and symposia in interdisciplinary topics of interest. Disciplines such as materials science, nanosciences, sustainability science, management sciences, computational sciences, mechanical engineering, industrial engineering, manufacturing, mechatronics, electrical engineering, environmental and civil engineering, chemical engineering, systems engineering and biomedical engineering are covered. Selected and peer-reviewed papers from events in these fields can be considered for publication in this series.

Jhareswar Maiti · Mamata Jenamani ·
Sri Krishna Kumar · Balagopal G. Menon ·
Anand Jacob Abraham · Abhishek Sharma
Editors

Recent Advances in Industrial and Systems Engineering

Selected Papers from International Conference on Industrial Engineering and Analytics (ICONIEA) 2024

Springer

Editors
Jhareswar Maiti
Department of Industrial and Systems Engineering
Indian Institute of Technology Kharagpur
Kharagpur, West Bengal, India

Mamata Jenamani
Department of Industrial and Systems Engineering
Indian Institute of Technology Kharagpur
Kharagpur, West Bengal, India

Sri Krishna Kumar
Department of Industrial and Systems Engineering
Indian Institute of Technology Kharagpur
Kharagpur, West Bengal, India

Balagopal G. Menon
Department of Industrial and Systems Engineering
Indian Institute of Technology Kharagpur
Kharagpur, West Bengal, India

Anand Jacob Abraham
Department of Industrial and Systems Engineering
Indian Institute of Technology Kharagpur
Kharagpur, West Bengal, India

Abhishek Sharma
Department of Industrial and Systems Engineering
Indian Institute of Technology Kharagpur
Kharagpur, West Bengal, India

ISSN 2522-5022 ISSN 2522-5030 (electronic)
Lecture Notes on Multidisciplinary Industrial Engineering
ISBN 978-981-96-8322-2 ISBN 978-981-96-8323-9 (eBook)
https://doi.org/10.1007/978-981-96-8323-9

This Springer imprint is published by the registered company Springer Nature Singapore Pte Ltd.
The registered company address is: 152 Beach Road, #21-01/04 Gateway East, Singapore 189721, Singapore

Preface

The Department of Industrial and Systems Engineering (ISE) at IIT Kharagpur, India, organized the *International Conference on Industrial Engineering and Analytics* (ICONIEA 2024) under the theme "Data to Decisions" during February 16–18, 2024. ICONIEA 2024 attracted a large number of full paper submissions, underscoring the global relevance and interest in data-centric solutions for industrial systems. Following a rigorous peer-review process, 19 papers were selected for inclusion in this book titled *Recent Advances in Industrial and Systems Engineering* as part of the *Lecture Notes on Multidisciplinary Industrial Engineering* series. These contributions highlight the innovative research organized around four major conference themes: *Supply Chain Innovations and Sustainable Business Strategies*, *Human-Centered Design and Ergonomics*, *Smart Manufacturing Systems and Industry 4.0*, and *Quality Assurance, Reliability, and Safety Engineering*.

Part I, *Supply Chain Innovations and Sustainable Business Strategies*, examines emerging models and frameworks that support sustainability and responsiveness in modern supply chains. Papers in this part discuss topics such as the "brick-to-click" evolution in retail, youth consumer behavior in social commerce, and take-back of end-of-life tires under Extended Producer Responsibility. Methodologies for sustainable supplier selection and green integration in supply chain design are also discussed here. Collectively, these studies emphasize the critical role of sustainability in today's supply chains and the ways in which data analytics can support strategic, evidence-based decision-making.

Part II, *Human-Centered Design and Ergonomics*, focuses on workplace design and ergonomics, with an emphasis on enhancing worker health, safety, and productivity. The papers in this part cover a range of applications, including eye-tracking for packaging analysis, health complaints among academic professionals, and whole-body vibration analysis in public transportation. The part also discussed ergonomic evaluation of employees in a pump manufacturing industry and ergonomic design considerations for school bags. By focusing on workplace improvements through ergonomics, particularly in industries that demand optimized physical conditions, this research underscores the importance of designing work environments that prioritize human well-being and performance.

Part III, *Smart Manufacturing Systems and Industry 4.0*, highlights innovations in manufacturing enabled by advanced applications of Industry 4.0 principles. Papers in this part include a case study on value stream mapping for lean manufacturing, approaches to integrating convergent technologies to enhance human-machine interaction, and the use of digital twins to streamline operations in smart manufacturing. These studies illustrate how digital transformation, and the integration of smart technologies can increase efficiency and flexibility in manufacturing systems, aligning them with the automated, data-driven paradigm of Industry 4.0.

The last part, *Quality Assurance, Reliability, and Safety Engineering*, addresses essential aspects of quality, reliability, and safety in industrial processes. This part includes research on reliability-centered maintenance using fuzzy fault tree analysis, cost analysis of extended warranties with imperfect repairs, intelligent hazard analysis in petrochemical enterprises, etc. Additionally, the part discusses interesting applications such as the use of regression for uncertainty quantification of artillery projectiles and also the use of image processing based techniques for quality inspection in the automotive industry. By quantifying risks and developing strategies for reliability and risk mitigation, these studies highlight the role of predictive maintenance, risk assessment, and quality control in enhancing operational safety and performance.

We extend our sincere gratitude to the authors, reviewers, and technical committee members for their valuable contributions and commitment to the quality of this volume. Special thanks are due to the editorial team at Springer Nature for their support in realizing this publication.

We hope this volume serves a valuable resource for researchers, students, and practitioners, offering insights and inspiration for future advancements in Industrial Engineering and Analytics. We look forward to the continued dialogue and progress inspired by ICONIEA 2024.

Kharagpur, India

Jhareswar Maiti
Mamata Jenamani
Sri Krishna Kumar
Balagopal G. Menon
Anand Jacob Abraham
Abhishek Sharma

Contents

About the Editors

Dr. Jhareswar Maiti, Suresh and Vidya Nair Chair Professor, is the former head and professor of the Department of Industrial and Systems Engineering at IIT Kharagpur. He is the Founder Chairman of the Center of Excellence on Safety Engineering and Analytics (CoE-SEA) at IIT Kharagpur since July 2020 and has developed its academics with Ph.D., MS and 2-year M.Tech. programs. He has also developed the Safety Analytics and Virtual Reality Lab at IIT Kharagpur. His research interests include safety engineering, management and analytics, human factors and ergonomics, risk assessment and uncertainty modeling, quality management, and AI/ML/VR applications in safety. He authored the text book on *Multivariate Statistical Modeling in Engineering and Management*. With more than 200 publications in the international journals and conferences of repute, Dr. Maiti is the recipient of several national and international awards such as P. C. Mahalanobis distinguished educator award, IEOM outstanding professor of occupational safety and health award, and Springer Nature Author Award.

Dr. Mamata Jenamani is a professor in the Department of Industrial and Systems Engineering, specializing in E-Business. With a research focus spanning auctions, recommender systems, social network analysis, consumer review analysis, and digital supply chain management, she has authored over 100 journal and conference publications in these areas. She has also led numerous sponsored and consulting projects, driving innovation and practical applications. As the head of an E-Business Center of Excellence, she has played a key role promoting research, industry collaboration, and technological advancements in the field.

Dr. Sri Krishna Kumar is an Associate Professor in the Department of Industrial and Systems Engineering. He received his Ph.D. from Wolfson School of Mechanical and Manufacturing Engineering at Loughborough University, UK in the area of virtual enterprise collaboration. His expertise includes, Supply Chain Management, Port Logistics, Knowledge Management, Game Theory and Optimization and published few good articles in the journal like *International Journal of Production Research,*

Computers in Industry, *Computers and Industrial Engineering*, *Annals of Operations Research*, and *Journal of Engineering Manufacture*.

Dr. Balagopal G. Menon is an Assistant Professor in the Department of Industrial and Systems Engineering at Indian Institute of Technology (IIT) Kharagpur, India. He obtained his B.Tech. degree in Production Engineering from University of Calicut, Kerala; Master's degree in Engineering Statistics from Cochin University of Science and Technology, Kerala; and Ph.D. in Industrial Engineering and Management from IIT Kharagpur. Dr. Menon was a postdoctoral fellow at Energy Institute, The University of Texas at Austin, USA, and Salford Business School, University of Salford, UK. His areas of interest include operations management, simulation, system dynamics and policy planning, applied multivariate statistical analysis, statistical design of experiments and econometric modeling of systems. His research has been published in prestigious journals such as *Expert Systems with Applications*, *Systems Research and Behavioural Science*, *Resources Policy* and *Energy Efficiency*. He is an editorial board member of the Springer journal *Energy Efficiency*.

Dr. Anand Jacob Abraham is an Assistant Professor at the Department of Industrial and Systems Engineering, IIT Kharagpur. He completed his B.Tech. in Mechanical Engineering from the University of Calicut, his M.Tech. in Industrial Engineering and Management from National Institute of Technology Calicut, and his Ph.D. in Management Studies from the Indian Institute of Science Bengaluru. Anand's main areas of research are resource economics, cooperative game theory, and pricing problems in supply chain management. Dr. Anand received the Indian Economics Association medal in 2022. His research has been published in prestigious journals such as *Mathematical Social Sciences* and *Economics Letters*.

Dr. Abhishek Sharma is an Assistant Professor at IIT Kharagpur in the Department of Industrial and Systems Engineering since December, 2021. Previously, he has served as faculty at IMT Business School Dubai and School of Management and Entrepreneurship, IIT Jodhpur in Operations and Supply Chain Management areas. He is Ph.D. from IIM Rohtak in operations area and Bachelor of Technology graduate in Chemical Engineering from NIT Srinagar. His research interests include behavioral operations, social preferences in supply chain management, channel coordination and supply contracts, multi-channel supply chains, agri-food supply chain, etc. His research work has been published in several reputed academic journals such as *Computers and Industrial Engineering*, *Journal of Business and Industrial Marketing*, *International Game Theory Review*, *Journal of Cleaner Production*, *Benchmarking: An International Journal*, *International Journal of Strategic Decision Sciences*, *International Journal of Indian Culture and Business Management*, *Operations and Supply Chain Management-An International Journal*, etc., and he has also presented his work in leading international conferences such as EuROMA conference 2019 held in Helsinki, ANZAM conference 2023 held in Wellington, New Zealand.

Part I
Supply Chain Innovations and Sustainable Business Strategies

Chapter 1
Brick to Click: Decoding the Dynamics of O2O Evolution in Brick-and-Mortar Stores in the Residential Neighborhoods of India

Anindita Priyadarsini and Sumana Gupta

Abstract Commercial centers are the repercussions of aggregated economic activities leading to the flow of capital and employment. The increasing digital agility among consumers has led to a shift in commerce. With the introduction of dark stores and brick-and-mortar stores entering the platform mode of commerce, the digital era is also reshaping traditional commerce. This research analyses the factors responsible for integrating online- to-offline (O2O) in the brick-and-mortar (B&M) stores in an Indian context. The evolution of commercial transactions over time was also explored to understand the shift trend. This study analyses the current situation of urban commerce and offers insightful information on the variables influencing the integration of online and offline sales in physical retail establishments. The interaction between the current dynamics at the nexus of digital and physical commerce leads to the extraction of components, which adds to a comprehensive knowledge of the dynamics of new-age urban commerce.

Keywords O2O · Platform-mode · Brick-and-Mortar · Dark stores · Residential neighborhoods

1.1 Introduction

A new age of cities is evolving to solve the challenge of urban functional flows and fixed locations and uncover a more detailed spatial structure [41]. With the meteoric urbanization in the urban and peri-urban areas, a vicious cycle of City Center creation followed by infrastructure continues in a loop to meet the rocketing demand of the

A. Priyadarsini (✉) · S. Gupta
Indian Institute of Technology, Kharagpur, India
e-mail: aninditap0405@kgpian.iitkgp.ac.in

S. Gupta
e-mail: sumana@arp.iitkgp.ac.in

J. Maiti et al. (eds.), *Recent Advances in Industrial and Systems Engineering*, Lecture Notes on Multidisciplinary Industrial Engineering,
https://doi.org/10.1007/978-981-96-8323-9_1

population. This results in a massive infiltration of migrants into the urban areas for better lifestyles and economic stability, thus increasing the demand for resources.

Commercial spaces have become crucial for land use planning to cater to the rising demand. Government-allotted commercial areas are provided in every neighborhood, with the sole purpose of catering to the adjacent catchment. Indian cities have suffered immensely due to the conventional planning policies and rules, which are far from ground reality. The gulf between the planned and ground realities makes the Indian cities suffer regarding infrastructure and management. The lack of a market area survey, results in stagnant commercial spaces with low footfall and economic turnover [11].

According to India Online Grocery Market Report 2021, online grocery sales experienced a growth rate of nearly 60% in 2020. By 2026, the market for online groceries is expected to develop at a CAGR of 28.99%. The COVID-19 pandemic catalyzed the shift in customer behavior and buying patterns from going outdoors and purchasing at home [16, 22].

Online delivery that targets efficiency in terms of time has led to the development of ghost kitchens and dark stores for quick last-mile delivery. Even though the online stores are at large, the existing brick-and-mortar stores in the urban areas still account for a surplus of the services. The emergence of online shopping has a major impact on existing retail, such as reduced profit margin, footfall, and customer pool [10]. The involvement of big companies in last-mile grocery delivery has also had a major impact on the B&M stores [28]. This research aims to comprehend the factors contributing to O2O shopping implementation within the existing B&M stores. The objectives are to understand the major indicators that influence the O2O participation of the B&M shops and identify their importance in initiating the trend shift.

1.2 Literature Survey

Commercial centers are the city centers or the business directs that generate economy and employment opportunities. According to [21], commercial areas focus on commerce and business and comprise settlements like offices, shopping centers, etc. The theory of commercial location suggests that the development of urban business districts is primarily influenced by the interplay of different rent structures, market supply and demand dynamics, and the mechanisms of aggregation and diffusion [8].

The highest intensity of commercial exchanges occurs when there is a significant difference, strong desire, and minimal Distance [13]. Commercial sectors' expansion and overgrowth have negative repercussions like encroachment, creation of informal sectors, pressure on the existing infrastructure, waste generation, and environmental issues like poor air quality, urban heat island effect, etc.

1.2.1 Evolution

Commercial interactions have existed since the dawn of civilization. The necessity for trading within or among groups arose when the social populations began to live in cohorts [5]. The growth of customers' options for products and services has been one of the most notable changes in commercial evolution over time. Trade liberalization, regulatory change, and information and communication technology developments contributed to this expansion [42].

From a stage of intra-household exchange in products and services, the exchange began to occur between families and households. However, a medium of exchange became necessary when the interaction occurred between various social communities. When money was introduced, all the issues with the barter trading system were eliminated, and trading became exceedingly easy [5]. The idea of commercial interaction has changed significantly over time. The concepts ranging from production, product, selling, profit, and social to holistic have been crucial to commercial evolution [26].

With technological development and the rapid influx of internet users, we have reached the era of digitized commerce. Online delivery that targets efficiency in terms of time is the new talk of the town. The various stages of Commercial development is shown in Fig. 1.1

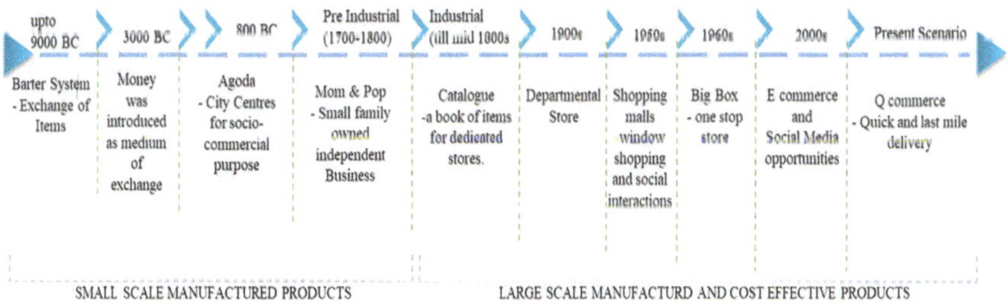

Fig. 1.1 Various stages of commercial development over time. *Source* Compiled by Author

1.2.2 Hierarchy of Commercial Centers

Various attempts have been made to sub-categorize the commercial activities happening at the nodes, which are done based on types of activity, economic turnover, scale, location, and the like. The segregation of commercial activities helps us better co-relate the interactions and flows within the urban structure.

- According to J L Berry, in 1967, a classification method using a hierarchical pattern was proposed for Chicago. The basic four classes are Retail Centers, Regional Centers, Other Shopping Goods, and Community and Neighborhood Centers.

- Garner also proposed a classification theory for Chicago, using the number and type of shop as the primary parameter. He classified the business districts under three categories: Regional, Community, and Neighborhood.
- A hierarchical classification of business centers into local, neighborhood, regional, and central business districts was done by Proudfoot in 1939.
- The modern economy has five market types: Resource, manufacturer, consumer, intermediary, and government [7].
- According to URDPFI, commercial centers are classified as a function of hierarchical planning units based on size. They are Convenience Shopping for Housing Clusters, neighborhood shopping centers, Community Centers, District centers, Sub City centers, City centers, Local Wholesale Markets/Mandi, and Weekly Markets.

1.2.3 Commercial Area in Residential Neighborhoods

The term neighborhood refers to the smallest unit of an urban or rural settlement, where people converge to share a place of shelter and form communities [32]. A neighborhood has a distinct physical and social attribute, distinguishing it from other areas. The clustering of neighborhoods results in towns, cities, and metropolitans.

Lewis Mumford defined neighborhood as a "fact of nature," which evolves when a group of individuals inhabits a shared space. Whittick, [44] characterizes a neighborhood unit as a harmonious, structured urban area connected to the bigger society, comprising residential areas, schools, commercial centers, places of worship, open spaces, and services.

Clarence Perry, in 1929, originally coined the term "Neighborhood Unit" in planned communities. The concept dates back to the fifteenth century [23], but Perry derived his inspiration from Ebenezer Howard's Garden City concept [3]. According to Perry, a Residential Neighborhood consists of major residential units with commercial areas, public semi-public areas, and 10% green and open space. Arterial and interior roads should be hierarchical to ensure easy understanding and safety.

Commercial areas in the residential neighborhoods are characterized by brick-and-mortar stores selling items comprising mostly daily, monthly, emergency, and impulse goods [40]. The most popular classification technique for commercial areas at national and local levels is based on the type of goods sold [13]. According to URDPFI, the government demarcates commercial areas known as neighborhood centers, which house shops for goods and services to cater to the needs of adjoining residential neighborhoods.

1.2.4 E-Commerce and Its Impact on Brick-and-Mortar

As we progress toward digital literacy, online platforms like Swiggy Instamart, Blinkit, Zepto, and Amazon Fresh also flourish among vivid internet users. These last-mile delivery platforms provide larger options and attract users with huge discounts and offers. The COVID-19 pandemic significantly shifted customer behavior and buying patterns from going outdoors to purchasing at home. The involvement of these big companies in the last mile of grocery delivery has a major impact on brick-and-mortar stores.

Even though there is a lack of human-led engagement and satisfaction of bargaining, the promptness in delivery, ease in availability, and flexibility of time for purchase have made online shopping the protagonist in the present Work from home or Lockdown situations. Even though the pandemic gave a major uplift to the introduction of Dark stores in India, [1] the major factors that have brought about the shift in shopping behavior among users can be mentioned below:

- Digital and technological agility.
- Better standard of living.
- The large variety of products are available at an effective price.
- Time efficiency and customer proximity for better delivery.
- Improving the payment options and making it more secure (Table 1.1).

The existing infrastructure or the brick-and-mortar stores are places of commerce where proprietors exhibit their goods in a tangible storefront; customers look around, interact, try on or test products, and make purchases. The intrusion of Online shopping has a major impact on various aspects of the existing retail as follows [10]:

- The Profit margin is affected by attractive offers, discounts, and limited inventory.
- Enhanced Customer Service and hassle-free purchases of e-commerce affect the customer pool.
- Increased window shopping and fewer transactions lead to a loss in footfall.

Table 1.1 Table showing the strengths and weaknesses of E-commerce

Strengths [22]	Weakness [39]
• Efficiency of time and money • Location is not the priority • Ease of availability and delivery • Hassle-free digital payments and cashback • Wide range of inventories • Comparative shopping and reviews • Discount coupons are more readily available online	• Physical examination • Chances of delay in delivery • Management of multiple inventories, expiration, and Substandard Goods, etc. • Returns of the goods are tricky • Chances of Payment frauds

1.2.5 O2O: Online-to-Offline

O2O was introduced to engage the local B&M shops in the digital presence [36]. It helped them expand their customer pool by providing consumers with a platform mode for online ordering and delivery systems [16]. Third-party and hybrid modes have been replacing the pure online and pure offline phenomena, giving rise to omnichannel retailing [37, 45]. The O2O model is centered on directing internet users to physical establishments to take advantage of offline offerings. The O2O method merges online and offline channels and builds a bridge between the virtual and real economies, instead of forging competitive links between them [19]. O2O platforms provide brick-and-mortar stores with two retailing modes: platform mode and mixed mode [16]. There are three platform modes: self-building, third-party mode, and mixed mode [20]. The crowd-shipping method has also gained momentum due to the introduction of O2O in B&M stores [48].

This research seeks to comprehend the factors contributing to O2O shopping implementation within the existing B&M stores. The major objectives are:

(a) Understanding the major indicators that influence the O2O participation of the B&M shops.
(b) Identifying the factors and their importance in initiating the trend shift using principal component Analysis.

1.3 Methodology and Materials

1.3.1 Study Area

The present study has been carried out in Bhubaneswar, the capital city of Odisha. It inhabits almost 2% of the total urban population of the state. Bhubaneswar, known as Ekamra Kshetra or the Temple City, is a perfect concoction of old charms and new-world comfort. The city has a planned precinct based on the neighborhood concept, providing ample opportunity to analyze the commercial spaces among the residential neighborhoods. The city has shown remarkable potential in the Tourism and Education sectors in the last decade. As per the 2011 census, the city has a commercial land use share of 2.17%, which has considerably increased over the last two decades. Commercial activities contribute to around 33% of employment in the city. As per the World Bank in 2014, Bhubaneswar became the best place for business in India as it has stood out as one of the rapidly growing, crucial trading and commercial centers in the state and eastern India.

Ward no. 8 was chosen for this survey as it is situated along a major road network axis, i.e., Nandan Kanan Road. It also consists of two major commercial hubs: planned residential and public and semi-public areas. The survey was carried out in the northwest zone, i.e., the growth direction of the city [33], thus making Ward 8 the most suitable study area.

1.3.2 Data Collection and Preparation

A literature survey was done to identify the variables responsible for introducing O2O mode in the B&M store. Cochran formula with a 90% confidence interval and a 5% precision level was used to find the sampling size. A physical survey was conducted, and the shopkeepers of the ward were approached with the questionnaire. A total of 287 usable samples were obtained from the data that was collected from the shopkeepers. The data collected for the present study was ordinal, i.e., on a Likert scale, where the shopkeepers were asked about their opinion on the factors amplifying the shift from B&M to O2O. Figure 1.2 shows the location of the surveyed shops in the Ward 8 of Bhubaneswar.

A focus group discussion was conducted before the main survey to determine the local parameters that catalyzed the shift among the retailers in the Indian context.

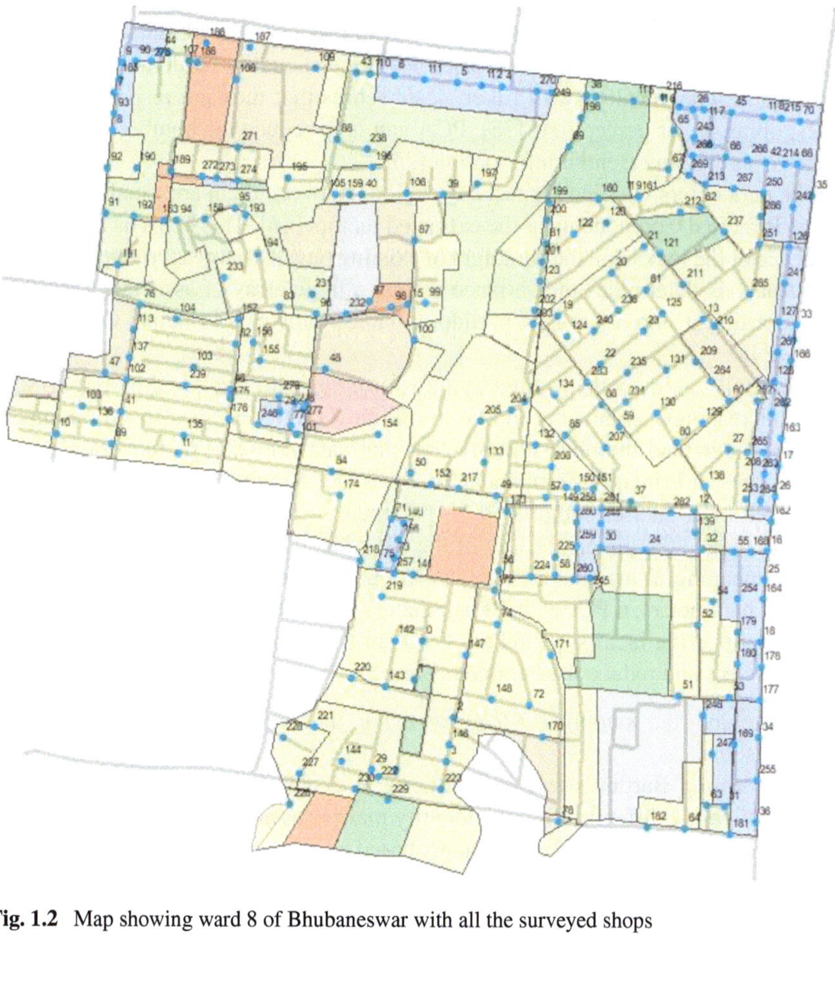

Fig. 1.2 Map showing ward 8 of Bhubaneswar with all the surveyed shops

An extensive discussion was conducted among 90 shopkeepers of the ward to understand their reasoning behind choosing the location of their shops, and a qualitative analysis of their transcript was conducted to find the indicators that influenced their decision the most. A total of 20 parameters that impact digital growth integration were identified from the Literature Survey and Focus Group Discussion.

The 20 parameters that were selected are Fast Internet, Usage of smartphones, Geolocation feature, User-friendly online platforms, and Smooth User interface of the application [2, 30], Catchment area, Income, Age, Gender, Educational Background, Occupation type of customers [12, 47], Distance of shop from the main road, presence of public transit stops, Road network connectivity, Employee wage, Profit and tax, Secure and hassle-free Payment, Rent/lease [10, 22, 27], Age of shop, Sustainable initiative and Crisis in the future [17, 34].

1. **Parameter Selection and Parameter Reduction**

Principal Component Analysis is a frequently used technique for parameter reduction of indices, as it effectively reduces information loss [14, 18, 29, 35, 43]. Principal Component Analysis is commonly used for dimensionality reduction in research, particularly when dealing with Likert scale scores that measure respondents' attitudes, opinions, or perceptions [38]. PCA can help reduce the number of variables while retaining the essential information [25].

The parameters with the highest correlations between them are usually clustered by the Principal Components or the extracted factors [46]. This helps us identify the influence of the factors on the paradigm of existing physical stores. Principal Component Analysis determines the variance among a large array of associated variables by converting it into a few new pseudo-variables known as Principal Components (PCs) [24].

The retail landscape is profoundly transforming, with brick-and-mortar (B&M) stores increasingly embracing online-to-offline (O2O) strategies to stay competitive. To understand the underlying factors that catalyze O2O participation among B&M stores, a Principal Component Analysis (PCA) with Varimax Rotation was conducted on a dataset comprising 20 parameters. This exploratory analysis was aimed to reduce the dimensionality of the data, identifying latent factors that could provide meaningful insights for retailers. By performing PCA on the obtained data set of 20 parameters, 6 PCs with 19 parameters out of 20 were identified that significantly defined the factors behind the O2O shift. The major procedures adopted in PCA to select and reduce the Parameters are:

(a) Assessing the sampling adequacy using the Kaiser–Meyer–Olkin (KMO) statistic.
(b) Performing Bartlett's test of sphericity to check if the correlation matrix is significantly different from the identity matrix for an appropriate PCA.
(c) Extracting the principal components based on the correlation matrix.
(d) Retaining components with eigenvalues greater than 1 and Scree plot analysis
(e) Orthogonally rotating the factors to simplify the structure and make it more interpretable using Varimax.

(f) Identifying the underlying factors and their weightages to understand the strength and direction of the relationship
(g) Interpreting the rotated factors loadings as each represents a latent variable and labeling each factor based on its parameters.
(h) Evaluating the retained factors' reliability or internal consistency using measures such as Cronbach's Alpha to guarantee that each factor's components measure the same underlying construct.

2. Results and Findings

The survey was conducted using a Likert Scale that measured shopkeepers' attitudes, opinions, or perceptions about the digital revolution in retailing. As suggested earlier, the parameters for the survey were taken from the Literature Survey and the Focus Group Discussion. As there is no significant data on the perception of Indian Retailers, the focus group helped understand their thought process and the local parameters that drove them to opt for the Hybrid mode of Retailing. A qualitative analysis of the discussion with the shopkeepers in the focus group discussion was done to develop some local parameters contributing to the shift.

The viability of the current data set was assessed using Kaiser–Meyer–Olkin (KMO) statistics and Bartlett's test of sphericity [43]. The Kaiser–Meyer–Olkin statistics are conducted to test for adequacy, while Bartlett's sphericity test is done to assess the relationship strength among the variables. The sampling is adequate if Kaiser–Meyer–Olkin is above 0.5 and within 1, and PCA can be conducted. For Bartlett's test of sphericity, the significance should be less than 0.05 to reject the Null Hypothesis that assumes the correlation Matrix to be an identity matrix. It ensures the data set is multivariate normal and accepted for component extraction [4].

The present data set has a Kaiser–Meyer–Olkin of 0.605 and Bartlett's test of sphericity significance of 0.000, which suggests that the dataset is significantly related among the parameters and adequate for Principal Component extractions. Principal Component Analysis was performed on the current data set using Varimax with Kaiser Normalization and factor extraction for parameters with Eigen Values of more than 1 using SPSS software.

Factor extraction is performed using Kaiser's criterion and Scree test to extract a few significant factors to define the relationship holistically [15]. Kaiser's criterion considers factors with Eigen value above 1, while Scree Plot considers factors above the elbow joint. PCA always assumes the variance associated with the parameters to be 1 before the factors are extracted.

Table 1.2 gives the communalities obtained using the Principal Component Method of all the parameters before and after the extraction. As Principal Component Analysis assumes the variance of the parameters to be 1 before extraction, the initial values are 1. Communalities represent the variance in each variable explained by the underlying components or factors [31]. The communalities after extraction show values above 50%, which implies a minimum of half the variance can be attributed to the identified factor or principal component. All the parameters except Geo Location (v1) and Employee Wage (v20) have less than 50% communality.

Table 1.2 Parameters with their labels and communalities

S. no.	Parameter	Label	Variance	
			Initial	Extraction
1	Geolocation feature	v1	1.000	0.499
2	Usage of smartphones	v2	1.000	0.525
3	Fast Internet	v3	1.000	0.708
4	User-friendly online platform and user interface	v4	1.000	0.661
5	Catchment area	v5	1.000	0.686
6	Gender of customers	v6	1.000	0.771
7	Age of customers	v7	1.000	0.533
8	Income of customers	v8	1.000	0.680
9	Occupation type	v9	1.000	0.658
10	Distance of shop from the main road	v10	1.000	0.616
11	Educational background	v11	1.000	0.632
12	Road network connectivity	v12	1.000	0.627
13	Profit and tax	v13	1.000	0.663
14	Presence of public transit stops	v14	1.000	0.759
15	Rent/lease	v15	1.000	0.611
16	Secure and hassle-free payment	v16	1.000	0.540
17	Crisis in the future	v17	1.000	0.617
18	Age of shop	v18	1.000	0.771
19	Sustainable initiative	v19	1.000	0.638
20	Employee wage	v20	1.000	0.406

Extraction method: principal component analysis

Table 1.3 gives all the factors with their Eigen values, individual Variance, and cumulative Variance. We are selecting till the 6th factor as their individual Eigen value is above 1, and the cumulative variance is 63.006%, which is deemed acceptable for explaining a substantial portion of the overall variability in the dataset. This decision is based on Kaiser's criterion, and this selection provides a meaningful and concise representation of the key factors influencing the phenomenon under study. It helps to achieve a compromise between preventing overfitting and preserving sufficient information to represent the dataset accurately. Adding more factors could cause overfitting, thus failing to identify underlying patterns.

Figure 1.3 shows the Scree plot with the Eigen values on the *y*-axis and the Principal Components on the *x*-axis. This also shows the number of factors or principal components to be selected is 6. This selection indicates that adding more components would have diminishing returns, explaining the variation in the dataset because it coincides with the point when the eigenvalues start to level off. The Scree plot gives a visual and systematic way to balance the complexity of the model with the amount of variation explained in a PCA analysis is to use.

Table 1.3 Variances of the components for extracted and rotated sums of squared loadings

Total variance explained									
PC	Initial Eigenvalues			Extraction sums of squared loadings			Rotation sums of squared loadings		
	Total	% of variance	Cumulative %	Total	% of variance	Cumulative %	Total	% of variance	Cumulative %
1	*3.244*	*16.220*	*16.220*	*3.244*	*16.220*	*16.220*	*2.725*	*13.623*	*13.623*
2	*2.914*	*14.569*	*30.789*	*2.914*	*14.569*	*30.789*	*2.633*	*13.166*	*26.790*
3	*1.910*	*9.549*	*40.338*	*1.910*	*9.549*	*40.338*	*1.893*	*9.463*	*36.253*
4	*1.756*	*8.778*	*49.115*	*1.756*	*8.778*	*49.115*	*1.886*	*9.429*	*45.682*
5	*1.521*	*7.607*	*56.722*	*1.521*	*7.607*	*56.722*	*1.809*	*9.047*	*54.729*
6	*1.257*	*6.284*	*63.006*	*1.257*	*6.284*	*63.006*	*1.655*	*8.277*	*63.006*
7	0.991	4.957	67.963						
8	0.815	4.074	72.037						
9	0.788	3.942	75.979						
10	0.724	3.622	79.601						
11	0.709	3.545	83.146						
12	0.605	3.027	86.173						
13	0.544	2.721	88.894						
14	0.446	2.232	91.126						
15	0.411	2.054	93.180						
16	0.340	1.700	94.880						
17	0.329	1.645	96.525						
18	0.256	1.280	97.805						
19	0.238	1.192	98.997						
20	0.201	1.003	100.000						

Extraction method: principal component analysis
Italics part indicate the six selected factors from the analysis that has Eigenvalues above 1, as per Kaiser's criterion, collectively explaining 63.006% of the cumulative variance, which is an acceptable level for representing key factors in the dataset

Table 1.4 gives the rotated Component matrix, which shows the loadings of the parameters and their clustering for the principal components extracted after the orthogonal rotation has been applied. It analyses and comprehends the relationship between the extracted major components and the parameters. The rotation converged after 10 iterations, giving the strength and direction of the relationship between the components and the parameters. The orthogonal rotation implies the common parameters have been removed, giving a clear interpretation of the structure of the components.

Table 1.5 gives the Component Transformation Matrix that illustrates the correlations between the components before and after rotation and the loadings of the retrieved Principal components. The loading helps us comprehend the contribution

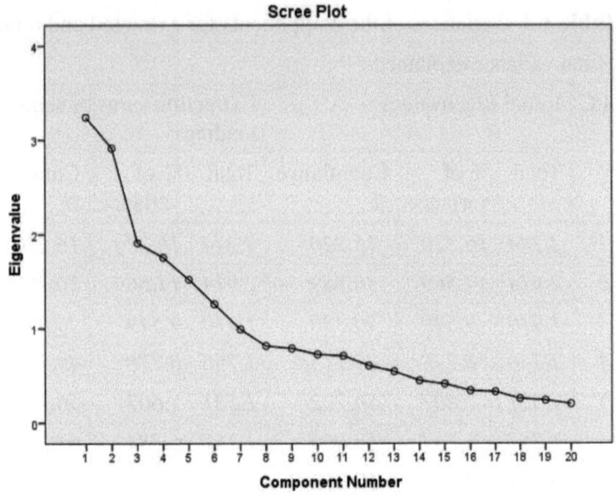

Fig. 1.3 Scree plot showing components and their Eigen values

Table 1.4 Rotated component matrix and their loadings

Rotated component matrix						
	Component					
	1	2	3	4	5	6
v14	0.818					
v12	0.724					
v10	0.683					
v1						
v13		0.723				
v15		0.704				
v17		0.614				
v21		0.558				
v5			0.751			
v6			– 0.712			
v7			0.628			
v9				0.776		
v11				0.652		
v8				– 0.575		
v20					0.754	
v19					– 0.717	
v18					0.695	
v4						0.727
v3						– 0.679
v2						0.624

Extraction method: principal component analysis
Rotation Method: Varimax with Kaiser Normalization[a]
[a] Rotation converged in 10 iterations

of the parameters in extracting the component, while the correlation implies the degree of association among the components. Moreover, studying the before and after loading values helps us fathom the impact of rotation on the factor structure and assess the relationships among variables and principal components in the analyzed data.

Reliability statistics were also conducted using Cronbach's Alpha, which evaluates the dependability of the data set. Cronbach's Alpha quantifies the correlation and the extent of internal consistency in a dataset. The major purpose of conducting a Reliability test is to understand the clustering within the data set and validate the scale so that consistent and dependable results are achieved. It is a quality control measure that helps us decide whether to retain or exclude a particular item in a data set so that the scale's reliability can be improved, thus improving the data quality overall. Cronbach's Alpha score should lie between 0 and 1. It suggests that the higher the scores obtained, the more likely the data set is consistent and reliable, thus increasing confidence in the validity of the results [6, 9] The Reliability Statistics were conducted using SPSS software, and the score was 0.707 for 20 items, suggesting that the data set is reliable and consistent.

Six Principal Components emerged from the initial 20 parameters, each representing a cluster of related factors. The 6 components, parameters, and loadings are given in Table 1.5. The loadings derived for the parameters suggest the contribution of each parameter to the component. A loading close to ± 1 indicates a strong influence, while a loading closer to 0 indicates a lower influence. The sign with the loading determines the direction of the influence. The extracted loadings are all above 0.5, suggesting a strong influence on the components. Parameters like Gender in the Demographic component, Educational background in the socio-economic component, age of shop in the Shop Image component, and fast Internet in the Digital Growth Component negatively influence, while all other components positively influence the components. Table 1.6 shows the identified components, their associated parameters, and their loadings.

Table 1.5 Component transformation matrix of the components before and after rotation

Component transformation matrix						
Component	1	2	3	4	5	6
1	0.308	0.794	0.355	0.384	– 0.015	0.022
2	0.855	– 0.045	– 0.317	– 0.304	0.190	0.196
3	– 0.216	– 0.018	– 0.201	0.388	0.630	0.604
4	0.163	– 0.313	0.594	0.013	0.607	– 0.394
5	0.302	– 0.518	0.249	0.566	– 0.434	0.264
6	0.096	– 0.027	– 0.564	0.538	0.105	– 0.609

Rotation method: varimax with kaiser normalization

Table 1.6 The labels, associated parameters, and loadings of the identified Six components

Component	Label of components	Parameters	Loadings
1	Spatial attributes	Distance of shop from the main road	0.818
		Enhanced consumer pool	0.724
		Road network connectivity	0.683
2	Operational cost	Profit and tax	0.723
		Presence of public transit stops	0.704
		Rent/lease	0.614
		Secure and hassle-free payment	0.558
3	Demographic profile of consumers	Age of customers	0.751
		Gender of customers	– 0.712
		Consumer catchment area	0.628
4	Socio-economic profile of consumers	Income of customers	0.776
		Occupation type	0.652
		Educational background	– 0.575
5	Shop image	Crisis in the future	0.754
		Age of shop	– 0.717
		Sustainable initiative	0.695
6	Digital growth	Usage of smartphones	0.727
		Fast Internet	– 0.679
		User-friendly online platform and smooth user interface	0.624

1.4 Discussions and Conclusion

Component Extraction using Principal Component Analysis imbibes the data with high objectivity. However, it fails to account for the local conditions. This present study has tackled the drawback by conducting interviews and discussions in the focus group discussion to understand better the local conditions for formulating the components. This strategic inclusion mitigates the limitation of overlooking the local conditions associated with the participation of brick-and-mortar stores in online-to-offline mode. Identifying six essential elements provides vital knowledge for physical retailers seeking to increase their online-to-offline (O2O) interaction. One parameter was excluded in the component extraction process due to its lower loading on any identified components to ensure that they were focused and robust, providing a clearer understanding of the factors influencing O2O participation.

The Spatial Attributes Component assesses the spatial factor and accessibility of the store, emphasizing the need for a placement strategy for maximum foot traffic. It includes elements that substantially impact the offline shopping experience, such as accessibility to main thoroughfares, road network connectivity, public transport access, and foot traffic. These metrics demonstrate the strategic importance of spatial qualities in influencing the O2O dynamics of brick-and-mortar establishments and become crucial factors for firms considering the integration of online channels. For instance, a shop not located on the main road can also enjoy a larger pool of consumers if it integrates an online mode of delivery. Thus, the hybrid mode of retailers can efficiently manage the competition regarding location and higher rent prices.

The Operational Cost component covers the financial factors associated with online-to-offline (O2O) integration, assisting retailers in prudent resource allocation. This component explores important financial factors such as staffing needs, maintenance costs, taxes and rent payments, and expenditures associated with adopting new technology. It also encompasses secure and hassle-free monetary transactions in a hybrid mode. Understanding the financial consequences through this component can reduce operational costs and increase sales throughout the year.

The Socio-economic and Demographic Profile Components shed light on the necessity of tailoring online-to-offline (O2O) strategies to the inclinations and actions of the intended audience. It explores the complex relationship between consumer traits and the transition from offline to online interaction as the second component. Understanding the socio-economic and demographic profile is essential, as customer demographics are crucial in decision-making processes. Variables including consumer age, income levels, educational background, occupation details, and catchment areas significantly shape this part. It provides insight into the unique preferences of their target market, allowing them to customize O2O tactics to suit these interests efficiently. A more detailed approach can be adopted by building a deeper relationship with their audience and recognizing the diversity among customer categories.

The Shop Image component highlights how important the image of the store is to draw in and keep consumers. It emphasizes how crucial creating a friendly and attractive shopping atmosphere is. It focuses on the shops' physical characteristics and surroundings, emphasizing elements like age, layout, aesthetics, sustainable approaches, and preparedness for a crisis. Generally, newer shops managed by the younger generations are keen to opt for a digital presence. Moreover, the recent pandemic has also played its part in motivating retailers to a year-round business.

The Digital Growth components emphasize the importance of digital technologies in improving client experiences and expediting operations. It covers the initial characteristics of incorporating digital technologies into conventional business structures. Key variables, including online inventory management, availability of smartphones and fast Internet services, and e-commerce platform integration, are significantly weighted on this component. In the dynamic world of online-to-offline connectivity, companies looking to maximize their efficiency and deliver a flawless customer experience must acknowledge the importance of digital transformation in these areas.

In conclusion, this study helps us understand the intricate details of the evolution of brick-and-mortar shops to platform mode and comprehensively understand the factors that catalyze the shift. It depicts the dynamic nature of the commercial industry and helps us identify the adaptations that commerce has undergone to enhance the customer experience. The Principal Component Analysis with Varimax rotation identifies 6 components that expedite the Brick-and-mortar retailers' online-to-offline participation. These components cover various elements, such as consumer profiles, digital procedures, physical shop attributes, and operating expenses. It will help retailers nurture a seamless integration of online and offline channels for improved consumer experiences and long-term business growth by creating focused strategies to manage the changing retail landscape. These factors help us better understand the commercial structure for strategic decision-making and policy-making to converge online and offline commerce for planning efficient urban space.

References

1. Abbas, A.: A review paper on mycoviruses. J Plant Pathol Microbiol **7** (2016). https://doi.org/10.4172/2157-7471.1000390
2. Asmare, A., Zewdie, S., Asmare, A.: Omnichannel retailing strategy: a systematic review Omnichannel retailing strategy: a systematic review. Int. Rev. Retail Distrib. Consum. Res. 1–22 (2021). https://doi.org/10.1080/09593969.2021.2016472
3. Baffoe, G.: Understanding the neighbourhood concept and its evolution: a review. Environ. Urban. ASIA **10**, 393–402 (2019). https://doi.org/10.1177/0975425319859115
4. Bah, S.: Discovering Statistics Using SPSS for Windows: Advanced Techniques for Beginners by A. Field Book Reviews (2019). https://doi.org/10.2307/2681235
5. Benton, R.A.: Unit I. Spok Pangasinan, pp. 1–34 (2019). https://doi.org/10.2307/j.ctv9hvspb.9
6. Bonett, D.G., Wright, T.A.: Cronbach's alpha reliability: interval estimation, hypothesis testing, and sample size planning. J. Organ. Behav. **36**, 3–15 (2015). https://doi.org/10.1002/job.1960
7. Carniel, A. What Marketing Is and How It Has Changed Over Time, pp. 1–16 (2021)
8. Chen, X.H., Jin, F.J., Zhang, Q., Yang, L.: Are investors rational or perceptual in P2P lending? Inf. Syst. E-bus. Manag. **14**, 921–944 (2016). https://doi.org/10.1007/s10257-016-0305-z
9. Cronbach, L.J.: Coefficient alpha and the internal structure of tests. Psychometrika **16**, 297–334 (1951). https://doi.org/10.1007/BF02310555
10. Dahiya, M.: Study on E-commerce and it's impacts on market and retailers in India. Adv. Comput. Sci. Technol. **10**, 1495–1500 (2017)
11. Ganesha, H.R., Aithal, P.S.: Theory of brick-and-mortar retailing in India **5**, 116–132 (2020)
12. Gujrati, R., Uygun, H.: Digital transformation has changed consumer behvoiur from digital transformation has changed consumer behvoiur from traditional market to digital market (2023)
13. Guy, C.M.: Classifications centres: some methodological issues. GeoJournal **45**, 255–264 (1998)
14. Gómez, N.: Epipelic diatoms from the Matanza-Riachuelo river (Argentina), a highly polluted basin from the Pampean plain: biotic indices and multivariate analysis Epipelic diatoms from the Matanza-Riachuelo river (Argentina). Aquat. Ecosyst. Health Manage. 4988 (2008). https://doi.org/10.1080/14634989908656966
15. Hadi, N.U., Abdullah, N., Lumpur, U.K., Ilham, S.: An easy approach to exploratory factor analysis: marketing perspective. Noor Ul Hadi (2016). https://doi.org/10.5901/jesr.2016.v6n1p215
16. He, B., Gupta, V., Mirchandani, P.: Online selling through O2O platform or on your own ? Strategic implications for local Brick-and-Mortar stores. Omega **103**, 102424 (2021a). https://doi.org/10.1016/j.omega.2021.102424
17. He, B., Gupta, V., Mirchandani, P.: Online selling through O2O platform or on your own? Strategic implications for local Brick-and-Mortar stores. Omega (United Kingdom) **103** (2021). https://doi.org/10.1016/j.omega.2021.102424
18. Helena, B., Vega, M., et al.: Temporal evolution of groundwater composition in an alluvial aquifer (Pisuerga River, Spain) by principal component analysis. Wat. Res. **34**, 807–816 (2000)

19. Hsieh, J.: The role of customers in in the O2O model (2015). https://doi.org/10.1108/JOSM-03-2016-0062
20. Hübner, A., Hense, J., Dethlefs, C.: The revival of retail stores via Omnichannel operations: a literature review and research framework. Eur. J. Oper. Res. **302**, 799–818 (2022). https://doi.org/10.1016/j.ejor.2021.12.021
21. Indianapolis, M.: Development (2012)
22. Jan. R.H., Lydia, E.L., Shankar, K., et al.: The increasing market of ecommerce and its impact on retailer. J. Crit. Rev. **6**, 122–127 (2019). https://doi.org/10.22159/jcr.06.05.21
23. Jenks, M., Dempsey, N.: Defining the neighbourhood: challenges for empirical research linked references are available on JSTOR for this article: defining the neighbourhood challenges for empirical research. Town Plan Rev. **78**, 153–177 (2007)
24. Jolliffe, A.I.T., Journal, S., Statistical, R., et al.: Discarding variables in a principal component analysis. I: artificial data. J. R. Statist. Soc. Ser. **21**, 160–173 (1972)
25. Karkra, R., Kumar, P., Bansod, B.K.S., Bagchi, S.: Classification of heavy metal ions present in multi-frequency multi-electrode potable water data using evolutionary algorithm. Appl Water Sci **7**, 3679–3689 (2017). https://doi.org/10.1007/s13201-016-0514-0
26. Keelson, S.: The evolution of the marketing concepts: theoretically different roads leading to practically same destination! Soc. Sci. Res. **1**, 35–41 (2012)
27. Kim, J.S., Ratchford, T.B.: Consumer choice and use of multiple information sources for automobile purchases. Int. J. Electron. Commer. **16**, 7–40 (2012). https://doi.org/10.2753/JEC1086-4415160301
28. Kiran, R., Jhamb, D.: A Strategic Framework for Consumer Preferences towards Emerging Retail Formats. J. Emerg. Knowl. Emerg. Mark. **3** (2011). https://doi.org/10.7885/1946-651x.1058
29. Lin, Y.: Firm heterogeneity and location choice of Chinese firms in Latin America and the Caribbean: corporate ownership, strategic motives and host country institutions. China Econ. Rev. **34**, 274–292 (2015). https://doi.org/10.1016/j.chieco.2015.03.001
30. Lokhande, J.S.: Online and offline shopping: a comparative study **10**, 775–786 (2022)
31. López del Val, J.A., Pérez, A., de Agreda, J.P.: Principal components analysis. Aten. Primaria **12**, 333–338 (1993). https://doi.org/10.5455/ijlr.20170415115235
32. Meenakshi: Neighborhood unit and its conceptualization in the contemporary urban context. India J. 81–87 (2011)
33. Pathy, A.C., Panda, G.K.: Modeling urban growth in Indian situation—a case study of Bhubaneswar City. Int. J. Sci. Eng. Res. **3**, 1–7 (2012)
34. Perlman, Y.: Pricing decisions of online and offline retailers, each offering a competing benefit. Oper. Res. Perspect. **9**, 100231 (2022). https://doi.org/10.1016/j.orp.2022.100231
35. Pesce, S.F., Wunderlin, D.A.: USE of Water Quality Indices to Verify the Â Rdoba City (Argentina) on Suquõâ a Impact of CO. 34 (2000)
36. Reinartz, W., Wiegand, N., Imschloss, M.: The impact of digital transformation on the retailing value chain. Int. J. Res. Mark. **36**, 350–366 (2019). https://doi.org/10.1016/j.ijresmar.2018.12.002
37. Ryu, D.H., Lim, C., Kim, K.J.: Development of a service blueprint for the online-to-offline integration in service. J. Retail. Consum. Serv. **54**, 101944 (2020). https://doi.org/10.1016/j.jretconser.2019.101944
38. Sanders, P., Zuidgeest, M., Geurs, K.: Liveable streets in Hanoi: a principal component analysis. Habitat Int. **49**, 547–558 (2015). https://doi.org/10.1016/j.habitatint.2015.07.001
39. Seidel, S.: One goal, one approach? A comparative analysis of online grocery strategies in France and Germany. Case Stud. Transp. Policy **9**, 1922–1932 (2021). https://doi.org/10.1016/j.cstp.2021.10.013
40. Smith, J.G., Clark, F.E.: Principles of marketing (1928)
41. Sui, D.: The new science of cities, by Michael Batty, Cambridge, MA, The MIT Press, 2013, 496 pp., $45.00/£31.95 (hard cover), ISBN 978-0-262-01952-1. Int. J. Geogr. Inf. Sci. **29**, 345–347 (2015). https://doi.org/10.1080/13658816.2014.937717

42. Toolkit, C.P.: The changing consumer and market landscape. Consum Policy Toolkit **15–30** (2010). https://doi.org/10.1787/9789264079663-4-en
43. Tripathi, M.: Use of principal component analysis for parameter selection for development of a novel water quality index: a case study of river Ganga. Ecol. Indic. **96**, 430–436 (2019). https://doi.org/10.1016/j.ecolind.2018.09.025
44. Whittick, A. (ed.) Encyclopedia of Urban Planning. McGraw-Hill Book Company, USA, pp. 714-715 (1974)
45. Wu, P.J.: O2O switching determinants and successful drivers in omnichannel retailing services. Serv. Bus. **17**, 771–788 (2023). https://doi.org/10.1007/s11628-023-00540-0
46. Xia, L.I. Integration of principal components analysis and cellular automata for spatial decision-making and urban simulation 45 (2002)
47. Xue, Y. Research on consumers' attitudes towards online and offline shopping 01018 (2020)
48. Yang, D., Zhang, X.: Omnichannel operations with ship-from-store. Oper. Res. Lett. **48**, 257–261 (2020). https://doi.org/10.1016/j.orl.2020.03.009

Chapter 2
Analysing Purchase Behaviour in Social Commerce Among India's Youth

Archiman Biswas and Subrata Chattopadhyay

Abstract Social Commerce is a fast-evolving trend that has turned social media from community platforms into online marketplaces. It is a subset of e-commerce involving online shopping driven by or integrated within social media platforms, allowing users to purchase products without leaving the platform. Several reports show that Social Commerce significantly contributes to the Indian GDP and employment generation. This study aims to understand the purchasing behaviour of young Indian consumers aged 18–29 in the context of social commerce. We have employed exploratory factor analysis and multiple linear regression to understand the role of income, lifestyle, attitude, Internet experience and perceived trust in shaping purchasing behaviour from Social Commerce. The exploratory factor analysis identified four factors that are significant in shaping the purchasing behaviour from Social Commerce in the Indian context. The factor scores were used in the multiple linear regression for understanding the relation between these factors and Social Commerce spending among the Indian youth. The findings of this study are instrumental in developing strategies to improve the Social Commerce experience by having a better understanding of different characteristics and how they affect the purchase of Social Commerce.

Keywords Social commerce · Exploratory factor analysis · Multiple linear regression · Young customers · Indian context

2.1 Introduction

In the recent years, the landscape of online commerce has undergone a profound transformation, with the advent and rapid ascent of social commerce (SC) [33]. Recognized as a subset of e-commerce, SC intertwines the realms of social media

A. Biswas (✉) · S. Chattopadhyay
Indian Institute of Technology Kharagpur, Kharagpur, West Bengal, India
e-mail: biswasarchiman9@kgpian.iitkgp.ac.in

S. Chattopadhyay
e-mail: schat@arp.iitkgp.c.in

J. Maiti et al. (eds.), *Recent Advances in Industrial and Systems Engineering*, Lecture Notes on Multidisciplinary Industrial Engineering,
https://doi.org/10.1007/978-981-96-8323-9_2

and commercial activities, presenting a novel paradigm in the digital marketplace [28]. Despite varying interpretations within the literature, SC can be fundamentally understood as a departure from conventional e-commerce models, seamlessly blending Web 2.0 technologies with social media platforms [6].

Essentially, SC represents a dynamic evolution, transcending traditional e-commerce by orchestrating a harmonious convergence of social networking and online retail [37]. Additionally, during and after COVID-19, numerous brick-and-mortar retailers have increased their online presence and harnessed social media live features, such as Facebook Live, Instagram Live, and YouTube Live, to sell their products [21, 38, 41]. This innovative approach reshapes the dynamics of business operations by capitalizing on influencers, celebrities, and individuals, thereby forging a distinctive synergy between social interactions and product transactions. The resulting amalgamation of marketing content and user-generated experiences gives rise to a novel and influential force in the commercial landscape [12]. The global surge in the SC market underscores its escalating significance. Recent estimations reveal a staggering global market size of $727.63 billion, with projections anticipating a trajectory to reach $1.3 trillion by the close of 2023 [31]. With an estimated 31.6% annual growth rate between 2023 and 2030, these figures highlight the exceptional rate at which SC is well-positioned for expansion [7].

Within the context of India, the SC arena is particularly dynamic, currently boasting a valuation of $5 billion in Gross Merchandise Value (GMV) [33]. Prospective estimates present a promising picture, envisioning the Indian SC market surging to an impressive $70 billion by 2030 [1]. This growth not only holds economic significance but also carries transformative potential for over 40 million small businesses and entrepreneurs across the nation. Recognizing the transformative power of SC, the Department for Promotion of Industry and Internal Trade, under the Government of India, envisions SC as a catalyst for employment generation. In this context, deciphering the intricate dimensions of SC becomes imperative, not only as a reflection of global market trends but as a strategic lever for economic empowerment and inclusive growth in the Indian context.

Very few previous studies have attempted to identify the factors that influence purchase behaviour from SC in the Indian context. Previous research has focused on comprehending how various factors, including social elements (such as informational support and community commitment), technical and socio-technical aspects, trust (towards members and community), and website quality, influence purchasing behaviour in the domain of SC in India. Limited research, with only one study focusing on Gen Z customers, has addressed the purchasing behaviour of young Indian consumers in the realm of SC. Hence, in this research, we have attempted to understand the purchasing behaviour of the young Indian population from the SC. Projections indicate that India is on track to become the third-largest online retail market by 2030 [15]. The age group chosen for the study is 18–29 years old. According to the Census of India in 2011, more than 55% of the population is under the age of 30, comprising a dynamic demographic open to experimenting with and potentially changing their shopping habits [5, 16]. It encompasses both working and non-working married and unmarried individuals, making it an intriguing segment for study. The major aims of this study are:

(i) To identify the factors that influence the purchasing behaviour of young Indian customers (aged between 18 and 29) in SC.
(ii) To gain insights into how these factors actively shape and determine the extent of participation in SC activities.

2.2 Literature Review

The advent of Web 2.0 technologies and the widespread adoption of social networking websites have ushered in new opportunities for online commerce. SC, a novel advancement in e-commerce, harnesses Web 2.0 technologies to facilitate customer interactions and enable them to become active content creators, aiding others in decision-making regarding product or service acquisition. Distinguishing itself from traditional e-commerce, SC operates on social technology platforms and fosters social media-supported commercial activities, leveraging online communities and encouraging customer engagement through reviews and recommendations. The surge of interest in SC is evident from the proliferation of studies exploring various facets of its impact on customer behaviour and purchase intentions.

The e-commerce sector in India has experienced significant growth, with projections indicating a rise to US$ 200 billion by 2026 [2]. Factors such as increasing smartphone penetration, the rollout of 4G networks, and rising consumer wealth contribute to this growth. Additionally, online retail sales in India are expected to increase by 31% to US$ 32.70 billion in 2018, and the online grocery market is projected to grow at a CAGR of 57% from $1.9 billion in 2019 to US$ 18.2 billion in 2024. Despite this growth, there remains a lack of comprehensive understanding regarding the purchasing behaviour of young Indian consumers in SC.

The role of social media in marketing has become increasingly significant, with businesses integrating social media techniques to enhance their relationships with customers [27]. Platforms like Facebook, YouTube, and Twitter have evolved into hubs for product information, brand awareness, and customer behaviour shaping techniques (Bai et al. 2015) [2]. Through extensive social networks, social media platforms offer a unique opportunity to amplify word-of-mouth marketing to a wider audience, enhancing consumer-to-consumer communications and expanding brand awareness [2, 12]. However, despite the widespread use of social media in marketing, there is still a need to understand how customer information and involvement influence the branding process, particularly in the context of SC.

Consumers frequently assess brands, compare them, and make purchasing decisions based on their brand affinity (Mou and Benyoucef 2021). Brand perceptions and social influences play a significant role in branding, with elements such as brand message and understanding determining the value associated with branding (Patwa et al. 2024). However, there is a research gap concerning the impact of branding on the purchasing behaviour of young Indian consumers in SC.

Previous studies have provided some insight into factors influencing purchasing behaviour from SC in India (Sharma et al. 2022), focusing on various factors including social elements, technical aspects, trust, and website quality. However, limited research, with only one study focusing on Gen Z customers, has addressed the purchasing behaviour of young Indian consumers in SC. Hence, this study aims to address these gaps by identifying the factors that influence the purchasing behaviour of young Indian consumers in SC and gaining insights into how these factors shape and determine their participation in SC activities. With the help of the literature review, we have identified the key indicators that significantly influence consumer purchasing behaviour in SC.

Through a thorough examination of existing literature, we have identified key indicators significantly shaping consumer purchase behaviour in SC. Various studies highlight the critical role of financial transaction security as a pivotal factor in influencing consumer's purchase decisions from the SC [17, 42]. Studies elucidate that a secure and trustworthy payment environment influences consumers' willingness to engage in SC purchases [2, 14]. Previous studies show that timeliness of delivery is an important determinant of consumer satisfaction, with efficient and prompt deliveries positively impacting the overall consumer experience and influencing repeat purchases [27, 47]. Lee et al. [18] highlighted that effective customer service mechanisms within SC platforms are essential for addressing issues such as damaged goods or late deliveries and positively impacting the purchase behaviour of the customers. Transparent pricing practices, including clear price information and the absence of hidden fees or charges, are crucial in building consumer trust on SC platforms [4, 13]. Several studies argued that the digital literacy and navigation skills of customers play a pivotal role in SC purchases [4, 10, 47]. The ease with which individuals can access and navigate these platforms influences their overall satisfaction and likelihood of engaging in transactions [22, 30]. Consumer confidence in making secure and satisfactory online purchases is closely tied to their trust in the overall online shopping experience [24, 44]. The literature also recognizes the importance of consumers' adaptability to new features, applications, or tools emerging on SC platforms [35, 45]. Those who are more adaptable and comfortable with technological advancements are likely to have a more positive experience [8, 29]. Several studies claimed that customers' purchase behaviour from e-commerce and SC depends on the customers' attitude towards technology and social media [19, 23, 32]. Based on these indicators the questionnaire for this survey was formulated.

2.3 Methodology

2.3.1 Questionnaire Preparation

Based on the literature review the questionnaire was prepared. The survey questionnaire was organized into three sections: customer characteristics, Internet experiences, social media usage, and spending in SC over the past year. Each primary section was further divided to enable a more in-depth exploration of the topics.

Respondents were asked to express their agreement with various statements related to social media usage, technology utilization, and the reliability of non-store shopping. A five-point Likert scale was employed for this purpose, with values ranging from 1 (strongly disagree) to 5 (strongly agree). To gather information on demographic characteristics, participants were directly queried about aspects such as age, gender, income level, and education level (Table 2.1).

2.3.2 Data Collection

The sample size for this study was determined using Cochran's formula. As per Cochran's Formula (Cochran 1977), with a 95% confidence level (corresponding to a z value of 1.96) and a 5% margin of error, the minimum required sample size is 384 or more. The survey questionnaire was distributed through a Facebook news feed and 67 Facebook groups, resulting in a total of 432 responses. Subsequently, the collected data underwent a thorough review to identify and eliminate unengaged responses, outliers, and error terms. Specifically, the review led to the exclusion of 14 responses due to missing values, 5 due to the respondents being outside the target age range, 3 for showing uniform answers across all Likert scale questions indicating disengagement, and 3 outliers identified through box plot analysis. After this cleaning process, a total of 407 samples remained for the analysis (Fig. 2.1). Table 2.2 elucidates the sample characteristics.

2.3.3 Data Analysis

Demographic information obtained through a questionnaire survey was utilized to construct a profile of SC shoppers in the Indian context, using simple numbers and percentages. To analyse data related to purchasing behaviour, a process of data reduction was implemented using exploratory factor analysis (EFA) within the SPSS software. Exploratory factor analysis (EFA) is a statistical approach for development of scale that is widely used to investigate the underlying dimensions and ideal structure of measuring items [26]. The collected data exhibited suitability for factor analysis, as indicated by correlation analysis, variable-wise measures of sampling

Table 2.1 Questions asked in primary survey (in 5-point Likert scale)

Indicator no	Indicator	Questions asked	Source
1.	Product quality confidence	How confident are you in the overall quality of products purchased through online shopping channels, considering factors such as accuracy in product descriptions and authenticity?	[17, 42]
2.	Transaction security	To what extent do you feel secure about the financial transactions you make on online shopping platforms?	[2, 14]
3.	Delivery reliability	How satisfied are you with the reliability of online shopping channels in terms of timely deliveries?	[27, 47]
4.	Customer service effectiveness	How well do you believe online retailers address and resolve customer service concerns such as damaged goods, late deliveries, or other issues?	[18]
5.	Price transparency	How transparent do you perceive online shopping channels to be in terms of price information, including any hidden fees or charges?	[4, 13]
6.	Internet navigation efficiency	How efficiently do you believe you navigate and utilize the Internet for various tasks?	[4, 10, 47]
7.	Purchase confidence	How confident are you in your ability to make secure and satisfactory purchases of goods online?	
8.	Adaptability to new platforms	To what extent do you feel comfortable and efficient in adapting to new online platforms, applications, or tools as they emerge and become part of your digital routine?	[35, 45]
9.	Tech improves life	To what extent do you feel comfortable and efficient in adapting to new online platforms, applications, or tools as they emerge and become part of your digital routine?	[9, 40]
10.	Influence of technology on life	Indicate the extent to which you agree or disagree with the following statement: "I believe that technology has significantly improved my overall quality of life."	[36, 37]
11.	Tech enhances productivity	Rate your agreement with the following statement: "I find technology to be a valuable tool in enhancing my productivity and efficiency."	[4, 13]

(continued)

Table 2.1 (continued)

Indicator no	Indicator	Questions asked	Source
12.	Importance of tech updates	To what extent do you agree or disagree with the statement: "I believe that staying updated with the latest technological advancements is important for personal and professional growth."	[4, 10, 47]
13.	Tech keeps us informed	Rate your agreement with the statement: "Technology keeps me more informed about current events, trends, and developments."	(Liang and Lai; [22, 30])
14.	Social media's effectiveness in maintaining personal connections	To what extent do you agree or disagree with the statement: "Social media enhances my overall experience of staying connected with friends and family."	[24, 44]
15.	Value of social media for news and current affairs	How much do you agree or disagree with the statement: "Social media is a valuable tool for staying informed about global news and current affairs."	[35, 45]
16.	Personal sense of happiness and fulfilment from social media use	Rate your agreement with the statement: "Social media positively contributes to my overall sense of happiness and fulfilment."	[8, 29]
17.	Social media for product research	Level of agreement with the following statement: "I consider social media as a reliable source for researching and discovering new products."	[19, 23, 32]

adequacy, Kaiser–Meyer–Olkin (KMO) test of sampling adequacy, and Bartlett's test of sphericity. The KMO value for the dataset is 0.855 (Table 2.3), well within the acceptable range. Factor extraction is carried out using the maximum likelihood method and Promax rotation. The determination of the number of dimensions relied on eigenvalues exceeding 1.0, following Kaiser's guideline.

Items with loadings below 0.4 are eliminated, ensuring that only significant contributors were retained. This elimination process is conducted iteratively, with items meeting the criteria being deleted one at a time. Subsequent to each deletion, individual KMO values and communalities are re-evaluated to confirm adherence to the established criteria [3, 39].

With the help of EFA, factor scores are extracted. Next, we have employed multiple linear regression to assess the impact of the extracted factor scores on annual SC spending, with factor scores as independent variables and annual SC spending as the dependent variable. This regression model aimed to elucidate the overall relationship between the identified factors and spending patterns. Before conducting the regression, we tested for the assumptions of multiple linear regression, including linearity, normality, homoskedasticity, and uncorrelated error terms. To assess the linearity

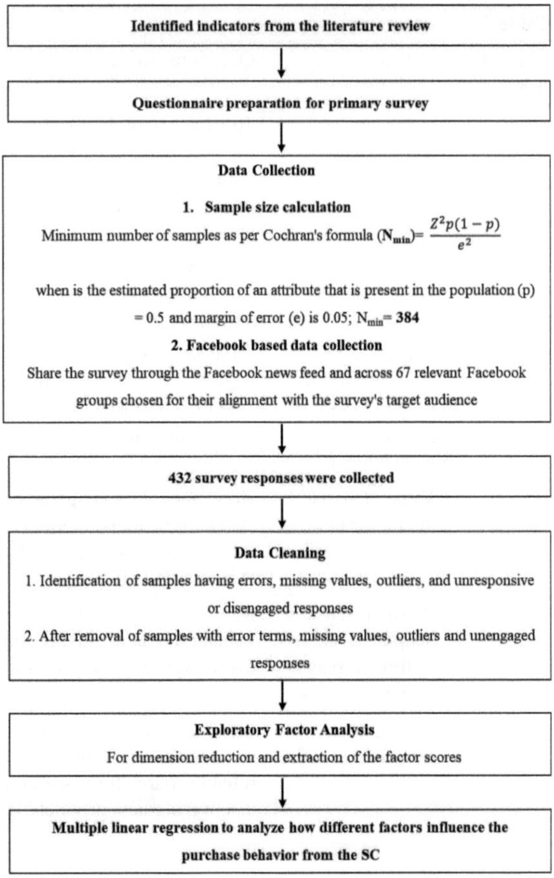

Fig. 2.1 Data collection and data analysis methodology

Table 2.2 Sample characteristics

Gender ($n = 407$)	Male = 181 Female = 226
Age	Mean =
Spending in SC in last 1 year (Dependent variable)	Mean = 5683.71 4876.08 (Male) 6330.53 (Female)
Level of income (Mean)	EWS = 79 LIG = 75 MIG 1 = 86 MIG 2 = 94 HIG = 71

Table 2.3 Sample size adequacy: KMO and Bartlett's test results

Kaiser–Meyer–Olkin measure of sampling adequacy		0.855
Bartlett's test of sphericity	Approx. Chi-Square	8042.660
	df	136
	Sig.	0.000

assumption, we performed an analysis of variance (ANOVA). The results, presented in Appendix 1, show that the relationship between the independent variables and the dependent variable is linear. The ANOVA table indicates that the linearity is significant ($p < 0.05$), confirming that the model captures the primary relationship without significant deviation from linearity (Appendix 1). The normality of the residuals was evaluated using a probability-probability (P-P) plot, illustrated in Fig. 2.2a. The P-P plot demonstrates that the residuals closely follow the 45-degree line, indicating that the residuals are approximately normally distributed. This visual inspection suggests that the normality assumption is satisfied. To check for homoskedasticity, we examined the scatter plot of the residuals against the predicted values. The scatter plot revealed that the residuals are randomly dispersed around the horizontal axis, without any discernible pattern (Fig. 2.2b). This randomness indicates that the variance of the residuals is constant across all levels of the independent variables, satisfying the homoskedasticity assumption. The Durbin–Watson (DW) statistic was calculated to test for autocorrelation in the residuals. The DW value obtained was 1.89, which is close to 2. This result suggests that the error terms are not significantly autocorrelated, thereby meeting the assumption of uncorrelated error terms. Based on the test results, the assumptions for multiple linear regression have been adequately met.

To understand the differences in purchase behaviour across genders, further separate regression analyses are conducted for male and female participants, allowing for

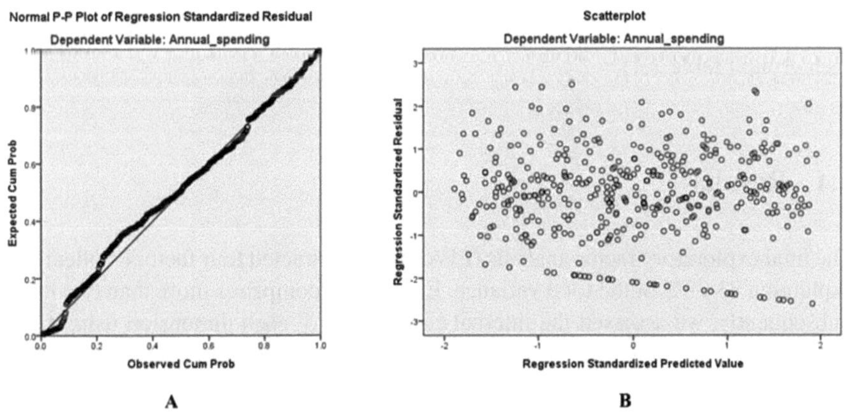

Fig. 2.2 Multiple linear regression assumption testing: **a** P-P plot, **b** Scatterplot

Table 2.4 EFA result: total variance explained

Factor	Initial eigenvalues			Extraction sums of squared loadings			Rotation sums of squared loadings[a]
	Total	% of Variance	Cumulative %	Total	% of Variance	Cumulative %	Total
1	4.399	25.878	25.878	3.395	19.972	19.972	4.129
2	3.882	22.835	48.713	4.143	24.373	44.344	3.494
3	3.521	20.709	69.422	3.557	20.922	65.266	3.413
4	3.087	18.158	87.580	3.182	18.715	83.981	3.390
5	0.302	1.779	89.359				
6	0.264	1.554	90.913				
7	0.216	1.268	92.181				
8	0.215	1.264	93.445				
9	0.192	1.130	94.575				
10	0.179	1.055	95.630				
11	0.162	0.955	96.585				
12	0.142	0.834	97.419				
13	0.123	0.723	98.142				
14	0.121	0.712	98.854				
15	0.101	0.593	99.446				
16	0.048	0.280	99.726				
17	0.047	0.274	100.000				

Extraction method: maximum likelihood
[a] When factors are correlated, sums of squared loadings cannot be added to obtain a total variance

a better understanding of how the identified factors influenced SC spending within each group (Table 2.4).

2.4 Results

The final exploratory factor analysis (EFA) model extracted four factors, collectively explaining 83.98% of the total variance. Each factor comprises more than two items. Subsequently, we assessed the internal consistency of each dimension using Cronbach's alpha, yielding values between 0.856 and 0.957 (Table 2.5). These results indicate reliability and internal consistency within each factor.

Following factor extraction, we assigned names to the dimensions based on the selected items. The first factor, encompassing five items, focuses on the reliability

Table 2.5 EFA result: pattern matrix

Pattern matrix[a]				
Indicator no	Factor			
	1	2	3	4
	α = 0.856	α = 0.956	α = 0.925	0.957
1	0.996			
4	0.927			
5	0.886			
3	0.862			
2	0.849			
10		0.954		
12		0.930		
13		0.928		
11		0.877		
9			0.933	
8			0.928	
6			0.924	
7			0.894	
15				0.999
18				0.933
17				0.863
16				0.862

Extraction method: maximum likelihood
Rotation method: promax with Kaiser normalization[a]
[a] Rotation converged in 4 iterations

of online channels, including aspects such as product quality, financial transactions, timely deliveries, and customer service. This factor is named ""Trustworthiness in Online Shopping" (TOS). The second factor, labelled "Attitude Towards Technology," (ATT) comprises four items gauging an individual's perspective on the positive impact of technology, social media, and the Internet on everyday life. The third factor, denoted as "Digital Proficiency" (DP), consists of four items related to customers' confidence in internet usage, encompassing aspects such as online payment, adapting to new online platforms, and making online purchases. The fourth factor, named "Attitude Towards Social Media" (ATSM), includes four items exploring attitudes towards social media's utility in communication, staying connected with others, contributing to overall well-being and happiness, and serving as a reliable source of news and updates.

The regression analysis yielded an R-square value of 0.481, indicating that 48.1% of the variance in annual SC spending could be explained by the model. Notably, TOS, ATT, DP, and ATSM emerged as significant influencers, all positively impacting

SC-based purchases. The regression weightage further highlighted that ATSM had the highest impact on SC purchase behaviour, followed by ATT (Table 2.6).

Upon gender-specific analysis, for the male population, the R-square value was 0.300 (Table 2.7). In the male dataset, similar to the overall results, TOS, ATT, DP, and ATSM were found to be important factors.

In contrast, for the female responders, the R-square value notably increased to 0.977, indicating that 97.7% of the variance in SC spending among females was accounted for by the model (Table 2.8). Interestingly, in the female dataset, the

Table 2.6 Multiple linear regression (entire dataset): coefficients

Model		Unstandardized coefficients		Standardized coefficients	*t*	Sig.
		B	Std. Error	Beta		
1	(Constant)	6548.863	560.352		11.687	0.000
	Age	– 27.022	20.596	– 0.047	– 1.312	0.190
	Monthly household income	– 0.001	0.002	– 0.018	– 0.512	0.609
	Trustworthiness_in_Online_Shopping	820.205	103.069	0.288	7.958	0.000
	Attitude_Towards_Technology	1060.981	105.441	0.367	10.062	0.000
	Digital_Proficiency	761.001	104.904	0.262	7.254	0.000
	Attitude_Towards_social_media	1359.601	102.924	0.478	13.210	0.000

[a] Dependent variable: Annual_spending

Table 2.7 Multiple linear regression (male population): coefficients

Model		Unstandardized coefficients		Standardized coefficients	*t*	Sig.
		B	Std. Error	Beta		
1	(Constant)	6097.626	1052.327		5.794	0.000
	Age	– 49.503	37.620	– 0.084	– 1.316	0.190
	Monthly household income	– 0.002	0.003	– 0.036	– 0.570	0.569
	Trustworthiness_in_Online_Shopping	661.567	192.565	0.222	3.436	0.001
	Attitude_Towards_Technology	916.794	201.430	0.295	4.551	0.000
	Digital_Proficiency	529.669	186.871	0.181	2.834	0.005
	Attitude_Towards_social_media	1150.460	185.530	0.395	6.201	0.000

[a] Dependent variable: Annual_spending

Table 2.8 Multiple linear regression (female population): coefficients

Model		Unstandardized coefficients		Standardized coefficients	t	Sig.
		B	Std. Error	Beta		
1	(Constant)	7001.777	198.301		35.309	0.000
	Age	− 15.071	7.428	− 0.030	− 2.029	0.044
	Monthly household income	0.002	0.001	0.042	2.862	0.005
	Trustworthiness_in_Online_Shopping	1096.482	36.979	0.433	29.652	0.000
	Attitude_Towards_Technology	1350.054	37.172	0.536	36.319	0.000
	Digital_Proficiency	945.305	38.638	0.357	24.466	0.000
	Attitude_Towards_social_media	1654.625	37.674	0.647	43.920	0.000

[a] Dependent variable: Annual_spending
[b] Selecting only cases for which Gender = 0.00

regression results revealed that in addition to TOS, ATT, DP, and ATSM, age and level of household income also emerged as significant factors influencing social media-based purchases.

2.5 Discussion

The results of this study provide significant information on the aspects that profoundly impact the purchasing decisions made by young people in India and explain the dynamics of their shopping behaviour in SC. The exploration of four major factors—TOS, ATT, DP, and ATSM—has provided a nuanced understanding of the underlying forces shaping the landscape of SC in the Indian context.

The results reveal that ATT has the highest percentage share of the total variance (24.373%). This underscores the impact of technology, social media, and the Internet in shaping their purchasing decisions. As SC operates at the intersection of technology and social interactions, a positive attitude towards these elements enhances users' willingness to participate actively in the platform. Result indicated that DP explains 20.922% of total variance. Followed by, TOS and ATSM explain 19.972 and 18.715% of total variance, respectively.

DP, encompassing aspects of confidence in internet usage and adaptability to online platforms, reflects the importance of users' comfort and familiarity with the digital landscape. On the other hand, TOS emphasizes the critical role of reliability in online channels, encompassing factors such as product quality, level of security in financial transactions, timely deliveries, and customer service. These findings align with the broader literature on e-commerce and SC, emphasizing the significance of trust and digital literacy in online consumer behaviour. The fourth factor, ATSM, suggests that while social media plays a crucial role in the SC ecosystem, other factors such as trustworthiness and DP exert a more pronounced impact on the purchasing decisions of the young Indian demographic.

Further, by utilizing the factor scores derived from factor analysis, a multiple linear regression is employed. The regression model shows the relationship between the identified factors and the SC spending pattern of the customers (Table 2.6). The r square value indicates that the model explains the purchase behaviour of females better than that of males. The descriptive statistics also show that SC purchase participation is more by females than males (Table 2.2). The data highlights the variables that affect the spending patterns of males and females, demonstrating the notable disparities in SC purchase behaviour across genders. Notably, the regression results indicate a noteworthy distinction in the impact of household income (HH income) on purchase behaviour between males and females. Specifically, the findings reveal that HH income emerges as a significant influencing factor for females, exerting a discernible effect on their SC purchase behaviour. In contrast, for males, the same variable is found to be not statistically significant, suggesting that other factors may play a more prominent role in shaping the spending patterns of male consumers in the realm of SC. This finding emphasizes the importance of gender-specific considerations in comprehending and predicting purchase behaviour in SC.

Young customers share a significant portion of the Indian market. These insights are invaluable for retail managers and business owners aiming to enhance their operations. Recognizing the significant impact of SC, the Department for Promotion of Industry and Internal Trade, under the Government of India, views SC as a key driver for creating employment opportunities. These findings could support the government in developing strategies to bolster the SC sector, thereby spurring economic development and generating more jobs.

Moreover, the understanding gleaned from this research isn't limited to SC alone but is also applicable to various other shopping channels. Future research could leverage the same indicators and factors to delve into purchasing behaviours across different retail settings. This wider relevance adds to the study's significance, offering a holistic approach for analysing consumer behaviour in diverse shopping contexts, both digital and traditional.

Businesses, policy makers, and other stakeholders can develop initiatives to improve the SC experience for Indian youth by having a better understanding of these characteristics and how they affect the purchase of SCs. To encourage increased involvement in SC platforms, e.g. interventions that focus on positive attitudes towards technology, enhancing DP, and supporting trustworthiness in online transactions can be essential. Moreover, recognizing the evolving landscape of SC in the Indian context provides a strategic roadmap for businesses to tailor their marketing strategies and user engagement initiatives to align with the preferences and concerns of the customer group.

2.6 Conclusion

In conclusion, our study revels into the intricate dynamics of SC among young Indians, shedding light on critical factors that shape purchasing decisions and analysing shopping behaviour within this evolving domain. The exploratory factor analysis extracted four major factors: TOS, attitude towards technology, DP, and attitude towards social media. Subsequent multiple linear regression analysis explains the substantial impact of these factors on annual spending within SC platforms.

The implications of our findings extend beyond academia, offering valuable insights for retail managers and business owners. Notably, our research underscores the importance of TOS. Results show that product quality, financial transactions, timely deliveries, and customer service improvements can enhance TOS. By prioritizing these elements, retailers can develop a more favourable environment, fostering trust and loyalty among consumers. This finding is in good agreement with many previous studies [25, 43].

Multiple linear regression results also show that ATSM plays a vital role in shaping SC-based purchase behaviour. This finding is also consistent with the results reported in various earlier researches [11, 33, 46]. This indicates the significance of carefully choosing influencers based on the demographic characteristics of their followers, ensuring alignment with the preferences and characteristics of the targeted customer groups. This targeted approach can maximize the impact of influencer marketing efforts and contribute to the overall success of a SC-based business.

In the Indian context, the Department for Promotion of Industry and Internal Trade, GOI, aims to promote SC as a means of employment generation [34]. Our research findings can offer valuable insights that can assist the government in supporting, promoting, and guiding business owners in SC.

However, it is important to acknowledge certain limitations in our research. Firstly, using social media platforms for data collection may introduce a sampling bias, as individuals not actively engaged in these platforms may be underrepresented. Additionally, the study focused solely on the age group between 18 and 35, limiting the generalizability of the findings to broader age demographics. Future research should include a more diverse sample, spanning different age groups, to obtain a comprehensive understanding of SC behaviour across various demographic segments.

Looking ahead, there are promising avenues for future research. Exploring the impact of cultural influences on SC behaviour could provide valuable insights, as could an in-depth examination of the role of trust and security in online transactions. Furthermore, investigating evolving trends in SC and adapting to changes in technology and consumer preferences will be crucial for staying abreast of this dynamic field. This research sets the stage for a more comprehensive exploration of SC dynamics, offering valuable insights for academics and practitioners in the ever-evolving digital commerce landscape.

Appendix: Test for Linearity

ANOVA Table

			Sum of Squares	df	Mean Square	F	Sig
Annual_ spending * Attitude_ Towards_ social_ media	Between Groups	(Combined)	3,225,403,923.342	399	8,083,719.106	1.122	0.487
		Linearity	642,375,578.258	1	642,375,578.258	89.148	0.000
		Deviation from Linearity	2,583,028,345.084	398	6,490,020.968	0.901	0.644
	Within Groups		50,440,000.000	7	7,205,714.286		
	Total		3,275,843,923.342	406			

References

1. Akram, U., Junaid, M., Zafar, A.U., et al.: Online purchase intention in Chinese social commerce platforms: being emotional or rational? J. Retail. Consum. Serv. **63**, 102669 (2021). https://doi.org/10.1016/J.JRETCONSER.2021.102669
2. Attar, R.W., Almusharraf, A., Alfawaz, A., Hajli, N.: New trends in E-commerce research: linking social commerce and sharing commerce: a systematic literature review. Sustain **14**, 16024 (2022). https://doi.org/10.3390/SU142316024/S1
3. Barann, B., Betzing, J.H.E., et al.: Exploring customers' likeliness to use e-service touchpoints in brick and mortar retail. Electron. Mark. **32**, 523–545 (2020). https://doi.org/10.1007/S12525-020-00445-0
4. Busalim, A.H., Hussin, A.R.C.: Understanding social commerce: a systematic literature review and directions for further research. Int. J. Inf. Manage. **36**, 1075–1088 (2016). https://doi.org/10.1016/J.IJINFOMGT.2016.06.005
5. Census of India.: Cities having population 1 lakh and above, Census 2011 (2011)
6. Chen, A., Lu, Y., Wang, B.: Customers' purchase decision-making process in social commerce: a social learning perspective. Int. J. Inf. Manage. **37**, 627–638 (2017). https://doi.org/10.1016/J.IJINFOMGT.2017.05.001
7. Chen, Y., Chen, L., Smith, R.: Linking passion to performance in the social commerce community: the role of collaborative information exchange. J. Bus. Ventur. Insights **18**, e00351 (2022). https://doi.org/10.1016/J.JBVI.2022.E00351
8. Dolata, U.: Technological innovations and sectoral change: Transformative capacity, adaptability, patterns of change: an analytical framework. Res. Policy **38**, 1066–1076 (2009). https://doi.org/10.1016/J.RESPOL.2009.03.006
9. Gatautis, R., Medziausiene, A.: Factors affecting social commerce acceptance in Lithuania. Procedia Soc. Behav. Sci. **110**, 1235–1242 (2014). https://doi.org/10.1016/J.SBSPRO.2013.12.970
10. Hajli, M.: A research framework for social commerce adoption. Inf. Manag. Comput. Secur. **21**, 144–154 (2013). https://doi.org/10.1108/IMCS-04-2012-0024/FULL/PDF
11. Hajli, M.N.: The role of social support on relationship quality and social commerce. Technol. Forecast Soc. Change **87**, 17–27 (2014). https://doi.org/10.1016/J.TECHFORE.2014.05.012
12. Handarkho, Y.D.: Impact of social experience on customer purchase decision in the social commerce context. J. Syst. Inf. Technol. **22**, 47–71 (2020). https://doi.org/10.1108/JSIT-05-2019-0088/FULL/PDF

13. Hanna, R.C., Lemon, K.N., Smith, G.E.: Is transparency a good thing? How online price transparency and variability can benefit firms and influence consumer decision making. Bus. Horiz. **62**, 227–236 (2019). https://doi.org/10.1016/J.BUSHOR.2018.11.006
14. Hossain, S.F.A., Xi, Z., Nurunnabi, M., Hussain, K.: Ubiquitous role of social networking in driving M-commerce: evaluating the use of mobile phones for online shopping and payment in the context of trust **10** (2020). https://doi.org/10.1177/2158244020939536
15. IBEF.: Indian Retail Industry Analysis Presentation (2021)
16. KPMG.: Indian Retail—The Next Growth Story (2014)
17. Lăzăroiu, G., Neguriță, O., Grecu, I., et al.: Consumers' decision-making process on social commerce platforms: online trust, perceived risk, and purchase intentions. Front. Psychol. **11**, 478039 (2020). https://doi.org/10.3389/FPSYG.2020.00890/BIBTEX
18. Lee, M.K.O., Turban, E.: A trust model for consumer internet shopping. **6**, 75–91 (2014). https://doi.org/10.1080/10864415.2001.11044227
19. Li, C.Y., Ku, Y.C.: The power of a thumbs-up: will e-commerce switch to social commerce? Inf. Manag. **55**, 340–357 (2018). https://doi.org/10.1016/J.IM.2017.09.001
20. Liang, T.-P., Lai, H.-J.: Effect of store design on consumer purchases: van empirical study of on-line bookstores
21. Liao, S.H., Widowati, R., Cheng, C.J.: Investigating Taiwan Instagram users' behaviors for social media and social commerce development. Entertain. Comput. **40**, 100461 (2022). https://doi.org/10.1016/J.ENTCOM.2021.100461
22. Lin, X., Li, Y., Wang, X.: Social commerce research: definition, research themes and the trends. Int. J. Inf. Manage. **37**, 190–201 (2017). https://doi.org/10.1016/J.IJINFOMGT.2016.06.006
23. Lin, X., Wang, X.: Towards a model of social commerce: improving the effectiveness of e-commerce through leveraging social media tools based on consumers' dual roles. Eur. J. Inf. Syst. **32**, 782–799 (2023). https://doi.org/10.1080/0960085X.2022.2057363
24. Lin, X., Wang, X., Hajli, N.: Building E-commerce satisfaction and boosting sales: the role of social commerce trust and its antecedents. Int. J. Electron. Commer. **23**, 328–363 (2019). https://doi.org/10.1080/10864415.2019.1619907
25. Lu, B., Fan, W., Zhou, M.: Social presence, trust, and social commerce purchase intention: an empirical research. Comput. Hum. Behav. **56**, 225–237 (2016). https://doi.org/10.1016/J.CHB.2015.11.057
26. Ma, X., Jiang, H., Liu, S., Zhang, S.: Development and validation of the intellectual property services scale in China. Heliyon **9**, e19892 (2023). https://doi.org/10.1016/j.heliyon.2023.e19892
27. Maia, C., Lunardi, G., Longaray, A., Munhoz, P.: Factors and characteristics that influence consumers' participation in social commerce. Rev. Gest. **25**, 194–211 (2018). https://doi.org/10.1108/REGE-03-2018-031/FULL/PDF
28. Mikalef, P., Sharma, K., Chatterjee, S., et al.: All eyes on me: predicting consumer intentions on social commerce platforms using eye-tracking data and ensemble learning. Decis. Support. Syst. **114039** (2023). https://doi.org/10.1016/J.DSS.2023.114039
29. Molinillo, S., Aguilar-Illescas, R., Anaya-Sánchez, R., Liébana-Cabanillas, F.: Social commerce website design, perceived value and loyalty behavior intentions: The moderating roles of gender, age and frequency of use. J. Retail. Consum. Serv. **63**, 102404 (2021). https://doi.org/10.1016/J.JRETCONSER.2020.102404
30. Osatuyi, B., Qin, H., Osatuyi, T., Turel, O.: When it comes to Satisfaction … It depends: an empirical examination of social commerce users. Comput. Hum. Behav. **111**, 106413 (2020). https://doi.org/10.1016/J.CHB.2020.106413
31. Qu, Y., Cieślik, A., Fang, S., Qing, Y.: The role of online interaction in user stickiness of social commerce: the shopping value perspective. Digit. Bus. **3**, 100061 (2023). https://doi.org/10.1016/J.DIGBUS.2023.100061
32. Rashid, R.M., Rashid, Q.ul.A., Pitafi, A.H.: Examining the role of social factors and mooring effects as moderators on consumers' shopping intentions in social commerce environments. SAGE Open **10** (2020). https://doi.org/10.1177/2158244020952073

33. Riaz, M.U., Guang, L.X., Zafar, M., et al.: Consumers' purchase intention and decision-making process through social networking sites: a social commerce construct. Behav. Inf. Technol. **40**, 99–115 (2021). https://doi.org/10.1080/0144929X.2020.1846790
34. Sarulatha, N., Sasirekha, V.: A study on the role of social commerce constructs for shopping fashion products. Int. J. Civ. Eng. Technol. **9**, 508–518 (2018)
35. Shakir Goraya, M.A., Zhu, J., Akram, M.S., et al.: The impact of channel integration on consumers' channel preferences: do showrooming and webrooming behaviors matter? J. Retail. Consum. Serv. **65**, 102130 (2022). https://doi.org/10.1016/J.JRETCONSER.2020.102130
36. Shen, X.L., Li, Y.J., Sun, Y., et al.: Understanding the role of technology attractiveness in promoting social commerce engagement: moderating effect of personal interest. Inf. Manag. **56**, 294–305 (2019). https://doi.org/10.1016/J.IM.2018.09.006
37. Shin, D.H.: User experience in social commerce: in friends we trust. Behav. Inf. Technol. **32**, 52–67 (2013). https://doi.org/10.1080/0144929X.2012.692167
38. Shirazi, F., Hajli, N., Sims, J., Lemke, F.: The role of social factors in purchase journey in the social commerce era. Technol. Forecast. Soc. Change **183**, 121861 (2022). https://doi.org/10.1016/J.TECHFORE.2022.121861
39. Tandon, A., Gupta, A., Tripathi, V.: Managing shopping experience through mall attractiveness dimensions: an experience of Indian metro cities. Asia Pacific J. Mark. Logist. **28**, 634–649 (2016). https://doi.org/10.1108/APJML-08-2015-0127/FULL/PDF
40. Um, N.H.: Antecedents and consequences of consumers' attitude toward social commerce sites. J. Promot. Manag. **25**, 500–519 (2019). https://doi.org/10.1080/10496491.2018.1448324
41. Vazquez, E.E., Patel, C., Alvidrez, S., Siliceo, L.: Images, reviews, and purchase intention on social commerce: the role of mental imagery vividness, cognitive and affective social presence. J. Retail. Consum. Serv. **74**, 103415 (2023). https://doi.org/10.1016/J.JRETCONSER.2023.103415
42. Williams, M.D.: Social commerce and the mobile platform: payment and security perceptions of potential users. Comput. Hum. Behav. **115**, 105557 (2021). https://doi.org/10.1016/J.CHB.2018.06.005
43. Wu, T., Zhang, R., Liu, X., et al.: A social commerce purchasing decision model with trust network and item review information. Knowl.-Based Syst. **235**, 107628 (2022). https://doi.org/10.1016/J.KNOSYS.2021.107628
44. Yeon, J., Park, I., Lee, D.: What creates trust and who gets loyalty in social commerce? J. Retail. Consum. Serv. **50**, 138–144 (2019). https://doi.org/10.1016/J.JRETCONSER.2019.05.009
45. Zhang, H., Lu, Y., Gupta, S., Zhao, L.: What motivates customers to participate in social commerce? The impact of technological environments and virtual customer experiences. Inf. Manag. **51**, 1017–1030 (2014). https://doi.org/10.1016/J.IM.2014.07.005
46. Zhang, K.Z.K., Benyoucef, M.: Consumer behavior in social commerce: a literature review. Decis. Support. Syst. **86**, 95–108 (2016). https://doi.org/10.1016/J.DSS.2016.04.001
47. Zhou, L., Zhang, P., Zimmermann, H.D.: Social commerce research: an integrated view. Electron. Commer. Res. Appl. **12**, 61–68 (2013). https://doi.org/10.1016/J.ELERAP.2013.02.003

Chapter 3
A Take-Back Mechanism for End-of-Life Tires Under Extended Producer Responsibility

Arup Ratan Paramanik and Biswajit Mahanty

Abstract This study proposes a take-back mechanism for collecting end-of-life tires (ELTs) from the consumers by considering their willingness to return. This mechanism involves the local tire repairing shop owners as take-back agents and their shops as local collection centers. A "multi-period mixed integer nonlinear programming (MP-MINLP)" model is also developed to maximize the total profit of the take-back agent by modeling its operations. The BONMIN solver in GAMS is used to run the developed MP-MINLP model. By using a case study of Indian scenario, it is found that the proposed mechanism has the capability to provide good profits to the take-back agents. This will boost the ELT collection from the consumers and thus, EPR implementation in tire industry.

Keywords End-of-life tires (ELTs) · Extended Producer Responsibility · Take-back · Nonlinear programming · Full factorial design

3.1 Introduction

In many countries, producers become responsible for ensuring environmentally sound management of their end-of-life products due to the implementation of Extended Producer Responsibility (EPR) [3, 5, 16, 17, 22, 23, 26, 27]. Recently, the Indian government has introduced EPR in tire industry and thus, the Indian tire producers become formally responsible for take-back and environmentally sound management of "end-of-life tires (ELTs)". Currently, landfilling or stockpiling of ELTs is one of the major ELT management practice in India, which leads to trapping

A. R. Paramanik (✉) · B. Mahanty
Department of Industrial and Systems Engineering, Indian Institute of Technology Kharagpur, Kharagpur, India
e-mail: aeroparamanik@gmail.com; aeroparamanik@kgpian.iitkgp.ac.in

B. Mahanty
e-mail: bm@hijli.iitkgp.ac.in

J. Maiti et al. (eds.), *Recent Advances in Industrial and Systems Engineering*, Lecture Notes on Multidisciplinary Industrial Engineering,
https://doi.org/10.1007/978-981-96-8323-9_3

valuable resources unutilized [18]. It should be noted that there are other important issues with the landfilling of ELTs, such as the stockpiled ELTs can act as a breeding ground for rats, mosquitoes, and several other disease vectors by retaining water after rain [9, 12]. In addition, they are at risk of catching fire and emitting hazardous gases, which is a blight on the public health and environment [2, 10, 12]. Moreover, since ELT wastes are non-biodegradable, they cause serious water, air, and soil contamination. Consequently, it is important to collect the ELTs from the consumers and recycle them to recover the secondary materials or resources.

Implementation of EPR in Indian tire industry may increase the take-back and recycling rates of ELTs. However, the take-back of ELTs from the consumers is not straightforward. Although there are some take-back schemes available in the literature [6, 11, 14, 15], the performance of these schemes has been far from satisfactory due to the lack of willingness to return of the consumers [7, 13, 19]. Since tires are essential for mobility in logistics and the prices of new tires, especially of truck tires, are very high, there is a tendency among the consumers not to change their tires even after the actual service life of the tires. This inappropriate practice reduces the safety of vehicles and increases the number of road accidents, since tires "provide grip between the road and vehicle for acceleration and braking, bear the weight of the vehicle, and act as vibration absorbers" [2]. Thus, use of ELTs in vehicle is risky and may reduce the overall efficiency of Indian logistics. Therefore, the main objective of our study is to propose a take-back strategy by considering the willingness to return of the consumers. In this study, we have suggested to use the local tire repairing shop owners as take-back agents. Additionally, we have modeled the operations of such take-back agents by developing a "multi-period mixed integer nonlinear programming (MP-MINLP)" model. This model maximizes the profit of the take-back agent over a planning horizon. The remainder of the current paper is organized as follows. Section 3.2 presents the problem description and formulation of the MP-MINLP model. Section 3.3 presents the solution approach for the MP-MINLP model. Section 3.4 presents the results and discussions through a case study. Lastly, Sect. 5 provides the conclusions.

3.2 Materials and Method

3.2.1 Problem Description

We propose to use the owners of tire repairing shops in a customer zone as take-back agents. They will collect the ELTs from the consumers by offering a take-back price and informing them about the negative impacts of ELTs on the environment and social health. Therefore, in this study, we consider the repairing shops as local collection centers (LCCs) and the operations at these centers are shown in Fig. 3.1.

It can be observed from Fig. 3.1 that whenever a consumer of tire (e.g., a truck owner) comes to the LCC, the take-back agent should check the condition of the tire.

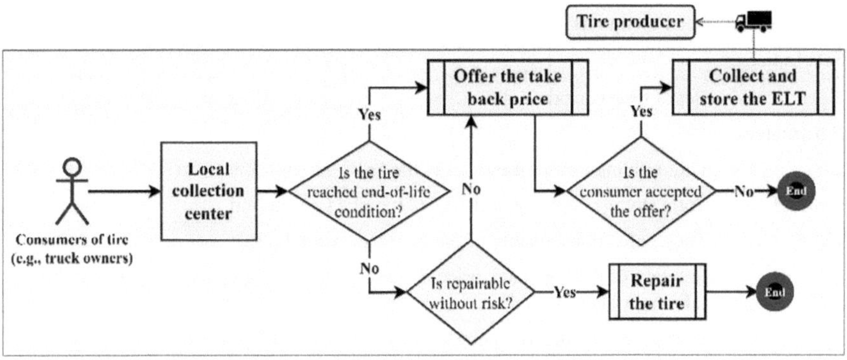

Fig. 3.1 Flowchart of the operations at an LCC for taking back the ELTs from the consumers

If the tire has reached its end-of-life condition, then the agent will offer the take-back price and collect the ELT based on the consumer's willingness. In contrast, if the tire has not reached its end-of-life condition but it is not repairable without risk, then also the agent will offer the take-back price to collect the damaged tire by considering it as an ELT; otherwise, the tire will be repaired. Note that for each collected ELT, these take-back agents will get a percentage of the offered take-back price as a commission or take-back reward. To model this take-back mechanism, a MP-MINLP model is developed in the next sub-section and the notations used are given in Table 3.1.

3.2.2 Problem Formulation

In this sub-section, the MP-MINLP model is developed. The corresponding objective function and the constraints are presented in the next two sub-sections, respectively.

Objective Function

The MP-MINLP model maximizes the profit of the take-back agent at the LCC. That is, the objective function maximizes the difference between the total revenue and total cost over a time horizon, as given in Eq. (3.1).

$$\text{Maximize } \Pi_{\text{TBA}} = R - TC \tag{3.1}$$

Here, the revenue, R comes from returning the ELTs to the producer, as shown in Eq. (3.2).

$$R = \sum_{t \in T} ((1 + f_R) \cdot PR_t \cdot Q_t) \tag{3.2}$$

On the other hand, the total cost, TC of the agent is the sum of the collection cost and inventory cost, as shown in Eq. (3.3).

Table 3.1 Notations used to develop the MP-MINLP model

Sets/Indices	
T	Set of time periods, $t \in T$
Parameters	
CH	Unit holding cost of the ELTs at the LCC per unit time
$CapELT$	Total storage capacity of the LCC for ELTs per period
$CapC$	Total collection capacity of the producer from the LCC per period
f_R	Fraction of the take-back price as a commission or take-back reward for the take-back agent
P_t^{ELT}	Probability that the tire entering the LCC is an ELT at period t
P_t^{NR}	Probability that the tire entering the LCC (which is not an ELT) is not repairable at period t
N_t	Quantity of tires entering the LCC at period t
a_t	Minimum price expected by the consumers for returning an ELT at period t
b_t	Maximum price expected by the consumers for returning an ELT at period t
$InvELT_0$	Initial inventory at the LCC
Integer variables	
Q_t	Quantity of ELTs collected by the producer from the LCC at period t
$InvELT_t$	Residual inventory of ELTs in the LCC at end of period t
Continuous variables	
PR_t	Offered unit take-back price for collecting the ELTs from the consumers at period t
RE_t	Probability of return of ELTs by the consumers by accepting the take-back price at period t

$$TC = \sum_{t \in T} \left(PR_t \cdot RE_t \cdot \left(P_t^{ELT} + P_t^{NR} \cdot \left(1 - P_t^{ELT}\right)\right) \cdot N_t \right) + \sum_{t \in T} (CH \cdot InvELT_t) \tag{3.3}$$

Constraints

Now, the objective function (3.1) is subjected to four different types of constraints, which are (i) capacity constraints, (ii) balance constraint, (iii) take-back constraints, and (iv) non-negativity constraints. These constraints are described below.

Capacity constraints

The inventory of ELTs is limited to the storage capacity at the LCC, whereas the returning quantity of ELTs is limited to the collection capacity of the producer in each period, as expressed in constraints (3.4) and (3.5), respectively.

$$InvELT_t \leq CapELT (\forall t \in T) \tag{3.4}$$

$$Q_t \leq CapC (\forall t \in T) \tag{3.5}$$

Balance constraint

The balance constraint determines the inventory in each period, which is equal to the inventory in the previous period and the difference between the inflow and the outflow in the present period, as expressed in constraints (3.6).

$$InvELT_t = InvELT_{t-1} + \left(RE_t \cdot \left(P_t^{ELT} + P_t^{NR} \cdot \left(1 - P_t^{ELT}\right)\right) \cdot N_t\right) - Q_t(\forall t \in T) \tag{3.6}$$

Note that $InvELT_{t-1}$ is equal to the $InvELT_0$ in the first time period, i.e., $t = 1$.

Take-back constraints

The take-back constraints determine the fraction of returns based on the offered take-back price by considering the willingness to return of the consumers, as expressed in constraints (3.7) and (3.8).

$$RE_t \leq (PR_t - a_t)/(b_t - a_t)(\forall t \in T) \tag{3.7}$$

$$a_t \leq PR_t \leq b_t(\forall t \in T) \tag{3.8}$$

To take-back the ELTs from the consumers, the take-back price, PR_t to be offered should be determined. However, based on PR_t, it is important to consider the willingness to return of the consumers. Since there is variation in consumer preferences on returning their ELTs, it is assumed that the expected price for each consumer may be modeled by a uniform distribution within the range $[a_t, b_t]$ [8, 19, 21], as expressed in constraint (3.7). Constraint (3.8) ensures that the PR_t is within the minimum and maximum prices expected by the consumers for an ELT, i.e., a_t and b_t, respectively.

Non-negativity constraints

All the decision variables are non-negative in each period, as expressed in constraint (3.9).

$$Q_t, InvELT_t, PR_t, RE_t \geq 0(\forall t \in T) \tag{3.9}$$

Moreover, the decision variables, Q_t and $InvELT_t$, are integer.

3.2.3 Solver and Software Used

In this study, we have used the "basic open-source MINLP (BONMIN) solver embedded in the general algebraic modeling systems (GAMS) software package" [1] to solve the MP-MINLP model. In the BONMIN solver,[1] different algorithms can be employed, such as "NLP-based branch-and-bound algorithm (B-BB),"

[1] https://github.com/coin-or/bonmin.

"outer-approximation decomposition algorithm (B-OA)," "Quesada and Grossmann's branch-and-cut algorithm (B-QG)," and "hybrid outer-approximation based branch-and-cut algorithm (B-Hyb)" for solving large-scale MINLP problems [4]. It is to be noted that we have used the B-BB algorithm in the BONMIN solver to obtain the solution of the proposed MP-MINLP model.

3.3 Case Study

In this section, we discuss a case study of the Indian scenario for demonstrating the effectiveness of the proposed take-back mechanism. We collect the required data for this case study based on a tire repairing shop located at the south western part of West Bengal state of India. In this study, we assume this shop as an LCC to run our simulation experiments. Based on the collected data, the values of N_t, P_t^{ELT}, P_t^{NR}, CH, $CapELT$, $CapC$, and $InvELT_0$ are set as $N(121, 34)$, $U(0.05, 0.2)$, $U(0.0005, 0.0015)$, 3.6 ₹/Week, 350, 300, and 0, respectively. The values of a_t and b_t are ₹800 and ₹2000, respectively, for each time period t. We have considered that the producer gives 5% of the take-back price to the agent as reward for collecting the ELTs from the consumers, i.e., $f_R = 0.05$. Based on these data, the total profits of the take-back agent are evaluated over four different time horizons of 26, 52, 78 and 104 weeks by using the MP-MINLP model, as plotted in Fig. 3.2.

Figure 3.2 displays that the owner of the tire repairing shop can get a profit of ₹1.29 lakh in 26 weeks (i.e., 6 months) by collecting the ELTs from the consumers as a take-back agent of the producer. Moreover, the total profit shows a positive increase with an increase in the time horizon. If we increase the time horizon to 52 weeks (i.e., 1 year) from 26 weeks, then the profit of the take-back agent becomes ₹2.54 lakh

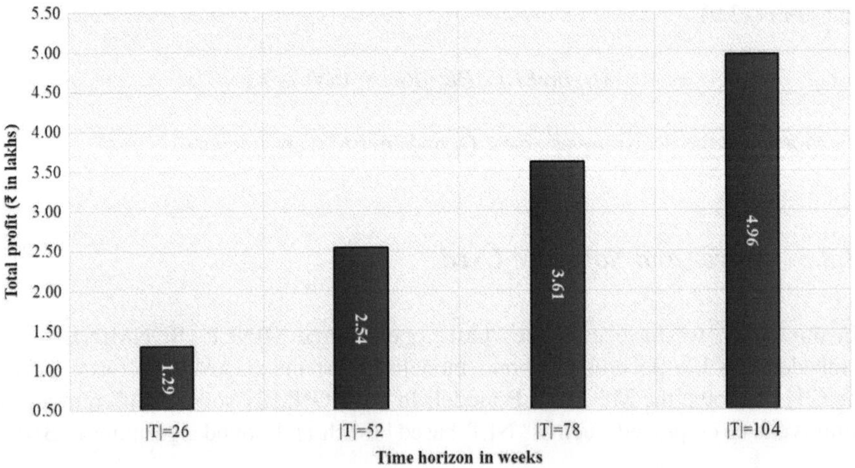

Fig. 3.2 Total profit of the take-back agent with 5% commission of the take-back price (i.e., $f_R = 0.05$) over four different time horizons

with an increase of 96.90%. Further, if we increase the time horizon to 78 weeks from 52 weeks, then the profit of the take-back agent becomes ₹3.61 lakh with an increase of 42.13%. Again, if we increase the time horizon to 104 weeks (i.e., 2 years) from 78 weeks, then the profit of the take-back agent becomes ₹4.96 lakh with an increase of 37.40%. Such positive trend infers that the proposed take-back mechanism has the potential to be profitable for the take-back agent over the time periods. It indicates toward economic sustainability.

3.3.1 Sensitivity Analysis

Since the expected profits are dependent on the commission percentage (which the agent get from the producer, i.e., f_R), we have also evaluated the profits over a time horizon of 52 weeks (i.e., 1 year) by considering different commission percentages from 1% (i.e., $f_R = 0.01$) to 10% (i.e., $f_R = 0.10$) as sensitivity analysis. The profits for different commission percentages are plotted in Fig. 3.3.

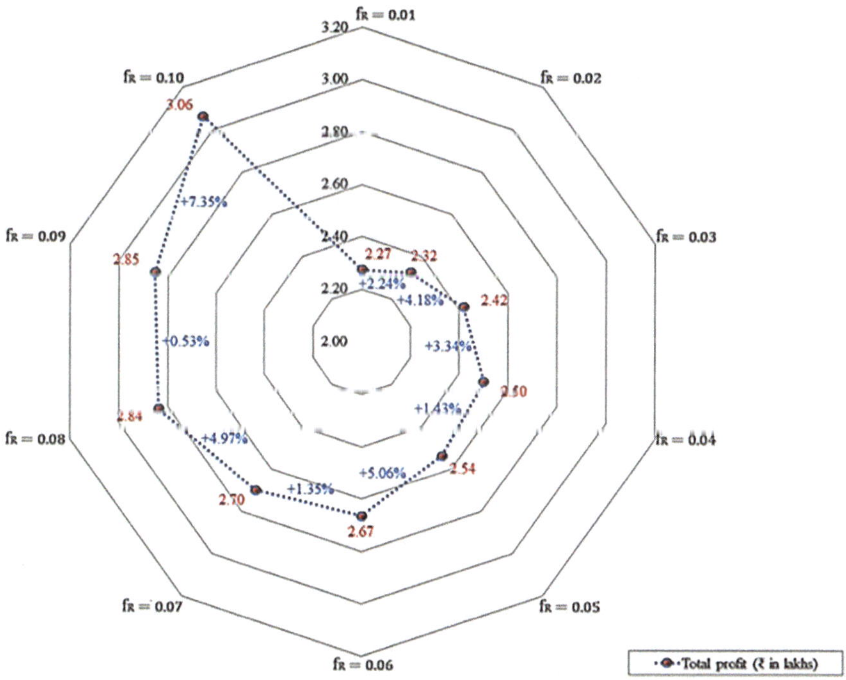

Fig. 3.3 Total profits of the take-back agent over a time horizon of 52 weeks (i.e., 1 year) for different commission percentages (i.e., different values of f_R) [Note: To ensure accuracy, percentage changes are derived from the absolute profit values without scaling, while the reported profit values are presented in lakhs of INR (₹) for improved readability and understanding.]

It can be seen from Fig. 3.3 that the total yearly profit of the agent is ₹2.27 lakh, when the agent gets a commission of 1% of the take-back price. When the agent receives a commission of 2% of the take-back price, the total yearly profit increases by 2.24%. When the commission is increased to 3% from 2%, the profit is increased by 4.18%. If the commission is increased to 4% from 3%, the profit is increased by 3.34%. Similarly, the total profit of the take-back agent increases positively with the increase in commission percentage. Overall, it can be inferred from Fig. 3.3 that, for an increase of 1% in the commission, the total profit of the take-back agent increases by 3.38%, on an average. This figure conveys that the owner of the LCC has a scope of getting a positive annual profit by acting as a take-back agent, even if the commission is 1% of the take-back price. This may attract the local tire repairing shop owners to be a part of the proposed take-back mechanism. It may enhance the collection rate of ELTs for successful implementation of EPR in the Indian tire industry.

3.3.2 Full Factorial Design and Analysis to Generalize the Findings

Since the data used in this study are collected from only a single tire repairing shop located in a particular area, data related to the model parameters, especially, the three different costs associated with the take-back agent (i.e., CH, a_t, and b_t), may significantly vary in different customer zones. Consequently, their potential impacts on the generalizability of the findings of this study should be examined. Therefore, we have conducted full factorial design (FFD) by following the strategy of Suard et al. [25] to show the effects of the three cost parameters on the total profit of the take-back agent. In this FFD analysis, we have discretized the range of each of these three parameters or factors into two levels, low and high. For the parameters, CH, a_t, and b_t, the lower levels are 2 ₹/Week, ₹800, and ₹2000; whereas the higher levels are 10 ₹/Week, ₹1200, and ₹4000, respectively. Thus, our experimental design needs 2^3 or 8 runs, which are feasible to perform. Based on these 8 runs, the total profits of the take-back agent are obtained over a time horizon of 52 weeks by considering $f_R = 0.05$. Since the other model parameters, i.e., N_t, P_t^{ELT}, are P_t^{NR} are determined using normal and uniform distributions, there is a chance of getting different profit values in different experimental runs without changing any parameter values. Hence, we have obtained the total profit of the take-back agent for 10 independent times and measured their minimum and maximum values for each of the 8 runs. These minimum and maximum profit values are considered as the response variables in the FFD analysis. It is to be noted that the FFD analysis is performed using the Minitab 19 software. The main effects and the interaction effects plots obtained by the FFD analysis are shown in Fig. 3.4. For evaluating the statistical significance of the effects, we also perform the "analysis of variance (ANOVA)" tests for both the minimum and maximum profits as response variables, as shown in Fig. 3.5.

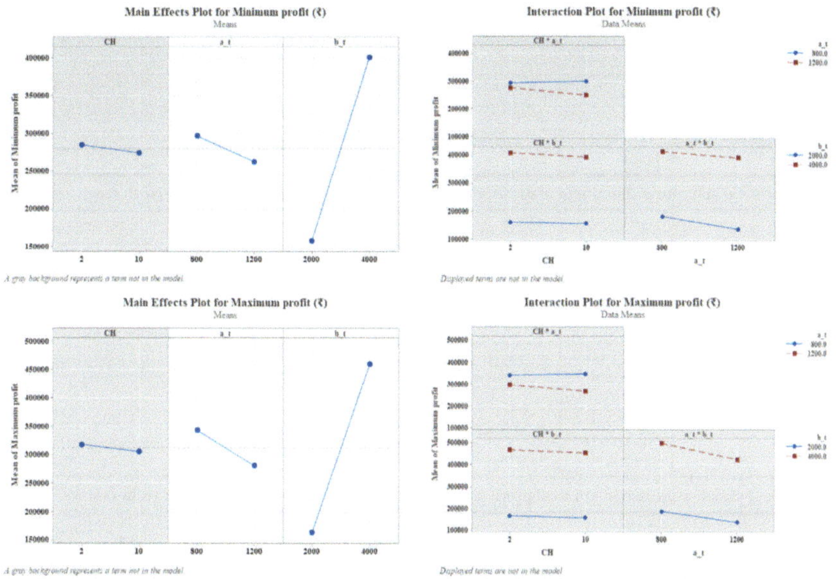

Fig. 3.4 Main effect and interaction effect plots for the minimum and maximum total profits of the take-back agents

From Fig. 3.4, it can be observed that we obtain an almost horizontal line across the reference line (i.e., the overall mean) for the parameter, CH, which indicates that CH does not cause significant variability in the total profit of the take-back agent. That is, each level of this parameter affects the total profit in the same way, and the response mean remains the same across all the factor levels. Since the higher level for this parameter is taken high enough, it is believed that this cost parameter may not be that significantly important in deciding the profitability of the agent, even in different customer zones. On the other hand, the lines are not horizontal for the parameters, a_t, and b_t, which indicates that significant main effects are present there and these two cost parameters affect the total profit of the take-back agent. Figure 3.4 displays that different levels of these two parameters affect the total profit differently. For lower level of the parameter, a_t, we have obtained higher total profit mean than its higher level. Surprisingly, higher level of the cost parameter, b_t, has resulted in higher total profit means than its lower level.

In the interaction plots, since we have obtained parallel lines, interaction effects are not significant. Thus, we do not need to consider the interaction effects while interpreting the main effects. The ANOVA results also support these main and interaction effects (Fig. 3.5). In general, steeper the slope of the line in a main effect plot, the greater the magnitude of the main effect. Therefore, it is evident from Fig. 3.4 that the parameter, b_t has more significant effect on the total profit than the parameter, a_t. To support this statement, further, we have also used the standardized regression coefficients [20, 24, 25] as sensitivity measures with respect to the individual

General Linear Model: Minimum and Maximum profits versus CH, a_t, and b_t					
Factor	*Type*	*Levels*	*Values*		
CH (₹/Week)	Fixed	2	2, 10		
a_t (₹)	Fixed	2	800, 1201		
b_t (₹)	Fixed	2	2000, 4000		
Analysis of variance for minimum total profit, considering backward elimination of terms (α to remove = 0.1)					
Source	*DF*	*Adj SS*	*Adj MS*	*F-Value*	*P-Value*
a_t	1	2268296520	2268296520	6.78	0.048
b_t	1	1.1800E+11	1.1800E+11	351.81	0.000
Error	5	1673498677	334699735		
Total	7	1.2200E+11			
Model Summary:					
S	*R-Sq*	*R-Sq (adj)*	*R-Sq (pred)*		
18294.8	98.62%	98.07%	96.48%		
Analysis of variance for maximum total profit, considering backward elimination of terms (α to remove = 0.1)					
Source	*DF*	*Adj SS*	*Adj MS*	*F-Value*	*P-Value*
a_t	1	7679229813	7679229813	21.28	0.006
b_t	1	1.7400E+11	1.7400E+11	482.96	0.000
Error	5	1804611466	360922293		
Total	7	1.8400E+11			
Model Summary:					
S	*R-Sq*	*R-Sq (adj)*	*R-Sq (pred)*		
18998	99.02%	98.63%	97.49%		

Fig. 3.5 ANOVA results

cost parameters, as plotted in Fig. 3.6. These standardized regression coefficients are non-dimensional and indicate the expected change in the total profit in response to an increase or decrease in the value of an individual parameter/factor by keeping all other input variables unchanged.

Figure 3.6 indicates that the parameter, CH has no significant influence on the total profit. The parameter, a_t, has a negative effect on the total profit and the profit may be decreased by 13.7–20.4% of its standard deviation for an increase of one standard deviation in the parameter, a_t. Similarly, the parameter, b_t, has a strong influence on the total profit with a positive effect and the profit may be increased by 97.4–98.4% of its standard deviation for an increase of one standard deviation in the parameter, b_t. Therefore, it is evident that the take-back agent may get best economic benefit when the minimum price expected by the consumers for returning an ELT (i.e., a_t) is low and the maximum expected price (i.e., b_t) is high. That is, the take-back agent may remain beneficial in those customer zone also, where the maximum price expected by the consumers for returning an ELT is very high.

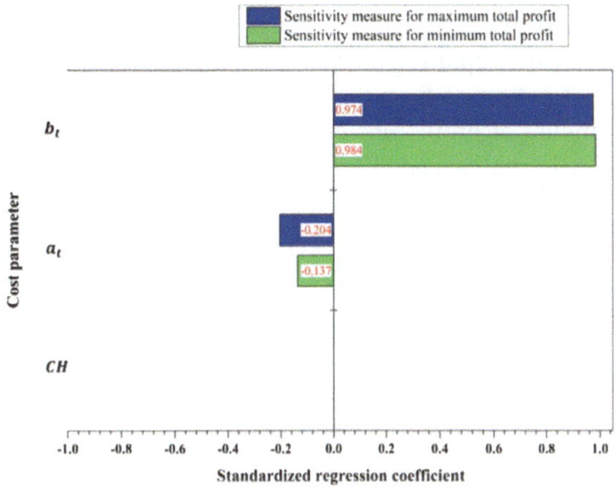

Fig. 3.6 Sensitivity analysis results based on the standardized regression coefficients

3.4 Conclusion

In this study, we have proposed a take-back mechanism for collecting ELTs from the consumers based on their willingness to return. Take-back of ELTs become necessary due to the implementation of EPR in many countries, including India. EPR ensures that the tire producers are actively involved in take-back and environmentally sound management of ELTs. However, take-back of ELTs from the consumers is not easy due to their unwillingness to return. Therefore, we incorporate the willingness to return of the consumers in the proposed take-back mechanism. In this mechanism, we propose to use the local tire repairing shops as LCCs for a customer zone and the owners of these centers as take-back agents. To model the operations of the LCC, a MP-MINLP model is developed to maximize the total expected profit and solved using the BONMIN solver in GAMS. We found that the proposed take-back mechanism has the potential to give positive profit for the take-back agents. We have also conducted a sensitivity analysis to investigate the effect of the commission percentages on the profit of the agent. It is found that the proposed mechanism can generate a good amount of positive profit for the take-back agent, even if the producer offers 1% of the take-back price as a commission to the agent. It indicates that the proposed take-back mechanism has the potential to help in the implementation of EPR in the Indian tire industry by encouraging the local tire repairing shop owners to collect the ELTs.

It is important to highlight that the major limitation of this work is that it is an initial step toward developing a take-back mechanism for the tire industry and only a single tire repairing shop is considered in this study to obtain the results. For generalizing the findings of this study, we have conducted a sensitivity analysis by adopting the FFD analysis, wherein the possible ranges of the values of the cost parameters involved in the proposed model are considered as the lower and upper levels. Through this FFD analysis, we have found that the take-back agent may achieve profitability even if the data related to the model parameters vary in those possible ranges. This clearly indicates toward the generalizability of our findings that approaching the local tire repairing shop owners as take-back agents may improve the scenario of ELT collection from the consumers in Indian context by providing the agents a good amount of profit. However, since these insights are drawn based on the single tire repairing shop and FFD analysis, as a future scope of research, the results of this study can be generalized in the Indian context through different case studies by considering different tire repairing shops located in different customer zones. Moreover, the operations of the producers should also be considered simultaneously with those of the take-back agents.

3.5 Competing Interests

The work is original in its contents and is not under consideration for publication in any other journals/proceedings. There is no potential conflict of interest. This work received funding from the Ministry of Education, India as a part of Prime Minister's Research Fellowship (PMRF) scheme (PMRF ID 2402361).

References

1. Ahmed, E.M., Rakočević, S., Ćalasan, M., et al.: BONMIN solver-based coordination of distributed FACTS compensators and distributed generation units in modern distribution networks. Ain Shams Eng. J. **13** (2022)
2. Araujo-Morera, J., Verdejo, R., López-Manchado, M.A., Hernández Santana, M.: Sustainable mobility: the route of tires through the circular economy model. Waste Manag. **126**, 309–322 (2021)
3. Banguera, L.A., Sepúlveda, J.M., Ternero, R., et al.: Reverse logistics network design under extended producer responsibility: the case of out-of-use tires in the Gran Santiago city of Chile. Int. J. Prod. Econ. **205**, 193–200 (2018)
4. Bonami, P., Biegler, L.T., Conn, A.R., et al.: An algorithmic framework for convex mixed integer nonlinear programs. Discret. Optim. **5**, 186–204 (2008)
5. Campbell-Johnston, K., Calisto Friant, M., Thapa, K., et al.: How circular is your tyre: experiences with extended producer responsibility from a circular economy perspective. J. Clean. Prod. **270**, 122042 (2020)
6. Eriksson, M., Ghosh, R., Mattsson, L., Ismatov, A.: Take-back agreements in the perspective of food waste generation at the supplier-retailer interface. Resour. Conserv. Recycl. **122**, 83–93 (2017)
7. Fattahi, M., Govindan, K.: Integrated forward/reverse logistics network design under uncertainty with pricing for collection of used products. Ann. Oper. Res. **253**, 193–225 (2017)
8. Fattahi, M. and Govindan, K.: Integrated forward/reverse logistics network design under uncertainty with pricing for collection of used products. Ann. Oper. Res. **253**(1), 193–225 (2017). https://doi.org/10.1007/s10479-016-2347-5
9. Fiksel, J., Bakshi, B.R., Baral, A., et al.: Comparative life cycle assessment of beneficial applications for scrap tires. Clean Technol. Environ. Policy **13**, 19–35 (2011)
10. Formela, K.: Sustainable development of waste tires recycling technologies—recent advances, challenges and future trends. Adv. Ind. Eng. Polym. Res. **4**, 209–222 (2021)

11. Gupta, V.K., Kaushal, R.K., Shukla, S.P.: Strategic policy modeling for stewardship program of spent portable batteries. J. Hazardous Toxic Radioact. Waste **22**, 1–12 (2018)
12. Jang, J.W., Yoo, T.S., Oh, J.H., Iwasaki, I.: Discarded tire recycling practices in the United States, Japan and Korea. Resour. Conserv. Recycl. **22**, 1–14 (1998)
13. Keyvanshokooh, E., Fattahi, M., Seyed-Hosseini, S.M., Tavakkoli-Moghaddam, R.: A dynamic pricing approach for returned products in integrated forward/reverse logistics network design. Appl. Math. Model. **37**, 10182–10202 (2013)
14. Klausner, M., Hendrickson, C.T.: Reverse-logistics strategy for product take-back. Interfaces (Providence) **30**, 156–165 (2000)
15. Manomaivibool, P.: Extended producer responsibility in a non-OECD context: the management of waste electrical and electronic equipment in India. Resour. Conserv. Recycl. **53**, 136–144 (2009)
16. Milanez, B., Bührs, T.: Extended producer responsibility in Brazil: the case of tyre waste. J. Clean. Prod. **17**, 608–615 (2009)
17. Niza, S., Santos, E., Costa, I., et al.: Extended producer responsibility policy in Portugal: a strategy towards improving waste management performance. J. Clean. Prod. **64**, 277–287 (2014)
18. Panagiotidou, S., Tagaras, G.: End-of-life tire recovery: the Thessaloniki initiative. In: Flapper, S.D.P., van Nunen, J.A.E.E., Van Wassenhove, L.N. (eds.) Managing closed-loop supply chains, pp. 183–193. Springer, Heidelberg (2005)
19. Paramanik, A.R., Mahanty, B.: A circular system for end-of-life tires under extended producer responsibility. Mater. Manuf. Process. **38**, 1964–1971 (2023)
20. Paramanik, A.R., Sarkar, S., Sarkar, B.: OSWMI: an objective-subjective weighted method for minimizing inconsistency in multi-criteria decision making. Comput. Ind. Eng. **169**, 108138 (2022)
21. Paramanik, A.R., Mahanty, B.: A collaborative circular supply chain network design framework for implementing individual producer responsibility under uncertainty. Transportation Research. Logist. Transp. Rev. **203**, 104332 (2025). https://doi.org/10.1016/j.tre.2025.104332
22. Park, J., Díaz-Posada, N., Mejía-Dugand, S.: Challenges in implementing the extended producer responsibility in an emerging economy: the end-of-life tire management in Colombia. J. Clean. Prod. **189**, 754–762 (2018)
23. Rodrigues, J.F.D., Lorena, A., Costa, I., et al.: An input-output model of extended producer responsibility. J. Ind. Ecol. **20**, 1273–1283 (2016)
24. Sarkar, S., Pramanik, A., Maiti, J., Reniers, G.: COVID-19 outbreak: a data-driven optimization model for allocation of patients. Comput. Ind. Eng. **161**, 107675 (2021)
25. Suard, S., Hostikka, S., Baccou, J.: Sensitivity analysis of fire models using a fractional factorial design. Fire Saf. J. **62**, 115–124 (2013)
26. Winternitz, K., Heggie, M., Baird, J.: Extended producer responsibility for waste tyres in the EU: lessons learnt from three case studies—Belgium, Italy and the Netherlands. Waste Manag. **89**, 386–396 (2019)
27. Zheng, X.-X., Chang, C.-T., Li, D.-F., et al.: Designing an incentive scheme for producer responsibility organization of waste tires: a MCGP cooperative game approach. Comput. Ind. Eng. **167**, 108009 (2022)

Chapter 4
q-Rung Orthopair Lance Distance TODIM Approach for Sustainable Supplier Selection

N. K. Vamsi Krishna and Divya Zindani

Abstract United nation's sustainable development goals (SDGs) play pivotal role in addressing social, economic, and environmental challenges, thereby ensuring sustainable future for all. Supplier selection plays a quintessential role in achieving SDGs since they directly impact the sustainable practices of business enterprises. However, choosing the best supplier to ensure environmental sustainability can be challenging for an organization given the uncertainties surrounding a business. To achieve this goal, this study introduces an innovative decision-making framework designed to assist organizations in the discerning choice of optimal suppliers. This newly proposed decision-making process leverages q-Rung Orthopair fuzzy sets to effectively model linguistic information. The utilization of Lance distance measures enables the comparison of alternatives, thereby assisting in the ranking of suppliers through the use of the Interactive Multicriteria Decision Making (TODIM) approach. The practicality and resilience of this approach have been demonstrated through a case study focused on the automotive sector. Its versatility enables its application for promoting thoughtful supplier selection across various industries, thereby serving as a pivotal instrument in advocating sustainable business practices.

Keywords Environment sustainability · TODIM · Lance distance measure · Supplier selection

---

N. K. Vamsi Krishna · D. Zindani (✉)
Department of Mechanical Engineering, Sri Sivasubramaniya Nadar (SSN) College of Engineering, Kalavakkam, India
e-mail: zindanidivya@gmail.com

N. K. Vamsi Krishna
e-mail: vamsi2010266@ssn.edu.in

J. Maiti et al. (eds.), *Recent Advances in Industrial and Systems Engineering*, Lecture Notes on Multidisciplinary Industrial Engineering,
https://doi.org/10.1007/978-981-96-8323-9_4

4.1 Introduction

Given the importance of sustainable development goals in ensuring a sustainable future for all, supplier selection has become important for organizations to promote sustainable innovations. Integrating sustainability into supplier selection involves evaluating and choosing suppliers based not only on traditional criteria like cost, quality, and reliability but also on their commitment to sustainable practices. The key aspects to this are environmental impact, ethical practices, social responsibility, regulatory compliance and certifications, innovation, and continuous improvement, etc. The selection of sustainable supplier can lead to enhanced brand reputation and customer trust, mitigation of supply chain risks related to environmental or ethical issues, cost savings through increased efficiency and reduced waste and improved relationships with stakeholders and regulators. The most important among all is that of environmental sustainability because of the stringent government regulations to circumvent the negative impacts of industrialization onto the environment. The organizations must adopt innovative approaches to promote effective utilization of natural resources and hence minimize the carbon footprint.

Setting critical sustainability criteria that align well with organization's values and goals is critical in assessing environmental sustainability of a supplier. Conducting thorough assessments of potential suppliers' sustainability practices based on environmental innovation criteria is however challenging owing to the vagueness and conflicting characteristics of these criteria. The presented work addresses this gap through the prospect of a novel decision-making framework that considers the various environmental innovation criteria. The vagueness in the judgment of experts, evaluating the suppliers, has been captured and modeled using q-Rung Orthopair fuzzy sets (q-ROFS) which are an extension of fuzzy sets that aim to address the limitations of traditional fuzzy sets by providing more flexibility in modeling uncertainty and vagueness. Various ranking approaches have been extended to include q-ROFS such as Technique for Order of Preference by Similarity to Ideal Solution (TOPSIS) [7, 13], Combinative Distance-Based Assessment (CODAS) [17, 25], Evaluation based on Distance from Average Solution (EDAS) [12, 22], COmplex PRoportional ASsessment (COPRAS) [24], Measurement of alternatives and ranking according to COmpromise solution (MARCOS) [2, 23], etc., to address wide array of decision-making problems such as solid waste management, selection of sustainable private health insurance policies, supplier selection in the defense industry, selection of unmanned aerial vehicles, cancer risk assessment, etc.

TODIM (an acronym in Portuguese for "Interactive and Multi-Criteria Decision Making") [9, 10] is one of the versatile multi-criteria decision-making (MCDM) methods to assist decision-makers in selecting the best alternative among several options when faced with multiple conflicting criteria. TODIM method considers the risk appetite in the decision-making process and ranks the alternatives based on overall prospect value. Owing to its idiosyncratic characteristic and versatility, TODIM has been applied to wide range of real-world decision-making problems such as portfolio selection [35], third-party logistics service providers selection [37], stock investment selection [38], green supplier selection [32], material selection [40], etc. To deal with unambiguity and vagueness of expert evaluation, TODIM approach has been combined with different decision-making environments such as intuitionistic fuzzy environment [34], interval-valued intuitionistic fuzzy environment (IVIF) [15], Pythagorean fuzzy environment [26], q-Rung orthopair fuzzy environment [2], linear diophantine uncertain linguistic fuzzy environment (2023), etc.

Although the TODIM approach has been extended to q-Rung orthopair fuzzy environment, the distance measures that have been used prominently to compare two q-Rung orthopair fuzzy numbers are Minkowski-type distance measures [2], Hausdorff metric [16], Hamming distance [33], distance measure based on herd mentality [30], etc. The said distance measures are sensitive to large, biased datasets and the same can be circumvented by Lance distance measure [19]. One of the lacunae which is prevalent in the existing TODIM approaches is the non-inclusion of such a versatile distance measure. The above discussion leads to the following contributions of the presented work:

- Proposition of a novel decision-making framework for sustainable supplier selection wherein the linguistic evaluations are modeled using q-rung orthopair fuzzy sets. Lance distance measures have been employed to compare two q-rung orthopair fuzzy numbers. The ranking phase has been accomplished using the TODIM approach.
- Applicability of the proposed approach to assess different suppliers in electric vehicle industry based on environmental sustainability innovation criteria. Five suppliers have been identified and evaluated based on ten influential environmental innovation criteria.

The remaining article has been structured into the following sections: basic definitions associated with q-rung orthopair fuzzy sets, Lance distance measures and the TODIM approach have been briefed in Sect. 4.2. Section 4.3 discusses the proposed approach. The applicability of the proposed approach has been demonstrated through a case study in Sect. 4.4 of the work. Conclusions and future recommendations have been shown in Sect. 4.5.

4.2 Literature Review

Numerous research studies have delved into supplier selection, specifically focusing on sustainability criteria that exert a substantial influence on organizational performance. Government policies and regulations have accelerated the pace at which the sustainable practices are being implemented by the organizations. The due consideration regarding the sustainability criteria has aided the organizations to enhance the performance of their supply chains and therefore gain competitive advantage in the market. Research investigations have shown the importance and the significant role that can be played by incorporating sustainability criteria in selection of the apt supplier in a supply chains (Muench et al. 2022; Barbanti et al. 2022; Ghosh et al.

2021). Moreover, the organizations want to position themselves in the market as being conservative about the environment and hence lay much emphasis on environmental sustainability [11]. However, the key is to select the apt supplier that can aid the organizations in realizing their environment sustainability objectives (Zhan et al. 2021). This has been addressed through the proposition of various novel decision-making frameworks out forth by varied researchers. Some of the recent studies in this regard have been summarized in Table 4.1 as follows.

As evident from the aforementioned recent literature, the most commonly employed decision-making frameworks are AHP, TOPSIS, VIKOR, COPRAS, DEMATEL, SAW, etc., and their fuzzy versions. However, the existing approaches fail to consider the psychological aspect of risk digesting capability of the experts while making such informed decisions. Moreover, the integration of the Lance distance measures that can aid in circumventing the issue of sensitiveness of other distance measures to large, biased datasets, within such a decision-making process is still a lacuna in the domain of decision science.

Table 4.1 also highlights that the industries considered for investigations by various investigators are electronics, steel, manufacturing, textile, automotive, logistics, etc. A wide array of environmental criteria such as green image, pollution control, financial capability, ISO standards, resource efficiency, etc., have been considered while carrying out the research investigations.

4.3 Preliminaries

This section of the work covers the fundamental definitions essential for constructing the decision-making framework.

Definition 1 A t-norm is a corollary T: $[0, 1]^2 \rightarrow [0, 1]$ that link up

T1. $T(\Pi_1, \Pi_2) = T(\Pi_2, \Pi_1)\ \forall\, \Pi_1, \Pi_2 \in [0, 1]$;
T2. $T(\Pi_1, \Pi_2) \le T(\Pi_3, \Pi_4)$ if $\Pi_1 \le \Pi_3, \Pi_2 \le \Pi_4\ \forall\, \Pi_1, \Pi_2, \Pi_3\ \Pi_4 \in [0, 1]$;
T3: $T(\Pi, 1) = \Pi\ \forall\, \Pi \in [0, 1]$;
T4: $T(\Pi_1, T(\Pi_2, \Pi_3)) = T(T(\Pi_1, \Pi_2), \Pi_3)$.

Definition 2 A t-conorm is a corollary S: $[0, 1]^2 \rightarrow [0, 1]$ that converge

S1. $S(\Pi_1, \Pi_2) = S(\Pi_2, \Pi_1)\ \forall\, \Pi_1, \Pi_2 \in [0, 1]$;
S2. $S(\Pi_1, \Pi_2) \le S(\Pi_3, \Pi_4)$ if $\Pi_1 \le \Pi_3, \Pi_2 \le \Pi_4\ \forall\, \Pi_1, \Pi_2, \Pi_3\ \Pi_4 \in [0, 1]$;
S3: $S(\Pi, 0) = \Pi\ \forall\, \Pi \in [0, 1]$;
S4: $S(\Pi_1, S(\Pi_2, \Pi_3)) = S(S(\Pi_1, \Pi_2, \Pi_2), \Pi_3)$.

Definition 3 The Aczel Alsina (AA) t-norm $\left(T_A^{\Lambda}\right)_{\Lambda \in [0,\infty]}$ is hypothesize as

$$T_A^{\Lambda}(\Pi_1,\ \Pi_2) = \begin{cases} T_D(\Pi 1,\ \Pi 2),\ \text{if}\ \Lambda = 0 \\ \min\{\Pi 1,\ \Pi 2\},\ \text{if}\ \Lambda = \infty \\ e^{-((-\ln \Pi 1)^{\Lambda} + (-\ln \Pi 2)^{\Lambda})^{1/\Lambda}},\ \text{otherwise} \end{cases}$$

Table 4.1 Recent studies

Authors	Criteria considered	Application domain	Method adopted
Nila and Roy (2023)	Reputation, Delivery reliability, Technological expertise, Geographical location, Resource consumption, Compliance with International Organisation for Standardization, (ISO) 14,000, Green distribution strategies, Environmental protection policies, Emission, effluents and waste generation, Local community influence, Compliance with International, Equity labor sources, Labour Organization (ILO) code, Health and safety practices, Staff training	Food manufacturing company	Triangular fuzzy numbers (TFN) LOgarithmic Percentage Change-driven Objective Weighting (LOPCOW), Full Consistency Method (FUCOM), and DOmbi Bonferroni (DOBI) methods
Ince et al. (2023)	Distance to Customer (km), Value-Added Time per Unit * (h/unit), Value-Added Cost Ratio per Unit * (%/unit), Customer Service Level (%), Rate of Returns on Costs (%), Net Profit Margin (%), Quality Level (%), Sustainable Sourcing (Wood Chips) (%/unit), Global Warming Potential ($kgCO_2$/unit), Water Footprint per Unit * (L/unit), Energy Footprint per Unit * (kwh/unit), Solid Waste Generation per Unit * (kg/unit), Recycled Raw Material Ratio per Unit * (%), Employee Satisfaction Rate (%), Contribution to society ($/year), Absenteeism Ratio (%), Gender Bias Ratio (%), Labor Practices (%)	Furniture industry	Fuzzy ARAS and Fuzzy TOPSIS methods
Yildizbasi and Arioz (2022)	Cost, Quality, Speed, Flexibility, After-sales service, Green Production, Green Technology, Green Logistics, Reduce, Reuse, Recycling(3R), Environmental Management System	Electronics industry	Fuzzy AHP, Fuzzy TOPSIS

(continued)

Table 4.1 (continued)

Authors	Criteria considered	Application domain	Method adopted
Nguyen et al. (2022)	Technological and financial capacity (TF), Time to Lead (LT), Adaptability (FL), Transport costs (TC), Timely delivery (DI), Product pricing (PR), Product Excellence (QA), Costs related to the environment (EC), Systems for Environmental Management (EM), Research and Development for Sustainability and Innovation (GI), Competencies in Environmental Sustainability (ES), Waste Reduction and Pollution Control (WP), Sustainable Manufacturing (GM), Sustainable Management (GR), Eco-friendly Packaging and Labeling (GL), Sustainable Design and Procurement (GP), Health and Security (HS), Customer Contentment (CS), Stakeholder Approval (SS), Employee Interests and Rights (IR)	Steel industry	Data Envelopment Analysis (DEA), spherical fuzzy analytic hierarchy process (SF-AHP), and spherical fuzzy weighted aggregated sum product assessment (SF-WASPAS)
Li et al. (2021)	Convenience of communication, Product price, Total cost of service, Date of delivery, Extra cost of payment, Technical level, Quality certification, Product qualification ratio, Service level, Enterprise scale, Market share, Management condition, R&D capability, Supply capacity, Management system, Reputation, Partner satisfaction, Resource consumption, Pollutant production, Resource recycling, Product versatility	Cloud manufacturing environment	Fuzzy BWM (Best–worst method) and TODIM approach
Karamaşa et al. (2021)	Technical Elements, Quality, Delivery, Green Elements, Cost, Service	Textile Enterprises	Range of value (ROV) method and Simple Additive Weighting (SAW)

(continued)

Table 4.1 (continued)

Authors	Criteria considered	Application domain	Method adopted
Shojaei and Bolvardizadeh (2020)	Transportation policies, experience in similar kind of project, reputation of the company, financial status of the company, environmental capital investment ratio, green technologies, environmental awareness, waste management during the process of green construction, environmental management system, equipment flexibility, green material design, plans on pollution control, reduced consumption for hazardous/harmful/toxic materials in construction process, analysis of materials' life cycle, legal compatibility of the organization with environment, appropriate site development, occupational health and safety, social responsibility, commitment to various regulatory standards such as ISO 14001, ODC, WEEE, etc., dealing with plants	Construction industry	Rough set, AHP, TOPSIS
Liu et al. (2019)	Cost of purchase, quality of product, capability regarding technology, environmental skill and performance regarding the delivery of the product	Watch manufacturing company	Best–worst method (BWM), Alternative queuing method (AQM), Interval-valued intuitionistic uncertain linguistic information

(continued)

Table 4.1 (continued)

Authors	Criteria considered	Application domain	Method adopted
Yazdani et al. (2017)	Financial solvency, environmental management protocols, waste management initiatives, managerial dedication, quality assurance systems, manufacturing processes, facility and reverse logistics, adaptability in quality, pricing, energy and natural resource efficiency, swift delivery, eco-conscious design, rates of re-use and recycling, and production scheduling	Dairy Company	Decision-making trial and evaluation laboratory (DEMATEL) method, Quality function deployment (QFD) model, Complex proportional assessment (COPRAS)
Wang et al. (2016)	Product cost, shipping expenses, policy terms and warranties, accolades for quality, implementation of ISO standards, product quality, delivery timeframes, component security in manufacturing, R&D proficiency, production capacity, technological prowess, adaptability, financial robustness, vendor reputation, organizational culture, new product introductions, innovation, communication transparency, interrelationship dynamics, pollution output, pollution management and mitigation, resource utilization, design for reusability and recycling, eco-conscious design, systems for environmental management, environmentally friendly reputation, proficiency in green practices, green product initiatives, employee training in environmental practices, dedication of management, green technology adoption	Luminance Enhancement Film (LEF) industry	Fuzzy analytic hierarchy process (AHP) and a fuzzy technique for order performance by similarity to ideal solution (TOPSIS)

Some special cases: $T_A^{\infty} = \min$, $T_A^0 = T_{D,} T_A^1 = T_{P.}\ T_{P.}$.

Definition 4 The AA t-conorm $\left(S_A^{\Lambda}\right)_{\Lambda \in [0,\infty]}$ is postulated as

$$S_A^{\Lambda}(\Pi_1, \Pi_2) = \begin{cases} S_D(\Pi 1, \Pi 2), \text{if } \Lambda = 0 \\ \max\{\Pi 1, \Pi 2\}, \text{if } \Lambda = \infty \\ 1 - e^{-((-\ln(1-\Pi 1)^{\Lambda} + (-\ln(1-\Pi 2)^{\Lambda})^{1/\Lambda}}, \text{ otherwise} \end{cases}$$

Some special cases: $T_A^{\infty} = \min$, $T_A^0 = T_{D,}T_A^1 = T_P.\ T_P.$.

Definition 4 A q-Rung Orthopair fuzzy set (Q) on X is defined as follows (Yager, 2017):

$$Q = \{x, \mu_Q(x), \nu_Q(x) : x \in X\}$$

In the above representation, the membership degree and non-membership degrees are denoted by μ_Q and $\nu_Q \rightarrow [0, 1]$. Following condition is required to be satisfied by the membership and non-membership degrees:

$$0 \leq \mu_Q^q + \nu_Q^q \leq 1$$

Definition 5: Let $\alpha^{\xi} = (\mu_{\alpha^{\xi}}, \nu_{\alpha^{\xi}})$, $\alpha^{\xi_1} = (\mu_{\alpha^{\xi_1}}, \nu_{\alpha^{\xi_1}})$, $\alpha^{\xi_2} = (\mu_{\alpha^{\xi_2}}, \nu_{\alpha^{\xi_2}})$ be any three q- ROFNs and $\aleph^h \geq 1$ and σ > 0, then the t-conorm operations on them and the AA t-norm are provided by the following formulations:

$$\alpha_1^{\xi} \oplus \alpha_2^{\xi} = \sqrt[q]{1 - e^{-\left(\left(-\log\left(1-\mu_{a\xi_1}^q\right)\right)\right)^{\aleph^h} + \left(-\log\left(1 - \mu_{a\xi_2}^q\right)\right)^{\aleph^h}\right)^{\frac{1}{\aleph^h}}},$$

$$\sqrt[q]{e^{-\left(\left(-\log\nu_{a\xi_1}^q\right)^{\aleph^h} + \left(-\log\nu_{a\xi_2}^q\right)^{\aleph^h}\right)^{\frac{1}{\aleph^h}}}}$$

$$\alpha_1^{\xi} \otimes \alpha_2^{\xi} = \sqrt[q]{e^{-\left(\left(-\log\left(1-\mu_{a\xi_1}^q\right)\right)\right)^{\aleph^h} + \left(-\log\left(1 - \mu_{a\xi_2}^q\right)\right)^{\aleph^h}\right)^{\frac{1}{\aleph^h}}},$$

$$\sqrt[q]{1 - e^{-\left(\left(-\log\nu_{a\xi_1}^q\right)^{\aleph^h} + \left(-\log\nu_{a\xi_2}^q\right)^{\aleph^h}\right)^{\frac{1}{\aleph^h}}}}$$

4.4 Methodology

This section of the work addresses the procedural steps related to the proposed methodology. A brief overview of the adopted methodology for the current study has been illustrated in Fig. 4.1. As per *stage 1* of the proposed framework, suppliers to be ranked and the criteria set based on which the identified suppliers are to be ranked are identified. Decision matrix is formulated next using the identified criteria and alternatives. *Stage 2* is about modeling linguistic evaluations. This stage incepts with the collection of the linguistic evaluations from the experts who assess the suppliers based on the identified criteria set. Subsequently, the linguistic evaluations are converted to q-Rung Orthopair Fuzzy numbers (q-ROFNs) [6]. Aggregation and normalization are processed in *stage 3* of the proposed approach. Aggregation of the collected linguistic evaluations is accomplished using q-Rung Orthopair Fuzzy Aczel-Alsina ordered weighted geometric (q-ROFAAOWG) operator [8] as depicted in Eq. (4.1).

$$q-ROFAAOWG\left(\alpha_1^{\xi}, \alpha_2^{\xi}, \ldots, \alpha_{\breve{T}}^{\xi}\right) = \sqrt[q]{e^{-\left(\sum_{\emptyset=1}^{\breve{T}} \breve{\varsigma}\,{}_{\phi}^{\varsigma}\left(-\log \mu^q_{\alpha^{\xi}_{\coprod(\emptyset)}}\right)^{\aleph\hbar}\right)}},\quad \sqrt[q]{e^{-\left(\sum_{\emptyset=1}^{\breve{T}} \breve{\varsigma}\,{}_{\phi}^{\varsigma}\left(-\log(1-\nu^q_{\alpha^{\xi}_{\coprod(\emptyset)}}\right)^{\aleph\hbar}\right)^{\frac{1}{\aleph\hbar}}}}, \tag{4.1}$$

where $\coprod(1), \coprod(2), \ldots, \coprod(\breve{T})$ are the permutation of $\left(\emptyset = 1, 2, \ldots, \breve{T}\right)$, including $\alpha^{\xi}_{\coprod(\emptyset-1)} \geq \alpha^{\xi}_{\coprod(\emptyset)} \quad \forall\, \emptyset = 1, 2, \ldots, \breve{T}$.

Once the aggregated decision matrix has been obtained, normalization is accomplished using the following Eq. (4.2):

$$\overline{r_{ij}} = \begin{cases} \overline{d_{ij}} = (\mu_{\alpha^{\xi}_{\emptyset}}, \nu_{\alpha^{\xi}_{\emptyset}}) & d_{ij} \in \text{Benefit criteria} \\ \overline{d_{ij}} = \left(\nu_{\alpha^{\xi}_{\emptyset}}, \mu_{\alpha^{\xi}_{\emptyset}}\right) & d_{ij} \in \text{Cost criteria} \end{cases} \tag{4.2}$$

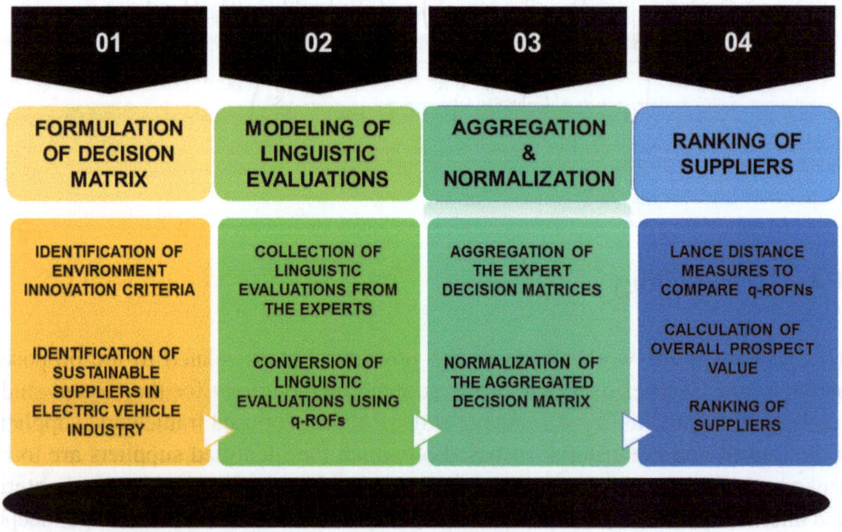

Fig. 4.1 The methodology adopted for the present research article

Stage 4 is the ranking phase wherein the identified suppliers are ranked using the TODIM approach. This stage begins with the calculation of the score values using Eq. (4.3). The score values are employed to determine the criteria (w_c) and relative criteria weights (w_{cr}). The criteria weights can be arrived at using the Shannon's entropy approach [39].

$$\mathrm{SV} = \frac{1 + \mu_{\alpha_{\emptyset}^{\xi}} - \nu_{\alpha_{\emptyset}^{\xi}}}{2} \tag{4.3}$$

Subsequently, the degree of dominance of one alternative over the other is computed using Eq. (4.4) in the following manner:

$$\delta(S_i, S_j) = \sum_{c=1}^{m} \rho_c(S_i, S_j) \quad \forall\ (i, j), \tag{4.4}$$

where

$$\rho_c(S_i, S_j) = \begin{cases} \sqrt{\frac{w_{cr}(\overline{r_{ic}} - \overline{r_{jc}})}{\sum_{c=1}^{m} w_{cr}}} \mathrm{if}(\overline{r_{ic}} - \overline{r_{jc}}) > 0 \\ 0 \mathrm{if}(\overline{r_{ic}} - \overline{r_{jc}}) = 0 \\ \frac{-1}{\theta}\sqrt{\frac{\sum_{c=1}^{m} w_{cr}}{w_{cr}(\overline{r_{ic}} - \overline{r_{jc}})}} \mathrm{if}(\overline{r_{ic}} - \overline{r_{jc}}) < 0 \end{cases}, \tag{4.5}$$

where $(\overline{r_{ic}} - \overline{r_{jc}}) > 0$ denotes the gain and $(\overline{r_{ic}} - \overline{r_{jc}}) < 0$ denotes the loss for one alternative over the other. These are derived using the Lance distance measure [31] through Eq. (4.6):

$$D_{\mathrm{Lance}}(\overline{r_{ic}}, \overline{r_{jc}}) = \frac{1}{3}\left(\frac{\left|\mu_{\alpha_{\emptyset}^{1}} - \mu_{\alpha_{\emptyset}^{2}}\right|}{1 + \mu_{\alpha_{\emptyset}^{1}} + \mu_{\alpha_{\emptyset}^{2}}} + \frac{\left|\nu_{\alpha_{\emptyset}^{1}} - \nu_{\alpha_{\emptyset}^{2}}\right|}{1 + \nu_{\alpha_{\emptyset}^{1}} + \nu_{\alpha_{\emptyset}^{2}}}\right) \tag{4.6}$$

The attenuation factor, denoted by θ in Eq. (4.5), represents the risk appetite of the experts participating in the decision-making process. Subsequently, the overall prospect value is then obtained through Eq. (4.7):

$$\varphi_i = \frac{\sum_{j=1}^{n} \delta(S_i, S_j) - \min \sum_{j=1}^{n} \delta(S_i, S_j)}{\max \sum_{j=1}^{n} \delta(S_i, S_j) - \min \sum_{j=1}^{n} \delta(S_i, S_j)} \tag{4.7}$$

The alternatives are assessed based on their overall prospect values, and the alternative with the highest prospect value score is considered the best among the options under consideration.

4.5 Case Study

The proposed approach's practicality has been demonstrated through a case study conducted in the electric vehicle industry. The suppliers cater to supply a critical component for the electric vehicle industry. As per *stage 1* of the proposed approach, five suppliers have been appraised based on their capability to innovate with due regard to environmental sustainability. The criteria have been listed in Table 4.2.

As per *stage 2,* the linguistic evaluations have been collected from three experts and modeled through q-Rung Orthopair Fuzzy numbers (q-ROFNs). Table 4.3 depicts the linguistic evaluations from expert 1.

The aggregation of the collected evaluations is processed in *stage 3,* and the aggregated decision matrix has been represented in Table 4.4. Subsequently, the aggregated matrix has been normalized by using Eq. (4.2).

Table 4.2 Environment sustainability innovation criteria set for evaluation of the suppliers

S. No.	Criteria	Reference/s
C_1	Degree of novelty, and level of sustainability	[14]
C_2	Financial availability for innovation	[1]
C_3	Technical expertise availability and investment in R&D for green practices	[18, 21]
C_4	Green manufacturing and operational capabilities development	[4, 18, 28]
C_5	Sustainable solutions to maximize society and environment	[14]
C_6	Implementation of socio-ecopolicies in organizations for sustainability	[11, 18, 29]
C_7	Reduction of resources in order to reduce waste	[5, 14]
C_8	Conducting regular environmental audits	[1]
C_9	Practices for improving energy efficiency	[5, 14, 20]
C_{10}	Reduced environmental accident	[20, 36]
C_{11}	Practices to improve water efficiency	[5, 14, 20]
C_{12}	Creating sustainable needs in order to change lifestyles	[14]
C_{13}	Production systems and suppliers to obtain environmental and social benefits	[14]

Table 4.3 Linguistic evaluation from expert 1

	C_1	C_2	C_3	C_4	C_5	C_6	C_7	C_8	C_9	C_{10}	C_{11}	C_{12}	C_{13}
S_1	H	H	N	H	H	VH	H	VH	VH	VH	VH	VH	H
S_2	MH	N	H	VH	H	VH	H	VH	MH	H	MH	VH	H
S_3	N	MH	EH	H	N	H	N	H	MH	H	MH	VH	MH
S_4	H	MH	VH	H	H	H	H	VH	VH	MH	H	MH	MH
S_5	H	H	H	H	VH	VH	H	MH	MH	MH	VH	VH	H

Table 4.4 Aggregated decision matrix

	C_1		C_2		C_3		C_4		C_5		C_6	
S1	0.96	0.86	0.96	0.86	0.90	0.82	0.96	0.86	0.96	0.86	0.97	0.86
S2	0.97	0.89	0.90	0.82	0.96	0.86	0.97	0.87	0.96	0.86	0.97	0.86
S3	0.91	0.84	0.97	0.88	0.97	0.67	0.96	0.86	0.90	0.82	0.96	0.86
S4	0.96	0.86	0.97	0.89	0.97	0.86	0.96	0.86	0.96	0.86	0.96	0.86
S5	0.96	0.86	0.96	0.86	0.96	0.86	0.96	0.86	0.97	0.86	0.97	0.86

In *stage 4* of the proposed framework, the determination of criteria weights and their relative weights follows the calculations of score values. Figure 4.2 provides a clear comparison of the weights assigned to each criterion. Subsequently, the normalized decision matrix has been employed to derive dominance degree, overall prospect value, and ranking results for the considered suppliers. The calculations of dominance degree entail calculating Lance distance measures using Eq. (4.6). Table 4.5 depicts the values of Lance distance measures for supplier 1. In Fig. 4.3, the overall prospect value and thereby the ranking results have also been depicted.

According to the illustration in Fig. 4.3, supplier 1 possesses the highest overall prospect value, establishing it as superior among the other suppliers under consideration. The application of the proposed approach has unveiled the following ranking order: $S_5 > S_4 > S_2 > S_3 > S_1$.

In Fig. 4.4, we see the sensitivity analysis results against the attenuation factor. Different values of the attenuation factor ($\theta = 0.1, 1, 2, 5$ and 10) have been considered while arriving at the overall prospect values. It is evident from the depiction that the ranking results are almost stable with different values of the attenuation factor barring at $\theta = 0.1$. This underscores the significance of the experts' risk appetite during the evaluation process.

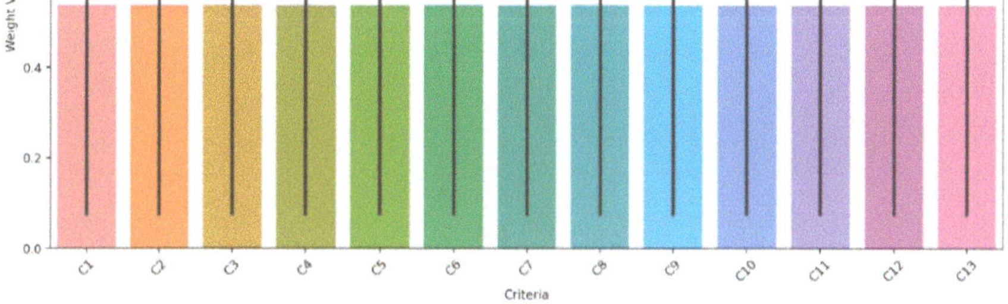

Fig. 4.2 Comparison of the weights assigned to each criterion

Table 4.5 Lance distance measurements for supplier 1

	C_1	C_2	C_3	C_4	C_5	C_6	C_7	C_8	C_9	C_{10}
S_2	−0.0036	0.0123	−0.0123	−0.0020	−0.1096	0.0000	0.0000	0.0000	−0.0026	0.0010
S_3	0.0083	−0.0028	0.0120	0.0000	−0.0993	0.0010	0.0123	0.0010	−0.0026	0.0010
S_4	0.0000	−0.0036	−0.0134	0.0000	−0.1096	0.0010	−0.0011	0.0000	0.0000	−0.0026
S_5	0.0000	0.0000	−0.0123	0.0000	−0.1102	0.0000	0.0000	−0.0012	−0.0013	−0.0026

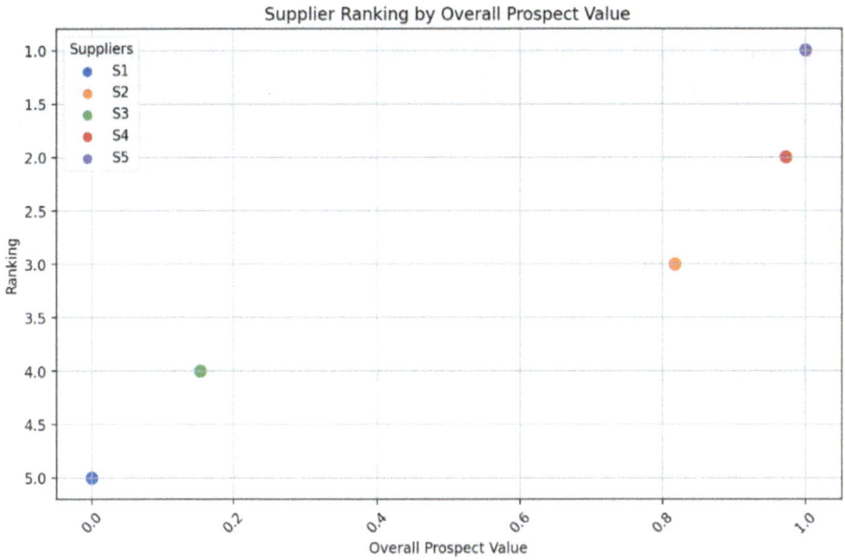

Fig. 4.3 Overall prospect value and the ranking results for different suppliers

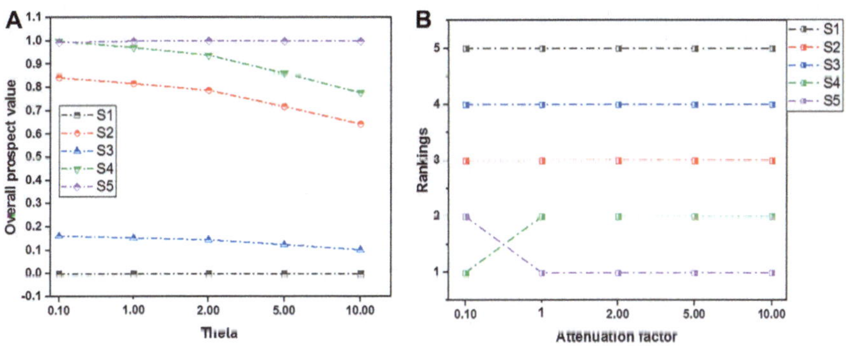

Fig. 4.4 Sensitivity analysis w.r.t attenuation factor for **a** overall prospect value of different suppliers **b** ranking results

4.6 Managerial Implications

The proposed decision-making framework can enable managers to systematically evaluate suppliers based on their capability to innovate with due regard to environmental sustainability, leading to more informed and balanced decisions. By considering these criteria, managers in the domain of electrical vehicle industry can better align supplier capabilities with organizational sustainability goals, mitigate risks, and enhance overall supply chain resilience. This holistic approach may also facilitate transparency and consistency in the decision-making process, fostering stakeholder confidence and support. Moreover, the use of the proposed framework can allow for the integration of both quantitative and qualitative data, aiding in the identification of the most suitable suppliers in complex and dynamic electrical vehicle market environments, ultimately contributing to improved operational efficiency and competitive advantage.

4.7 Conclusion

The present article proposed a novel decision-making framework to assess various suppliers based on the environmental sustainability innovation criteria and hence their willingness to meet the sustainability targets. The proposed decision-making process uses q-Rung Orthopair fuzzy sets to model the linguistic information collected from the experts. Lance distance measures are employed to compare the alternatives and therefore aid in ranking the suppliers through the TODIM approach. Five suppliers have been identified and evaluated based on thirteen influential environmental sustainability innovation criteria. Supplier 5 has been revealed to be the superior among the considered suppliers. A comprehensive evaluation of the prosed approach reveals its robustness regarding the obtained ranking results.

The existing approaches fail to consider the psychological aspect of risk digesting capability of the experts while making such informed decisions. Moreover, the integration of the Lance distance measures that can aid in circumventing the issue of sensitiveness of other distance measures to large, biased datasets, within such a decision-making process, is still a lacuna in the domain of decision science. However, the employability of q-Rung Orthopair fuzzy sets results in complexity of the decision-making process. Moreover, the additional parameters inherent in q-ROFS models reduce their interpretability compared to simpler fuzzy sets, thereby further complicating the process. As a future research direction, the applicability of the proposed approach can be explored to encourage judicious selection of suppliers in any sector such as such as the automotive and electronics industries, e-commerce and fashion, healthcare sector, the food and beverage industry and hence play a pivotal role in advocating sustainable practices.

References

1. Ahmadi, H.B., Lo, H.W., Gupta, H., Kusi-Sarpong, S., Liou, J.J.: An integrated model for selecting suppliers on the basis of sustainability innovation. J. Clean. Prod. **277**, 123261 (2020)
2. Ali, J.: A q-rung orthopair fuzzy MARCOS method using novel score function and its application to solid waste management. Appl. Intell. **52**(8), 8770–8792 (2022)
3. Ali, J., Bashir, Z., Rashid, T.: A cubic q-rung orthopair fuzzy TODIM method based on Minkowski-type distance measures and entropy weight. In: Soft Computing, pp. 1–25 (2023)
4. Ansari, Z.N., Kant, R.: A state-of-art literature review reflecting 15 years of focus on sustainable supply chain management. J. Clean. Prod. **142**, 2524–2543 (2017)

5. Cho, Y.S., Linderman, K.: Resource-based product and process innovation model: theory development and empirical validation. Sustainability **12**(3), 913 (2020)
6. Ecer, F., Ögel, İ.Y., Krishankumar, R., Tirkolaee, E.B.: The q-rung fuzzy LOPCOW-VIKOR model to assess the role of unmanned aerial vehicles for precision agriculture realization in the Agri-Food 4.0 era. In: Artificial intelligence review, pp. 1–34 (2023)
7. Erdebilli, B., Gecer, E., Yılmaz, İ, Aksoy, T., Hacıoglu, U., Dinçer, H., Yüksel, S.: Q-ROF fuzzy TOPSIS and VIKOR methods for the selection of sustainable private health insurance policies. Sustainability **15**(12), 9229 (2023)
8. Farid, H.M.A., Riaz, M.: Q-rung orthopair fuzzy Aczel-Alsina aggregation operators with multi-criteria decision-making. Eng. Appl. Artif. Intell. **122**, 106105 (2023)
9. Gomes, L.F.A.M., Lima, M.M.P.P.: TODIMI: basics and application to multicriteria ranking. Found. Comput. Decis. Sci **16**(3–4), 1–16 (1991)
10. Gomes, L.F.A.M., Lima, M.M.P.P.: From modeling individual preferences to multicriteria ranking of discrete alternatives: a look at prospect theory and the additive difference model. Foundations Comput. Decis. Sci. **17**(3), 171–184 (1992)
11. Govindan, K., Muduli, K., Devika, K., Barve, A.: Investigation of the influential strength of factors on adoption of green supply chain management practices: an Indian mining scenario. Resour. Conserv. Recycl. **107**, 185–194 (2016)
12. Güneri, B., Deveci, M.: Evaluation of supplier selection in the defense industry using q-rung orthopair fuzzy set based EDAS approach. Expert Syst. Appl. **222**, 119846 (2023)
13. Gurmani, S. H., Garg, H., Zulqarnain, R.M., Siddique, I.: Selection of unmanned aerial vehicles for precision agriculture using interval-valued q-Rung orthopair fuzzy information based TOPSIS method. Int J Fuzzy Syst 1–15 (2023)
14. Gomes, S., Lopes, J.M., Travassos, M., Paiva, M., Cardoso, I., Peixoto, B., Duarte, C.: Strategic organizational sustainability in the age of sustainable development goals. Sustainability **15**(13), 10053 (2023)
15. Huang, G., Xiao, L., Zhang, G.: An integrated design concept evaluation method based on best-worst best–worst entropy and generalized TODIM considering multiple factors of uncertainty. Appl. Soft Comput. **140**, 110165 (2023)
16. Hussain, Z., Abbas, S., Yang, M.S.: Distances and similarity measures of Q-Rung orthopair fuzzy sets based on the Hausdorff metric with the construction of orthopair fuzzy TODIM. Symmetry **14**(11), 2467 (2022)
17. Kausar, R., Farid, H.M.A., Riaz, M., Gonul Bilgin, N.: Innovative CODAS algorithm for q Rung orthopair fuzzy information and cancer risk assessment. Symmetry **15**(1), 205 (2023)
18. Kusi-Sarpong, S., Gupta, H., Sarkis, J.: A supply chain sustainability innovation framework and evaluation methodology. Int. J. Prod. Res. **57**(7), 1990–2008 (2019)
19. Lance, G.N., Williams, W.T.: Computer programs for hierarchical polythetic classification ("similarity analyses"). Comput. J. **9**(1), 60–64 (1966)
20. Le, T.T., Ikram, M.: Do sustainability innovation and firm competitiveness help improve firm performance? Evidence from the SME sector in vietnam. Sustain Prod Consumption **29**, 588–599 (2022)
21. Li, Y., Mathiyazhagan, K.: Application of DEMATEL approach to identify the influential indicators towards sustainable supply chain adoption in the auto components manufacturing sector. J. Clean. Prod. **172**, 2931–2941 (2018)
22. Naz, S., Akram, M., Fatima, A., Nadeem, A.: q-Rung orthopair fuzzy 2-tuple linguistic Hamy mean operators for MAGDM with modified EDAS method. In: Real Life Applications of Multiple Criteria Decision-Making Techniques in Fuzzy Domain. Springer Nature Singapore, Singapore, pp. 369–415 (2022)
23. Peng, X., Garg, H., Luo, Z.: When content-centric networking meets multi-criteria group decision-making: optimal cache placement policy achieved by MARCOS with q-rung orthopair fuzzy set pair analysis. Eng. Appl. Artif. Intell. **123**, 106231 (2023)
24. Pinar, A.: Supplier evaluation with Q-Rung orthopair fuzzy-based COPRAS method. In: Multiple Criteria Decision Making with Fuzzy Sets: MS Excel® and Other Software Solutions, pp. 13–26. Springer International Publishing, Cham (2022)

25. Ramírez-Ochoa, D.D., Pérez-Domínguez, L., Martínez-Gómez, E.A., Torres-Argüelles, V., Garg, H., Sansabas-Villapando, V.: Supplier selection process based on CODAS method using q-rung orthopair fuzzy information. In: q-Rung Orthopair Fuzzy Sets: Theory and Applications. Springer Nature Singapore, Singapore, pp. 219–240 (2022)
26. Sarkar, B., Biswas, A.: TODIM-based pythagorean fuzzy multicriteria group decision making through similarity measure. Int J Adv Intell Paradigms **22**(1–2), 184–199 (2022)
27. Singh, S., Kanwar, N., Zindani, D.: Linear diophantine uncertain linguistic-based prospect theory approach for performance evaluation of islanded microgrid-system scenarios. Clean Energy **7**(2), 263–282 (2023)
28. Somsuk, N., Laosirihongthong, T.: Prioritization of applicable drivers for green supply chain management implementation toward sustainability in Thailand. Int J Sust Dev World **24**(2), 175–191 (2017)
29. Tariq, A., Badir, Y.F., Tariq, W., Bhutta, U.S.: Drivers and consequences of green product and process innovation: a systematic review, conceptual framework, and future outlook. Technol. Soc. **51**, 8–23 (2017)
30. Tian, X., Niu, M., Zhang, W., Li, L., Herrera-Viedma, E.: A novel TODIM based on prospect theory to select green supplier with q-rung orthopair fuzzy set. Technol. Econ. Dev. Econ. **27**(2), 284–310 (2021)
31. Wang, H., Xu, T., Pamucar, D., Li, X., Feng, L.: A lance distance-based MAIRCA method for q-Rung orthopair fuzzy MCDM with completely unknown weight information. In: Informatica, pp. 1–24 (2023)
32. Wang, Z., Cai, Q., Wei, G.: Modified TODIM method based on cumulative prospect theory with type-2 neutrosophic number for green supplier selection. Eng. Appl. Artif. Intell. **126**, 106843 (2023)
33. Wang, Y., Tang, X., Pan, Y., Cai, M.: Chebyshev distance entropy combined with TODIM method in Lq-ROFS multiattribute group decision making. Discrete Dynamics in Nature and Society (2023)
34. Wu, C., Zhang, D.: Ranking products with IF-based sentiment word framework and TODIM method. Kybernetes **48**(5), 990–1010 (2019)
35. Wu, Q., Liu, X., Qin, J., Zhou, L., Mardani, A., Deveci, M.: An integrated generalized TODIM model for portfolio selection based on financial performance of firms. Knowl. Based Syst. **249**, 108794 (2022)
36. Xue, M., Boadu, F., Xie, Y.: The penetration of green innovation on firm performance: effects of absorptive capacity and managerial environmental concern. Sustainability **11**(9), 2455 (2019)
37. Zhang, D., Su, Y., Zhao, M., Chen, X.: CPT-TODIM method for interval neutrosophic MAGDM and its application to third-party logistics service providers selection. Technol. Econ. Dev. Econ. **28**(1), 201–219 (2022)
38. Zhao, M., Wei, G., Wei, C., Wu, J.: Improved TODIM method for intuitionistic fuzzy MAGDM based on cumulative prospect theory and its application on stock investment selection. Int. J. Mach. Learn. Cybern. **12**, 891–901 (2021)
39. Zindani, D., Maity, S.R., Bhowmik, S.: Excogitating material rankings using novel aggregation multiplicative rule (AMR): a case for material selection problems. Arab. J. Sci. Eng. **45**, 5631–5646 (2020)
40. Zindani, D., Das, G., Bhagwat, V., Singh, G.: Performance analysis of Mg-Al-Zn (AZ) anodic alloys for metal-air batteries. Mater. Today Proc. **72**, 2315–2321 (2023)

Chapter 5
Green Integration as a New Dimension of Supply Chain Classification for Strategic Analyses and Emission Control: A MILP Approach

Santanu Banerjee, Amit Kumar Singh, Pranav Sanjay Gore, Thiruchitrambalam B., and Ankit Patel

Abstract In this article, we design a multi-stage supply chain consisting of Suppliers, Distribution Centres, Branches and Customers, allowing multi-commodity product transfer to satisfy the end consumer demands. We take into account both fixed and variable emissions due to the operation and amount of goods transferred through the intermediary echelon-facilities of Distribution Centres and Branches, as well as during transportation. The decisions include whether to keep

Authors Ankit Patel, Thiruchitrambalam B., and Pranav Sanjay Gore are equal contributors.

S. Banerjee (✉) · A. K. Singh · B. Thiruchitrambalam · A. Patel
Department of Industrial and Systems Engineering, Indian Institute of Technology Kharagpur, Kharagpur 721302, West Bengal, India
e-mail: santanu@kgpian.iitkgp.ac.in

A. K. Singh
e-mail: amit2021@kgpian.iitkgp.ac.in

B. Thiruchitrambalam
e-mail: thirubaskkaran@gmail.com

A. Patel
e-mail: ankitsinghpatel24@gmail.com

P. S. Gore
Terex India Pvt. Ltd, Bangalore, India
e-mail: pranavgoremsd@gmail.com

B. Thiruchitrambalam
Logistics and Materials Handling, Production Engineering Department, TVS Motor Company Pvt. Ltd., Krishnagiri, India

A. Patel
Fair Isaac India Software Private Limited (FICO), Bengaluru, Karnataka, India

P. S. Gore
Department of Mechanical Engineering, Indian Institute of Technology Kharagpur, Kharagpur 721302, West Bengal, India

J. Maiti et al. (eds.), *Recent Advances in Industrial and Systems Engineering*, Lecture Notes on Multidisciplinary Industrial Engineering,
https://doi.org/10.1007/978-981-96-8323-9_5

the intermediary echelon-facilities running during the planning horizon, and the product flow path from the Suppliers to the Customers through the functional intermediary echelon-facilities. We consider capacity constrained facilities and homogeneous vehicles, with the optimization goal of the green supply chain being the overall emission minimization. In addition to the green objective of emission reduction, we separately model fixed and variable supply chain costs similar to emissions, and estimate the emission trade-off when such cost minimization objectives are used. Further, we also consider the impact of supply chain integration on both of these objectives, with the same MILP being used on integrated supply chains for optimization; the process of integration of two or more supply chains involves coalescing each individual stage of the supply chains separately (e.g. all Branches across the supply chains being integrated are merged together to form a single set of Branches, and likewise) with additional new transportation links being provided for allowing the new transport pathways. We find that trade-off in emissions when using the cost objective (i.e. for efficient supply chains) is proportional to the size of the supply chain, and propose the introduction of a new classification strategy for environmentally friendly supply chains, that being of green supply chains with the fundamental objective being to minimize wastes and emissions. All codes to replicate the computational results, and synthetic datasets generated by us are hosted online.

Keywords Green supply chain design · Supply chain integration · Emission reduction · Multi-Commodity flow · Mixed integer linear programming · Supply chain classification

5.1 Introduction

Supply chains (SC) form an integral part of present-day mass-consumerism and the need for sustainability within SCs is ever growing. Developing individual green emission-less components within existing SCs through retrofitting, upgradations or replacements will be a time-taking process, but this is the most important and necessary step towards achieving a carbon-negative planet and developing a sustainable consumer–supplier ecosystem. During this steady transition towards building individual green sustainable SC components, the effective management of SCs becomes an important decision problem; this includes integration of SCs to decrease the total emissions. We compare the reduction in emissions when performing the general industry norm of cost minimization separately on the individual SCs. In this paper, we model a SCI problem considering decisions on the flow of commodities from main sources (Suppliers) to final sinks (Customers), through other intermediate facilities. We consider the fixed and variable emissions of these intermediate facilities (consisting of Distribution Centres and Branches), allowing the decision of closing them during the considered operational period. The fixed and variable emissions for each unique transportation link (unique connection from one vertex to another) are also taken into account. This presents an offshoot of the facility location problem that has long been studied with primary focus on the cost aspect of location, transportation and inventory management. Our green SC design problem focusses on the usage of an intermediate facility as being a decision contributing to emissions within a SC, such that non-usage of a particular intermediate facility would also be considered a feasible option if it decreases the overall emissions. We synthetically generate five SC datasets to compare their overall emissions when using a cost minimization objective of these separate SCs, with the results when using an emission minimization objective.

The definition of Supply Chain Integration (SCI) although being incomplete [1], internal integrations and integrating customers are more strongly related to improved performance compared to integrating suppliers. SCI is inherently difficult due to hampering competitiveness through the collaborative efforts necessary for achieving SCI [2], due to which few companies actually engage in extensive SCI practices, and theoretically idealistic principles are seldom incorporated within real-world SC practices. Critical reviews on SCI [3] analyse existing performance measures while suggest other subjective measures. Although green SC is a well-studied topic, through an extensive literature search we hardly find a handful of studies having been done on the problem of optimizing emissions within integrated SCs. We delve into the emission-tradeoff analysis comparing SCs separately and when they are integrated, showing progressive increase in savings in the emissions for integrated SCs versus when they operate separately.

5.2 Literature Review

Road transportation accounts for 35% of total liquid commercial fuel consumed by all Indian sectors [4]. The mitigation of greenhouse gas (GHG) emissions is challenging [5] especially during the actual implementation of plans. The dimensions of policy-making revolve on the aspects of energy security, local environment and climate change [6], with feasible approaches to achieving developmental objectives with carbon dioxide (CO_2) co-benefits. The current dependence on oil for security and defensive applications could be surely shifted to electrical vehicles with wireless power transfer (WPT) technologies enabling charging while on the road. Similar technological applications of using WPT [7] in unmanned vehicles [8] for defensive [9] or civilian applications would require moving vehicles on powered roadways.

The emissions from industries and facilities set up to ease the commodity flow towards the end consumer is also a key factor during the development of a sustainable SC, with decisions being whether to keep a specific product at a facility, the routing of vehicles loaded with goods to specific facilities in an hierarchical process, as well as whether to keep a facility partly functional or closed during a certain period depending on the product flow amount to reduce emissions and resources usage. The emissions of volatile organic compounds (VOCs) have "*unfair manifestation under the background of economic development and ecological civilization construction*" [10] monitoring which require hyperspectral. Multiple conflicting objectives including ensuring sustainable and effective conveyances have been considered [11] comparing with other Pareto-based approaches. The impact of loading and unloading operations of Stock Keeping Units (SKUs) in warehouses on the environmental CO_2 emissions has also been dealt with [12]. It is necessary to implement innovative decarbonization policies supporting prosperous low-carbon SCs [13]; among available net-zero carbon technologies across multiple sectors, Hydrogen as fuel will be a key technological changemaker.

We identify a major gap in literature regarding SCI, and modelling the facility-operation and transportation emissions. Few papers in literature investigate the global warming potential (GWP); a case for a treatment plant with Life Cycle Assessment (LCA) approach [14] found the main GWP contributions coming from electricity needs (more than 70% both in summer and in winter). Complex disruptions pre and post the coronavirus disease of 2019 (COVID-19) necessitates aspects of decentralization for resilience in SCs [15]. Similar to a research on determining minimum cost routes from factory warehouses to distribution centres for a multi-echelon dairy SC [16], in this research, we consider the objective of minimizing total emissions (which we propose should be enhanced into GWP minimization when doing real case studies) and further show the requirement of integration of independent SCs for a much lower overall emission.

One of the major environmental concerns within the focus area of cold SCs is that it supporting infrastructure consumes heavy energy to maintain the refrigeration temperatures ultimately resulting in increased emissions. Various technologies have been implemented in SCs to save energy consumed and reduce carbon emissions contributing to sustainability. Methods to evaluate the environmental impacts of food cold SCs have been studied considering LCA [17], a bottom-up approach was used to evaluate GHG emissions and estimate cold-chain activities on perishable food products to constitute about 1–3% of the overall emissions from China in 2018. Cold logistics must be optimized in order to achieve carbon neutrality [18] because of its enormous energy usage and waste. Utilizing energy-efficient refrigeration technologies, minimizing greenhouse gas emissions, reducing waste, and conserving natural resources to mitigate environmental impact is the primary objective and among the prominent strategies in the environmental dimension within sustainable cold chains [19]. The literature survey conducted on green SCs in the context of food cold chains is presented in Table 5.1, highlighting the methodology used in each case.

In the context of the SC, profit sharing refers to a cooperative financial agreement between the different SC participants, including distributors, retailers, suppliers, and manufacturers. Under this arrangement, the various partners split up a share of the profits made throughout the SC according to predetermined agreements or criteria. The aim of profit sharing is to synchronize the interests of the parties involved in the SC, promoting collaboration and coordinated endeavours to improve overall efficiency. Collaborative performance is one of the key aspects in profit sharing in SC. Profit sharing and the overall performance of the SC are frequently connected. Profits from a successful overall SC are distributed among the involved parties, fostering a sense of shared accountability and cooperation. Research on the profit distribution of SCs in e-commerce environments is uncommon, with the majority of

Table 5.1 Literature survey on environmental sustainability in cold chains

Ref. Paper	Sustainability focus			Research study		Objective and strategy
	Environmental	Economical	Social	Empirical	Modelling	
Bag, 2016 [20]	✔	✔		✔		Modelling barriers of Green Cold SC using Interpretive Structure Model
Chiang *et al.*, 2019 [21]	✔	✔		✔		Route optimization of unmanned aerial vehicles through MILP mathematical model and genetic algorithm
Hu *et al.*, 2019 [22]	✔			✔		Study of GHG emission reductions using non-linear optimization
Zhang *et al.*, 2019 [23]	✔			✔		Low-carbon cold chain using ant colony optimization algorithm
Neusel and Hirzel, 2022 [24]	✔	✔		✔		Energy efficiency in food SCs using thermal management
Jeong and Lee, 2023 [25]	✔	✔		✔		Cost optimization for rapid delivery of perishable food SC using optimization by multimode vehicles
Wu *et al.*, 2019 [26]	✔	✔		✔	✔	Environmental trade-offs in fresh-fruit SCs with life cycle assessment

(continued)

Table 5.1 (continued)

Ref. Paper	Sustainability focus			Research study		Objective and strategy
	Environmental	Economical	Social	Empirical	Modelling	
Chkanikova, 2016 [27]	✔			✔		Study of sustainable purchasing in green product retailing
Arora *et al.*, 2023 [28]	✔	✔	✔	✔		Performance assessment of perishable food SC using structural equation modelling

studies focussing on SCs at a single level, but this is an essential factor for practical operation of green SCs.

Profit sharing in SC among multiple suppliers is relatively less explored area in literature and this has a great scope in future academic work; very few academicians have laid emphasis on this. In one such effort [29], an evolutionary game model for distribution of profit in SCs is formulated, the idea being merging enterprises of the same SC to reduce production cost. Authors have also considered a two-level and four-party SC that includes a demander, an e-commerce platform, a supplier, and third-party logistics [30]. Although studies on profit sharing for the case of multiple suppliers is scare, studies performed in the same context for multiple buyers are available. Managing SCs of any product is a not a short-term process and decisions to make profit among the stake holders should consists of diverse eco-friendly long-term strategies; this concept is analysed using game theory for profit distribution [31].

Every SC targets all three dimensions of sustainability, i.e. social, economic and environmental dimensions [32, 33]. The integration of sustainable practices within SCs has become a key focus for businesses due to increasing awareness of environmental, social and economic impacts. Several researchers have also focussed on a particular dimension of sustainability [34] keeping in mind the social aspects for micro, small and medium enterprises. Green practices in SCs are inevitable in today's world and profit distribution in SCs should not be done without incorporation of green practices. Today several policies like cap and trade are used for limiting emissions; such limitations significantly affect companies' decisions about reducing their carbon emissions, their production processes and how profits are distributed [35–37]; suppliers coming together for common SC of their products can be also studied in detail in diverse sectors.

5.3 Problem Description and Mathematical Formulation

Based on the literature review, we find that the considered problem of optimizing emissions within a SC, and comparing independent SCs with integrated ones has not been dealt with extensively. The few papers that have worked on similar problems include the consideration of a network design problem for SCs [38] accounting for CO_2 emissions and modelling the relationship of these emissions with vehicle weight using a concave function. The impact of green technologies and carbon policies on integrated inventory management has also been considered [39] for a two-echelon SC with a single vendor and buyer. A two-echelon SC network of the battery manufacturing industry with a green objective function has also been considered in the literature [40] with a carbon emission cost. Carbon trading integration within SC network design has also been considered [41] allowing decision-makers to understand the trade-off between impacts of GHG reduction and logistics costs. A review on green SC network design [42] found four most successfully incorporated policies that have been able to substantially reduce emission with minor cost increase, sensing a dearth of models that capture the complex nature of emissions, and with the effect of emissions on demand being hardly accounted for.

In this paper, we design our problem focussing on the requirement of a facility being open (remain functional for the duration of the considered objective) as a decision. We then compare individual SCs separately, as well as integrated SCs, highlighting reduction in emissions after the integration. We consider that after integration, a new integrated SC will have its own holistic singular decision-making capability, and can be optimized using this same formulation. Some other, more complex integration variants could be the incorporation of inter-connected decisions necessitating new additional constraints after every new SCI, which we do not consider for our study.

We model our variant of the problem considering SCI as the separate merging of each of the sets (stages of the problem) as indicated in Fig. 5.1, i.e. all sets of the same type from individual SCs are merged in order to obtain the corresponding set of the integrated SC. We consider the problem having the following features:

i. The requirement of a facility to being open is a decision (to save on its emissions).
ii. We consider a four-level network consisting of the following stages (sets):

    1. Suppliers (S)
    2. Distribution Centres (R)
    3. Branches (B)

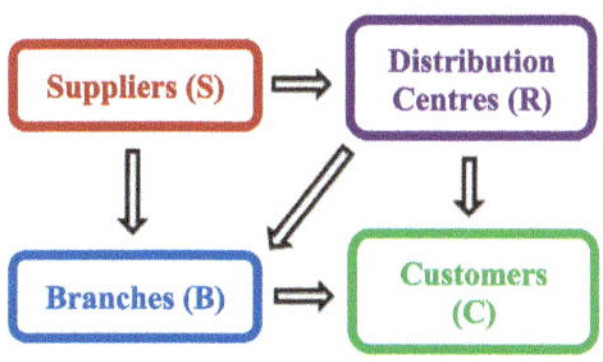

Fig. 5.1 Inter-set connections available for transportation

4. Customers (C)

iii. All commodity flow happens in similar sequence as indicated in Fig. 5.1, i.e. from S to R, from R to B, and from B to C. We also allow two other transportation routes, from S to B, and from R to C. Elaborating on the intricacies of the commodity flow, this flow of commodities through the network as described in Fig. 5.1 is driven by the diverse demand generated at the end customers (indicated by the set C) to deliver some quantity of various types of products at their respective locations. These demands could be based on actual requests, but could as well be forecasted predictions, in which case the Customer set C would contain end customers as well as the smallest of stores; any demand can be satisfied by any of the appropriate Supplier (depending on its product availability). During this transfer of products from any Supplier to a Customer, the products must pass through either a Distribution Centre (R) or a Branch (B), or both.

iv. We consider both fixed as well as load-based variable emissions at each R and B.

v. We consider a single type of vehicle on the transportation network, and further consider any inter-stage transportation to be done by a unique vehicle (we do not consider any intra-stage transportations, thereby doing away with the complexity of modelling complex vehicle route plans, to clarify further each of the four stages have been modelled into the four sets as described in Fig. 5.1). We assume all possible transportation links as show in the Fig. 5.1 are present, and each transportation link is being catered to by a single vehicle of sufficient capacity (Q).

vi. We model both fixed and load-and-distance-based variable emission during transportation.

The notations of the sets and parameters used for modelling the problem are detailed in Table 5.2, with Table 5.3 describing the decision variables used in the Mixed Integer Linear Programming (MILP) formulation.

We model the Objective Function as in Eq. 5.1 to represent the total emission, and minimize it. The first term represents the fixed emission and variable (as per load and distance) emission due to transportation of the loads. The fixed emission, which represents the emission of the empty vehicle per unit distance travelled, is doubled due to vehicles' (empty) return journey. The second term represents the fixed emission from those Distribution Centres and Branches which are made to remain operational. The last two terms represent the variable emissions (as the sum of emissions for handling each product) to maintain Distribution Centres and Branches, respectively.

$$\begin{aligned} \text{minimize} \sum_{(i\rightarrow j)\in E} T_{i,j} \cdot \left(2 \cdot x_{i,j} \cdot F^0 + V^0 \cdot \sum_{k\in K} y^k_{i,j}\right) + \sum_{i\in R,B} z_i \cdot F_i \\ + \sum_{i\in R} V_i \cdot \sum_{k\in K}\sum_{j\in S} y_{j,i,k} + \sum_{i\in B} V_i \cdot \sum_{k\in K}\sum_{j\in C} y_{i,j,k}, \end{aligned} \quad (5.1)$$

Table 5.2 Sets and parameters used

Notation	Sets and parameters
S	Set of all suppliers
R	Set of all distribution centres
B	Set of all branches
C	Set of all customers
K	Set of all products that are in demand by the customers
E	Set of all edges connecting the different types of nodes/vertices $(i \rightarrow j)$. The considered connections between the elements from the different sets are: $S \rightarrow R \cup B, R \rightarrow B$, and $R \cup B \rightarrow C$; as shown in Fig. 5.1
$T_{i,j}$	Travelling distance from vertex i to vertex j for the transportation, where $(i \rightarrow j) \in E$
U_i^k	Refers to the amount of product k, $(k \in K)$, available to be supplied by supplier i, where $i \in S$
H_i^k	Inventory holding and handling capacity of product k, $(k \in K)$, at the facility at the vertex i, where $i \in R \cup B$
D_i^k	Demand refers to the amount of product k, $(k \in K)$, required by customer i, where $i \in C$
F_i	Fixed emission for location i, where $i \in R \cup B$
V_i	Variable emission for location i, where $i \in R \cup B$
F^0	Fixed emission of an empty vehicle per unit distance travelled
V^0	Variable emission per unit quantity of product being transported, per unit distance (this is calculated over and above the fixed emission)
Q	Capacity of each homogeneous vehicle
M	An arbitrary high-value positive real number
G_i	Fixed cost for location i, where $i \in R \cup B$
W_i	Variable cost for location i, where $i \in R \cup B$
G^0	Fixed cost of an empty vehicle per unit distance travelled
W^0	Variable cost per unit quantity of product being transported, per unit distance (in addition to the fixed cost)

Table 5.3 Notations and description of the decision variables used in the MILP model

Notation	Variable description
$x_{i,j}$	Binary decision variable referring to whether the path from vertex i to vertex j is selected to be used during the transportation
$y_{i,j}^k$	Continuous decision variable referring to the amount of product of category k being transported from vertex i to vertex j
z_i	Binary decision variable referring to whether the (facility at) vertex i is used in the solution

We minimize Eq. 5.1 subject to the constraints in the Eqs. 5.2–5.12.

$$\sum_{j\in R,B} y_{i,j}^k \le U_i^k, \quad \forall k \in K, \ \forall i \in S, \tag{5.2}$$

$$\sum_{j\in B,C} y_{i,j}^k \le \sum_{j\in S} y_{j,i}^k, \quad \forall k \in K, \ \forall i \in R, \tag{5.3}$$

$$\sum_{j\in C} y_{i,j}^k + \sum_{j\in S} y_{j,i}^k \le \sum_{j\in R} y_{j,i}^k, \quad \forall k \in K, \ \forall i \in B, \tag{5.4}$$

$$\sum_{i\in S} y_{i,j}^k \le H_j^k, \quad \forall k \in K, \ \forall j \in R, \tag{5.5}$$

$$\sum_{i\in S,R} y_{i,j}^k \le H_j^k, \quad \forall k \in K, \ \forall j \in B, \tag{5.6}$$

$$\sum_{i\in R,B} y_{i,j}^k = D_j^k, \quad \forall k \in K, \ \forall j \in C, \tag{5.7}$$

$$\sum_{k\in K} y_{i,j}^k \le Q \cdot x_{i,j}, \quad \forall (i \to j) \in E, \tag{5.8}$$

$$\sum_{j\in S} x_{j,i} + \sum_{\text{if}(i\notin R)\to(j\in R)} x_{j,i} + \sum_{if\,(i\notin B)\to(j\in B)} x_{i,j} + \sum_{j\in C} x_{i,j} \le M \cdot z_i, \quad \forall i \in R \cup B, \tag{5.9}$$

$$x_{i,j} \in \{0, 1\}, \quad \forall (i \to j) \in E, \tag{5.10}$$

$$y_{i,j}^k \ge 0, \quad \forall (i \to j) \in E, k \in K, \tag{5.11}$$

$$z_i \in \{0, 1\}, \quad \forall i \in R \cup B, \tag{5.12}$$

The constraint in Eq. 5.2 ensures the supply of each product from the Suppliers to the Distribution Centres and Branches should be less that the amount of that product available for supply. Equation 5.3 ensures the supply of each type of product from the Distribution Centres to the Branches and Customers should be less that the amount of inventory available at the respective R for that specific product. Equation 5.4 restricts the supply of each product from Branches to Customers to less that the available inventory at the respective B. In Eq. 5.5, the aggregation of the same products sent to a R from Suppliers should be less that the inventory holding and handling capacity for that specific product at the concerned R. Equation 5.6 similarly restricts the aggregate of a product type sent to each element in B from facilities in S and R to be less than the respective H. Equation 5.7 ensures that all requested customer demands are met, and Eq. 5.8 ensures that the singular vehicle plying between each unique connection is not overloaded. Equation 5.9 populates the variable z, indicating which facilities should remain functional (stay open) to satisfy the given set of customers.

Table 5.4 Results from generated datasets show the possible rise in emissions when the supply chains are meant to be efficient rather than green

Supply chains	$\lvert S\rvert$	$\lvert R\rvert$	$\lvert B\rvert$	$\lvert C\rvert$	$\lvert K\rvert$	No. of vehicles allowed to be used $\lvert E\rvert$	When minimizing the emission objective of Eq. 5.1				Minimizing the cost objective of Eq. 5.13				Increase in emission when minimizing cost instead of emissions $(c)^{*}$ (%)
							Emission value (a)	Open distribution centres	No. of open branches	Vehicles used	Emission value (b)	No. of open distribution centres	Open branches	No. of vehicles used	
1	2	3	5	10	2	111	511.11	1	0	12	630.50	2	0	13	23.36
2	2	3	5	10	2	111	704.04	1	0	12	797.01	2	0	13	13.21
3	2	3	5	10	2	111	645.26	2	0	14	692.98	2	0	12	7.4
4	1	1	2	5	3	20	350.95	1	0	6	350.95	1	0	6	0
5	1	2	3	5	3	36	253.91	1	0	6	253.91	1	0	6	0

$^{*}c = 100 \cdot (b - a)/a$

To generate the dataset for each unique original Supply Chain in Table 5.4 (i.e. for the Supply Chains 1, 2, 3, 4, 5), the necessary input was obtained randomly from within a range:

1. Latitude and Longitude (for all elements in the sets S, R, B and C) were obtained from a uniform distribution from within [− 50, 50]
2. The Fixed Emission for each Distribution Centre ($F_i, \forall i \in R$) was obtained from a uniform distribution from within [0, 50]
3. The Variable Emission for each Distribution Centre ($V_i, \forall i \in R$) was obtained from a uniform distribution from within [0, 1]
4. The Fixed Emission for each Branch ($F_i, \forall i \in B$) was obtained from a uniform distribution from within [0, 10]
5. The Variable Emission for each Branch ($V_i, \forall i \in B$) was obtained from a uniform distribution from within [0, 1]
6. All product availability at suppliers ($U_i^k, \forall k \in K, \forall i \in S$, varied within [0, 1000]), inventory holding and handling capacities ($H_j^k, \forall k \in K, \forall j \in R$ varied within [0, 750] at R and $H_j^k, \forall k \in K, \forall j \in B$ varied within [0, 250] at B) and demands at each customer ($D_j^k, \forall k \in K, \forall j \in C$, varied within [0, 50]) were obtained accordingly
7. All cost parameters $\left(G^0, W^0, G_i, W_i\right)$ were obtained similarly to their respective emission parameters $\left(F^0, V^0, F_i, V_i\right)$
8. Vehicle specifications of capacity ($Q = 500$), Fixed Emission ($F^0 = 100$) and Variable Emission ($V^0 = 5$) were kept unchanged over all instances.

5.4 Results and Discussion

To compare the emission values when the optimization objective is the minimization of total emissions, versus the emission values when the minimization of cost is being performed, we use the parameters G_i, W_i, G^0 and W^0 from Table 5.2 to develop the Cost Minimization objective function of Eq. 5.13; when using the Emission Minimization objective, we may also use only the expression of Eq. 5.13 to calculate the Cost of the entire operation.

We find considerable increase in emissions when the cost objective from Eq. 5.13 is used instead of minimizing the emissions (Eq. 5.1).

$$\begin{aligned} \text{minimize} \sum_{(i \to j) \in E} T_{i,j} \cdot \left(2 \cdot x_{i,j} \cdot G^0 + W^0 \cdot \sum_{k \in K} y_{i,j}^k\right) + \sum_{i \in R,B} z_i \cdot G_i \\ + \sum_{i \in R} W_i \cdot \sum_{k \in K} \sum_{j \in S} y_{j,i,k} + \sum_{i \in B} W_i \cdot \sum_{k \in K} \sum_{j \in C} y_{i,j,k}, \end{aligned} \tag{5.13}$$

We generate data for five instances, creating five different SCs as indicated in Table 5.4, and further integrate them in Table 5.5 (the last row in Table 5.5 is the

Table 5.5 Progressive decrease in emissions as supply chains coalesce while using the green objective of emission reduction

Integrated supply chains	|S|	|R|	|B|	|C|	|K|	No. of homogeneous vehicles available (|E|)	Minimizing the emission objective of Eq. 5.1				Minimizing the cost objective of Eq. 5.13				Rise in emission while minimizing cost instead of emissions [100·(b − a)/a] (%)	Emission savings of this integrated SC w.r.t. its individual SCs* (%)
							Emission value (a)	Open distribution centres	No. of open branches	Vehicles used	Emission value (b)	No. of open distribution centres	Open branches	No. of vehicles used		
1, 2	4	6	10	20	3	444	1170.95	2	0	27	1494.5	3	0	33	27.63	3.64
4, 5	2	3	5	10	3	111	416.89	2	0	12	440.1	2	0	13	5.57	31.08
1, 2, 3	6	9	15	30	3	999	1756.48	3	0	44	2030.33	4	0	50	15.59	5.59
3, 4, 5	4	6	10	20	3	444	740.41	4	1	28	910.02	3	0	26	22.91	40.78
1, 2, 4, 5	6	9	15	30	3	999	1002.63	3	1	42	1826.75	4	1	54	82.2	44.91
1, 2, 3, 4, 5	8	12	20	40	3	1776	1268.9	4	1	58	2229.25	5	0	64	75.68	48.53

*For each instance (row), the sum of emission values of its constituent SCs (labelled a in Table 5.4) is compared with the total emission value of this integrated SC (column a of Table 5.5)

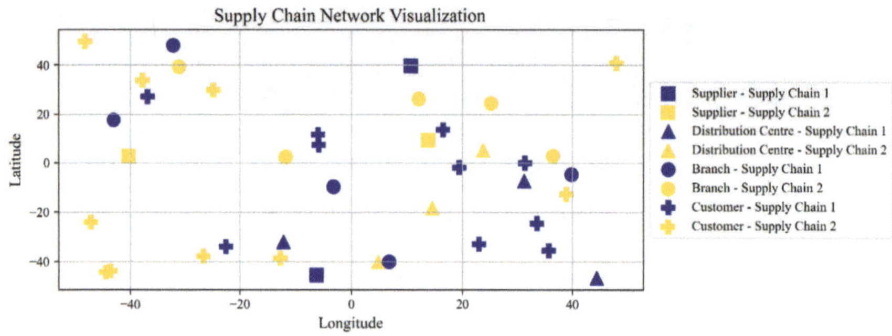

Fig. 5.2 An integrated supply chain is shown (first instance of Table 5.5); its two original supply chains (1 and 2 from Table 5.4) are depicted using different colours

integration of all the five original SCs essentially forming a large supply web) to show the improvement in terms of reduction in the modelled emissions. One such integrated supply chain is shown in Fig. 5.2. The dataset for the original five SCs in Table 5.4 is synthetically generated, and we use this same data after SCI in the Table 5.5.

We observe from our computations in Tables 5.4 and 5.5 that there is a dramatic increase in the difference in emissions with respect to (w.r.t.) changing objectives of Emission Reduction and Cost Minimization when the size of the SC increases. Thus, larger SCs play a bigger role towards ensuring sustainable practices and, optimizing larger SCs with emission minimization intent would prove more beneficial just in terms of the percentage improvement through reduced emissions (and definitely in terms of the actual quantity of emissions avoided).

5.5 Findings and Contributions

In this problem, initially we consider a few separate Maintenance, Repair and Operations (MRO) Supplies-based SCs for individual corporate entities; and show how integration of these SCs would help reduce the total emission, w.r.t. the separate individual SCs. The integration in this case is defined to be performed by separately merging each of the corresponding sets of Suppliers, Distribution Centres, Branches and Consumers present within the original SCs; the parameters and characteristics of all the original elements are retained after integrations. Our results indicate that minimizing the individual costs of the five SCs separately increases emissions by about **114.78%** w.r.t. the case when minimizing the total emissions in the integrated SC (when all the five independent SCs in Table 5.4 have been integrated into a single SC in Table 5.5, with any Suppliers being able to cater to any Customer through the similarly available inter-connections among the newly integrated stages as shown in Fig. 5.1). We believe that instead of further developing random variables for the parameters considered for our and similar problems to develop and compare stochastic programming approaches towards emission reduction, real-world case studies would reveal the actual potential of clubbing independent SCs into a single green entity; this should result in magnitudes of emission reduction, much more than estimated by us for the single holistic green integration with random parameter values. SCI is generally complex to implement due to multiple profit-sharing methodologies available, favouring one-player over the other and the lack of a general consensus on the same. The aspect of profit sharing in integrated SCs is poorly dealt with in the literature, being a complicated subject in itself. We therefore need to extensively dwell on developing the best profit-sharing approaches promoting SCI, focussing especially on those which improve green sustainable SC performance, envisioning a future government mandate towards integration of SCs for developing a universal green sustainable supply web.

5.5.1 Policy Implications

Metamorphosing our current economy-driven society into being climate conscious, and taking the path of sustainability for our civilization's longevity, would involve drastic policy implementations including:

- Mandating such SCI and promoting self-integration through collaboration of market players by introducing higher GWP Tax on the product/components: The underlying aspects of SCI for the case of European firms [43] indicate that comprehensive SCI is not near reality and therefore it is necessary to introduce mechanisms that model the actual consumer requirements w.r.t. a future direction of elongating our existence by mitigating climatic threats. Appropriate policies would therefore require simulation and introspection; implementing a GWP Tax would automatically require industry and corporate collaborations for a sustainable SCI. In this regard, wastes (like plastics) and effluents would also need to be categorized and weighted for assigning of appropriate ecosystem destruction potentials through LCA and environmental impact assessment.
- Mandating corporate decision support systems as well as any independent decision-making to be focussed on minimizing GWP as the primary objective: For any commercial entity currently, the main optimization criteria is being massively profit making, with the sustainability issue being secondary. We show in this paper, that such a secondary approach towards climate consciousness is able to more-than-at-least double our emissions, which is unfortunately shocking. Mandating GWP minimization to be the primary objective will be a win–win for all life-forms and this is extremely easy to implement, especially since this requires only small software tweaks impacting the subsequent internal policy and decision-making of large corporates.

The soul of a decision-making entity can be determined by the real optimization approach being considered, i.e. cost minimization versus emission minimization. An international mandate of corporate decision intelligence focussed on sustainability instead of profit maximization (i.e. shifting the primary decision objective from being Efficient or Responsive towards being Green and Sustainable) would cascade down towards a healthier cooler planet Earth, an urgent need of the hour; current government mandates do not bind internal corporate decision-making algorithms to mandatorily use emission minimization objectives and therefore a governing mechanism needs to be established. Individual countries and states would soon follow suit as changing simple decision intelligence software with sustainability mandates would require simple algorithm changes, and this would also help in driving more research into designing and modelling green sustainable integrated SCs with zero-waste and zero-emissions; this would also ultimately promote prosperity of the various interacting elements within our complex supply–demand web.

5.6 Conclusions and Future Work

In this study, we designed a four-stage SC consisting of Suppliers, Distribution Centres, Branches and Customers (Fig. 5.1), for multi-commodity flow via homogeneous vehicles, to satisfy end customer demands. Variable and fixed emissions are modelled for the functional operations and product handling in Distribution Centres and Branches, as well as during transportation. We focus on minimizing the emissions with the decisions being the usage of certain Distribution Centres and Branches, and necessary transportation links. We also develop a cost objective (modelling it similar to emissions, with both fixed and variable components for transportation costs, and for the facility usage costs of *R*s and *B*s) and minimize it for the same problem instances to compare the trade-off in emissions in this case.

Further comparison is done after the SCs in our considered instances coalesce and integrate; we find higher rise in emissions when the cost objective is used instead of the emission minimization objective the larger the SC. Expanding on the fundamental classical ideas of classification of SCs being efficient and responsive only, the dimension of sustainability needs to be used for assessing performance in SCs as well. In this progressive era necessitating enormous climate action, previous performance analysis parameters are insufficient and evolution of each type of SC (improving over the classical approaches) is shown in Table 5.6.

As a future research area, our current modelling assumption of each transportation link (unique connection from one vertex to another) being catered to by a single vehicle of sufficient capacity (Q) could be expanded to allow multiple heterogeneous vehicles to ply on the network. Parallel approaches to compare the difference in emissions from different objectives like total cost minimization or minimizing the total time could yield deeper insights and provide a problem-case-specific understanding; this would help towards the requirement of developing a Green Sustainable SC instead of optimizing for other goals, through elaborate case studies using real-world parameter values. Availability of any existing inventory can also be incorporated by altering the existing model. Another appropriate direction for further investigation would be to also consider the temporal causality, especially during

Table 5.6 Increasing scope of supply chain classification to enable achievement of optimal strategic fit

		Types of Supply Chain		
		Efficient	Responsive	**Green & Sustainable (with necessary Integration)**
Primary Goal →		Lowest Cost	Quick Response	Lowest Emissions and least impact on future organisms of Earth, by reusing raw materials, reducing GHG emissions and generating lesser wastes
STRATEGY	Product Design	Min. Product Cost	Modularity to allow postponement	Including sustainable practices and innovation allowing reduced emissions as well as minimum waste generation. The generated waste after useful life cycle of the product should be biodegradable or collected back for recycling
STRATEGY	Pricing	Lower Margins	Higher Margins	Existing customers are willing to pay more for sustainable products and practices
STRATEGY	Manufacturing	High Utilization	Capacity Flexibility	High utilization of intelligent manufacturing techniques including less waste generation and upcycling
STRATEGY	Inventory	Minimize Inventory	Buffer Inventory	Green Warehouses use energy efficient lighting sources like LED lights; reduces wastes through efficient packaging using bio-degradable materials; opting for cool roofs thereby increasing cooling efficiency of the warehouse
STRATEGY	Lead Time	Reduce but not at expense of greater cost	Aggresively reduce even if costs are significant	Reduce but not at the cost of emissions, greater waste generation or unsustainable practices
STRATEGY	Supplier Selection	Cost and Low Quality	Speed, Flexibility, Quality	Green production channel and sustainable practices including waste management standards
STRATEGY	Transportation and Distribution	Greater reliance on low cost modes	Greater reliance on responsive (fast) modes	Reliance on Green Hydrogen Vehicles, Bio-Fuels, CNG, E-Vehicles for transportation thereby developing a Green Supply Chain. Developing back-channels for waste collection preventing wastes to be dumped to degrade the environment
Introducing Penalty Factors		Transitioning into **Sustainable Efficient** supply-chain	Transitioning into **Sustainable Responsive** supply-chains	Including penalties for products and supply chains which don't follow green, clean, sustainable practices and proving government subsidies to promote eco-friendly, products, services, and supply-chains.
End-of-Life product management through Reverse Logistics		Essential to introduce and effectively integrate End-of-Life Product Management, thereby establishing services for back-propagation of wastes ensuring Up-Cycling and Re-Cycling		End-of-Life Product Management is included and should be maintained as well as scrutinized by external authorities periodically
Metrics to measure Sustainability of Supply Chains:				Impact on the Planet, People Engagement, Profits
		Social, Financial and Environmental Responsibilities need to be properly addressed through innovative labour practices and CSR		

consideration of the inventory holding and handling capacities. Currently the model assumes that the entire process of obtaining the loads from appropriate transportation networks into the facilities happens at once. The capacity constraint for the facilities, (constraining capacities to not exceed above their respective value of H), could therefore be improved since all the loading unloading process will not take place at the same time in reality, and therefore, the facilities may not get overwhelmed easily.

Multiple emission parameters should be further considered (instead of a single type of emission) and weighing these different types of emissions with their respective GWPs to obtain a cumulative objective function would enable minimization of GWP for the integrated SC. A government mandate on the integration is essential with further elaborate research on profit sharing in such extensive hierarchical integrated supply webs.

Future research could also focus on finding easy practically implementable pareto-optimal fronts of the two clashing objectives (profit-vs-emissions); another idea is to estimate the profit lost and the positive climate impact achieved due to the pareto-optimal solution implementation and weighing the benefits, government tax benefits in this regard would help shift pareto-fronts towards the green sustainable domain.

Data Availability Statement The dataset along with the necessary code is publicly accessible and hosted online on GitHub; they may be obtained from the below link to replicate the results:

https://github.com/SanTanBan/DataSet-ICONIEA-2024-Emission-Reduction-through-Green-Sustainable-Supply-Chains

CRediT Author Statement

Santanu Banerjee: Conceived and Designed the Study; Investigation and Methodology: Devised the Mathematical Models, Created Datasets for the Computational Analyses; Original Draft Writing and Editing, Compilation and Manuscript Revisions.

Amit Kumar Singh: Literature Review of Existing Studies to Identify Research Gaps; Original Draft Writing, Manuscript Review.

Pranav Sanjay Gore, **Thiruchitrambalam B**, and **Ankit Patel**: Manuscript Review.

References

1. Flynn, B.B., Huo, B., Zhao, X.: The impact of supply chain integration on performance: a contingency and configuration approach. J. Oper. Manag. **28**(1), 58–71 (2010)
2. Fawcett, S.E., Magnan, G.M.: The rhetoric and reality of supply chain integration. Int. J. Phys. Distrib. Logistics Manag. **32**(5), 339–361 (2002)
3. van der Vaart, T., van Donk, D.P.: A critical review of survey-based research in supply chain integration. Int. J. Prod. Econ. **111**(1), 42–55 (2008)
4. Singh, A., Gangopadhyay, S., Nanda, P., Bhattacharya, S., Sharma, C., Bhan, C.: Trends of greenhouse gas emissions from the road transport sector in India. Sci. Total. Environ. **390**(1), 124–131 (2008)
5. Rahman, S.M., Khondaker, A.N., Hasan, M.A. and Reza, I.: Greenhouse gas emissions from road transportation in Saudi Arabia—a challenging frontier. Renew. Sustain. Energy Rev. **69**, 812–821 (2017)
6. Dhar, S., Shukla, P.R.: Low carbon scenarios for transport in India: co-benefits analysis. Energy Policy **81**, 186–198 (2015)

7. Huang, C.-H.: Design and Development of Wireless Power Transmission for Unmanned Air Vehicles (2012)
8. Rosu, F., Badescu, A.: Electric and magnetic design of a deployable WPT system for industrial and defense UAV applications. Electronics **10**(18), 2252 (2021)
9. Shin, J., Shin, S., Kim, Y., Ahn, S., Lee, S., Jung, G., Jeon, S.-J., Cho, D.-H.: Design and implementation of shaped magnetic-resonance-based wireless power transfer system for roadway-powered moving electric vehicles. IEEE Trans. Industr. Electron. **61**(3), 1179–1192 (2014)
10. Xing, C., Liu, C., Lin, J., Tan, W., Liu, T.: VOCs hyperspectral imaging: a new insight into evaluate emissions and the corresponding health risk from industries. J. Hazard. Mater. **461**, 132573 (2024)
11. Das, S.K., Yu, V.F., Roy, S.K., Weber, G.W.: Location–allocation problem for green efficient two-stage vehicle-based logistics system: a Type-2 neutrosophic multi-objective modeling approach. Expert Syst. Appl. **238**, 122174 (2024)
12. Sadeghi, M., Nikfar, M., Rad, F.M.: Optimizing warehouse operations for environmental sustainability: a simulation study for reducing carbon emissions and maximizing space utilization. J. Future Sustain. **4**(1), 35–44 (2024)
13. Mani, A., Budd, T., Maine, E.: Emissions-intensive and trade-exposed industries: technological innovation and climate policy solutions to achieve net-zero emissions by 2050. RSC Sustain. (2024)
14. Lu, T., Lü, F., Liao, N., Chai, H., Zhang, H., He, P.: Material flow analysis and global warming potential assessment of an industrial insect-based bioconversion plant using housefly larvae. J. Environ. Sci. **139**, 483–495 (2024)
15. Adana, S., Manuj, I., Herburger, M., Cevikparmak, S., Celik, H., Uvet, H.: Linking decentralization in decision-making to resilience outcomes: a supply chain orientation perspective. Int. J. Logistics Manag. **35**(1), 256–280 (2024)
16. Wofuru-Nyenke, O.: Routing and facility location optimization in a dairy products supply chain. Future Technol. **3**(2), 44–49 (2023)
17. Dong, Y., Miller, S.A.: Assessing the lifecycle greenhouse gas (GHG) emissions of perishable food products delivered by the cold chain in China. J. Clean. Prod. **303**, 126982 (2021)
18. Bin, L., Jiawei, L., Aiqiang, C., Theodorakis, P.E., Zongsheng, Z., Jinzhe, Y.: Selection of the cold logistics model based on the carbon footprint of fruits and vegetables in China. J. Clean. Prod. **334**, 130251 (2022)
19. Adams, D., Donovan, J., Topple, C.: Sustainability in large food and beverage companies and their supply chains: an investigation into key drivers and barriers affecting sustainability strategies. Bus. Strateg. Environ. **32**(4), 1451–1463 (2023)
20. Bag, S.: Modeling barriers of green cold chain management in India. J. Supply Chain Manag. Syst **5**(1) (2016)
21. Chiang, W.-C., Li, Y., Shang, J., Urban, T.L.: Impact of drone delivery on sustainability and cost: realizing the UAV potential through vehicle routing optimization. Appl. Energy **242**(0306–2619), 1164–1175 (2019)
22. Hu, G., Mu, X., Xu, M., Miller, S.A.: Potentials of GHG emission reductions from cold chain systems: case studies of China and the United States. J. Clean. Prod. **239**, 118053 (2019)
23. Zhang, L.-Y., Tseng, M.-L., Wang, C.-H., Xiao, C., Fei, T.: Low-carbon cold chain logistics using ribonucleic acid-ant colony optimization algorithm. J. Clean. Prod. **233**, 169–180 (2019)
24. Neusel, L., Hirzel, S.: Energy efficiency in cold supply chains of the food Sector: an exploration of conditions and perceptions. Clean. Logistics Supply Chain **5**, 100082 (2022)
25. Jeong, H.Y., Lee, S.: Drone routing problem with truck: optimization and quantitative analysis. Expert Syst. Appl. **227**, 120260 (2023)
26. Wu, W., Beretta, C., Cronje, P., Hellweg, S., Defraeye, T.: Environmental trade-offs in fresh-fruit cold chains by combining virtual cold chains with life cycle assessment. Appl. Energy **254**, 113586 (2019)
27. Chkanikova, O.: Sustainable purchasing in food retailing: interorganizational relationship management to green product supply. Bus. Strateg. Environ. **25**(7), 478–494 (2016)

28. Arora, M., Kumar, R., Bangwal, D., Ahmad, V., Negi, S.: An empirical study of cold chain issues and performance: applying structural equation modeling. J Food Process Eng **46**(12)
29. Zhang, Y., Geng, H.: An analysis on distribution of cooperative profit in supply chain based on multi-agent. Phys. Proc. **33**, 698–704 (2012)
30. Zhang, J.-H., Yin, R.-R., Zhang, J., Nie, M.-J.: The profit distribution of supply chain under e-commerce. Discrete Dyn Nat Soc **2014**, 287925 (2014)
31. Hosseini-Motlagh, S.-M., Choi, T.-M., Johari, M., Nouri-Harzvili, M.: A profit surplus distribution mechanism for supply chain coordination: an evolutionary game-theoretic analysis. Eur. J. Oper. Res. **301**(2), 561–575 (2022)
32. Ageron, B., Gunasekaran, A., Spalanzani, A.: Sustainable supply management: an empirical study. Int. J. Prod. Econ. **140**(1), 168–182 (2012)
33. Dubey, R., Gunasekaran, A., Childe, S.J., Papadopoulos, T., Fosso Wamba, S.: World class sustainable supply chain management: critical review and further research directions. Int. J. Logistics Manag. **28**(2), 332–362 (2017)
34. Uttam, N., Dutta, P., Singh, A.: Micro, small, and medium suppliers' perspectives on supply chain social sustainability: new evidence from India. J. Clean. Prod. **379**, 134473 (2022)
35. Jiang, W., Lu, W., Xu, Q.: Profit distribution model for construction supply chain with cap-and-trade policy. Sustainability **11**(4), 1215 (2019)
36. Jiang, W., Wei, M., Wu, W.: Profit distribution strategy of cooperation construction supply chain under carbon tax policy. In: IOP Conference Series: Earth and Environmental Science, vol. 304, no. 3, p. 032110 (2019)
37. Jiang, W., Yuan, L., Wu, L., Guo, S.: Carbon emission reduction and profit distribution mechanism of construction supply chain with fairness concern and cap-and-trade. PLoS ONE **14**(10), 1–23 (2019)
38. Elhedhli, S., Merrick, R.: Green supply chain network design to reduce carbon emissions. Transp. Res. Part D: Transp. Environ. **17**(5), 370–379 (2012)
39. Marchi, B., Zanoni, S.: Inventory management in supply chains with consideration of logistics, green investment and different carbon emissions policies. Comput. Ind. Eng. **175**, 108870 (2023)
40. Sherif, S.U., Asokan, P., Sasikumar, P., Mathiyazhagan, K., Jerald, J.: Integrated optimization of transportation, inventory and vehicle routing with simultaneous pickup and delivery in two-echelon green supply chain network. J. Clean. Prod. **287**, 125434 (2021)
41. Ramudhin, A., Chaabane, A., Kharoune, M., Paquet, M.: Carbon market sensitive green supply chain network design. In: 2008 IEEE International conference on industrial engineering and engineering management, Singapore (2008)
42. Waltho, C., Elhedhli, S., Gzara, F.: Green supply chain network design: a review focused on policy adoption and emission quantification. Int. J. Prod. Econ. **208**, 305–318 (2019)
43. Bagchi, P.K., Chun Ha, B., Skjoett-Larsen, T., Boege Soerensen, L.: Supply chain integration: a European survey. Int. J. Logistics Manag. **16**(2), 275–294 (2005)

Chapter 6
A Descriptive Study on the Growth of Unorganized Independent Pharmaceutical Retail Sector in India

Koushik Mondal, Balagopal G. Menon, and Sunil Sahadev

Abstract For over a century, the Indian pharmaceutical retail market has been dominated by the unorganized pharmaceutical retail sector, which not only makes essential drugs available to the huge diversified population of India but also provides millions of jobs. Presently, these unorganized pharmacies are facing tremendous challenges in attractive prices and the large variety of products offered by organized and e-pharmacies. The survival of these independent pharmacies is of prime concern in this research. A field survey of Indian unorganized pharmaceutical retailers and customers is carried out in the present study to analyze the influence of organized retail and e-pharmacies on unorganized pharmaceutical retail sectors. According to the study, unorganized retailers in the catchment area of organized retailers suffered a reduction in sales and profit after the entry of organized retailers and e-commerce. The unorganized pharmaceutical retailers away (more than 2 km) from the organized retailers also suffer in sales due to the presence of predatory pricing policy of the organized retailers and e-commerce.

Keywords Independent pharmacy · Unorganized and organized retailing · E-pharmacy · Semi-structured interview · Emerging economy

---

K. Mondal (✉) · B. G. Menon
Department of Industrial and Systems Engineering, Indian Institute of Technology, Kharagpur, West Bengal, India
e-mail: koushik.phd21@gmail.com

S. Sahadev
Department of Management, Sheffield Hallam University, Sheffield, United Kingdom
e-mail: s.sahadev@shu.ac.uk

J. Maiti et al. (eds.), *Recent Advances in Industrial and Systems Engineering*, Lecture Notes on Multidisciplinary Industrial Engineering,
https://doi.org/10.1007/978-981-96-8323-9_6

6.1 Introduction

A pharmaceutical retailer (in literature referred to as a "community pharmacy") delivers medications to patients as prescribed by their doctors. Pharmacy accessibility is key for the emerging role of community pharmacies as providers of patient-centered medication management services in addition to traditional dispensing roles [1]. The tasks of a community pharmacist include processing prescriptions, checking for drug interactions, dispensing and disposing of medication, providing advice, and promoting a healthy lifestyle. The type of pharmaceutical retails ranges from small, individually owned, unorganized pharmacies in isolated rural towns to large organized pharmaceutical chains in the main hubs. The term "organized retailing" refers to the business activity carried out by registered or licensed shops (e.g., supermarkets, privately operated large retail chains, hypermarkets, etc.). On the contrary, unorganized retailing refers to the traditional small neighborhood shops such as mom-pop shops, grocery shops, general stores, etc. In literature, these types of stores are known by several names like "nanostores" [2], "small unorganized retailers" [3], "micro-retailers" [4], etc. In the context of developing countries, these unorganized retail outlets are also known by various region-specific native names, such as "kirana" in India, "changarro" in Mexico, "sari sari" in the Philippines, "bodega" in Peru, "warung" in Indonesia, or "spaza" in South Africa [5]. These independent shops are mainly run by the owner or family of the owner or managed by them, have cheap real estate and labor costs, and low taxes. The major benefits of unorganized retail stores are proximity and consumer familiarity, which may be passed down through generations. These unorganized or informal stores operate on a small scale, serving only a few hundred consumers around their vicinity and having storage and cash constraints [6]. They also offer relationship-based credits, personalized services, and even delivery services to become competitive in the market [7]. In developed countries, organized retailing dominates the market, while unorganized retailing is more prevalent in emerging economies [8]. Blanco and Fransoo [2] estimate that there will be around 50 million nanostores in emerging economies. There are about twelve million nanostores in India [9], six million in China [10], and millions more across other emerging economies like Mexico, Brazil, Nigeria, etc. In the majority of developing economies, traditional or unorganized retail is the major distribution channel for consumer-packaged goods (CPG), with a market share of more than 85% in South Asia and Africa and approximately 50% in Latin America [6].

According to the Indian Brand Equity Foundation [11], the pharmaceutical retail industry is projected to grow from an estimated INR 1720 billion in 2020 to INR 3078 billion by 2026. The e-pharmacy industry in India has been growing rapidly and is expected to reach INR 317 billion by 2026 from a mere INR 55 billion in 2021 [12]. Organized pharmacies are also growing substantially, with a projected growth of INR 370 billion by 2026 from INR 134 billion in 2020. The market share of both organized and e-pharmacies is growing rapidly [12].

Despite this growth, the traditional unorganized pharmaceutical retail industry, which constitutes approximately 88% of Indian retail trade, faces significant challenges [13]. Big organized chain pharmacies and e-pharmacies, have entered the market with predatory pricing like a minimum of 20% off on all medicines, customer loyalty programs, and extensive product variety that the unorganized pharmaceutical retail sector is struggling to compete with. The unorganized sector has seen a drastic decline in market share, falling from 90% in 2020 to an estimated 77.63% in 2026 due to high competition from organized and e-pharmacies [12] (Fig. 6.1).

In the pharmaceutical science and retailing, the struggle between organized chain pharmacies and small unorganized independent pharmacies [14], the importance of independently owned pharmacies providing access to community pharmacies in rural areas [1], and the role of healthcare provider from a retailer in chronic disease management [15] are discussed in great details but the problem was looked from a pharmaceutical angle rather than from a management angle. Despite the significance of this phenomenon, the impact of organized retailers and e-commerce on unorganized retailers in emerging economies is still not fully understood [8]. In this present study, the impact of organized chain pharmacies and e-pharmacies on independent small, unorganized pharmacies is investigated.

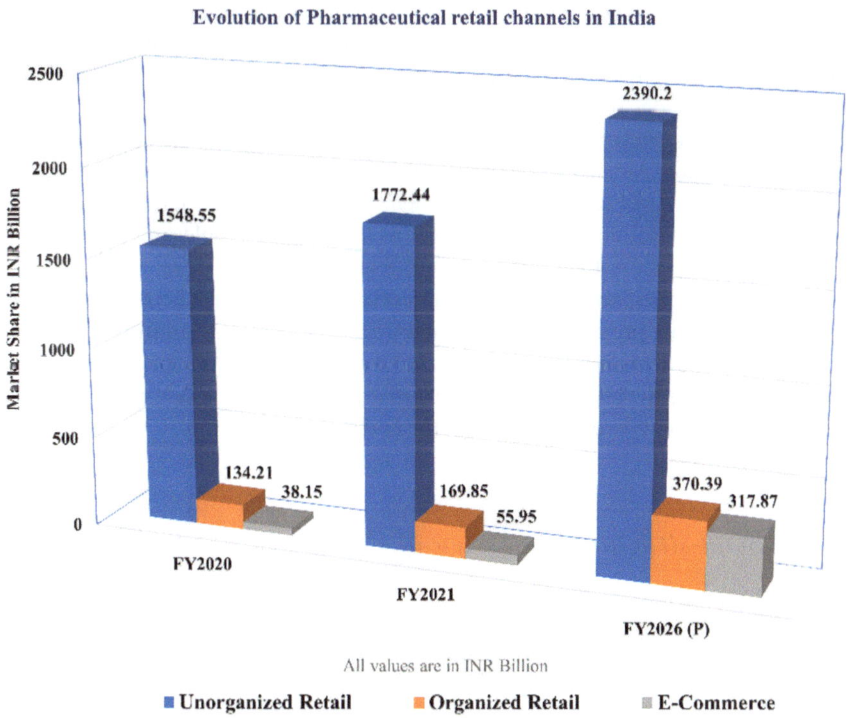

Fig. 6.1 Evolution of pharmaceutical retail channels in India

For that purpose, a block in the state of West Bengal, India, is chosen for the study area. Several field studies, including semi-structured interviews of unorganized pharmaceutical retailers, customers, and stakeholders of the pharmaceutical supply chain, are carried out to validate our hypothesis.

6.2 Methodology

This section briefly describes the methodology and sampling design followed for collecting primary data through semi-structured interviews with unorganized retailers, consumers, and stakeholders in the pharmaceutical sector in India.

6.2.1 Semi-structured Interview of Unorganized Pharmaceutical Retailers

It is observed from the review of the literature and general interaction with unorganized pharmaceutical retailers that those retailers are more prone to the adverse effect of organized chain pharmacies if they fall inside the catchment area of the organized pharmacy retail outlets [16]. The catchment area is defined as the area (radius of distance in km) from where the organized retailers are expected to draw potential customers. In the present study, these locations of the organized retail outlets are referred to as hubs. A distance criterion is used to distinguish between the treatment sample and the control sample. The treatment samples are those unorganized stores located within two kilometers of the hubs, and the control samples are those that are located far away (more than 2 km) from the hubs. This was done to compare the two scenarios of the unorganized pharmaceutical retail sector where the organized pharma retailers are present (treatment) and where they are not (control). The difference between the two samples will help to find out the impact of organized retail. In the present study, twenty unorganized pharma retailers were randomly selected as treatment, and twenty were selected for the control sample.

6.2.2 Semi-structured Interview of Customers

A consumer survey is intended as an integral aspect of the current impact assessment study because consumers constitute a major stakeholder in the expansion of organized retail, and consumer purchasing behavior is affected by the presence of organized retail (Joseph et al. 2008). A random sampling of one hundred customers was carried out and interviewed.

Table 6.1 Size and age of the sample unorganized stores

Sample size		Average size of store (sq. ft.)		Average age of the outlets (years)	
Treatment sample	Control sample	Treatment sample	Control sample	Treatment sample	Control sample
20	20	142	128	23	19

6.2.3 Scope of Research Work

1. The study is limited to traditional or unorganized pharmaceutical retail sector in India.
2. The relevant primary data is collected from the eastern states of India.

6.3 Findings and Contributions

This section describes the survey results and findings from the interviews of unorganized retailers, consumers, and stakeholders in the pharmaceutical sector in India.

6.3.1 Survey Results: Unorganized Pharmaceutical Retailers

6.3.1.1 Size and Age of the Unorganized Stores

The minimum size of a pharmaceutical shop governed by the Drug Control Department is 120 square feet [17]. The sampled unorganized retail outlets have an average size between 128 and 142 sq. ft (Table 6.1)

6.3.1.2 Impact on Customer Footprint, Turnover, and Profit

Table 6.2 brings out the comparative situation of the customer footprint of both treatment and control samples in two scenarios, with and without the presence of organized retailers in the market. Both of the samples have reported a steady decline in customer footprint as well as turnover and profit over the last ten years.

6.3.1.3 Reasons for Decline in Overall Turnover

The unorganized pharmaceutical retailers are asked about the major reason for the decline in overall turnover. Table 6.3 shows that retailers in the treatment sample (in

Table 6.2 Impact on customer footprint: control versus treatment sample

	Before organized retail present		After organized retail present	
	Treatment sample	Control sample	Treatment sample	Control sample
Customer visit per day	200–250	70–100	100–150	30–50

Table 6.3 Reasons for decline in overall turnover: treatment versus control sample (in % of sampled retailers)

	Treatment sample	Control sample
1. Direct competition from organized retail	95	45
2. Direct competition from e-commerce	55	25
3. Direct competition from unorganized retail	20	60
4. Predatory pricing of organized retail and e-commerce	95	80
5. Government regulations/license problem	25	55
6. Other factors	15	45

the catchment area) faced the most direct competition from organized retail and e-commerce, while the control sample retailers faced more indirect effects of organized retail, predatory pricing of organized retail and e-commerce, frequent stockouts, strict government regulation like Goods and Services Tax (GST), pharmacist license problem, etc.

According to the study, unorganized retailers in the catchment area of organized retailers suffered a reduction in sales and profit after the entry of organized retailers. This result is in line with previous studies of Kohli and Bhagwati [9] and Joseph et al. [16].

6.3.2 *Customer Survey Results*

The purpose of the customer survey is to understand the behavior and travel patterns of the customers when purchasing from unorganized retail vs. organized retail vs. e-commerce.

6.3.2.1 Income Level of Customers

As anticipated, individuals with higher incomes typically prefer shopping at organized retail stores, whereas the middle-class demographic, encompassing aspirational individuals with monthly household incomes ranging from Rs.10,000 to Rs.1,00,000, emerges as a vital customer segment for all three retail sectors. Notably, this middle-class cohort demonstrates a nuanced consumer behavior by engaging in

Table 6.4 Average monthly household income of customers (% share)

		Share of avg monthly spending shopping from		
Income group (Rupees)	Population	Unorganized	Organized	E-commerce
Up to 10,000	11	90.91	9.09	0.00
10,001–20,000	55	74.55	21.82	3.64
20,001–50,000	21	42.86	47.62	9.52
50,001–1,00,000	9	22.22	55.56	22.22
1,00,001–10,00,000	4	8.25	76.50	15.25

shopping activities across organized and unorganized retail outlets and e-commerce, as evidenced by the data presented in Table 6.4. This triple retail engagement highlights the dynamic nature of consumer preferences within this influential income stratum.

6.3.2.2 Spatial Advantages for the Unorganized Retailers

The advantage of location is evident in the case of unorganized retailers, as indicated by Fig. 6.3, which shows that the average distance from consumers' residences to the nearest unorganized outlets is 0.88 km, while it is 2.04 km for organized outlets (Fig. 6.2). Predictably, the majority of consumers travel to organized retail locations by personal vehicle or public transport, whereas the majority of consumers walk to the traditional unorganized retailers.

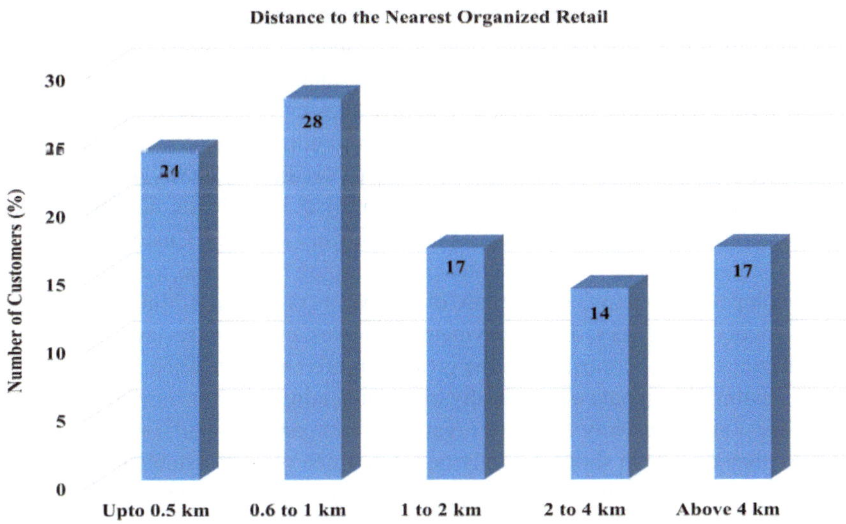

Fig. 6.2 Distance to the nearest organized retail (mean distance = 2.04 km)

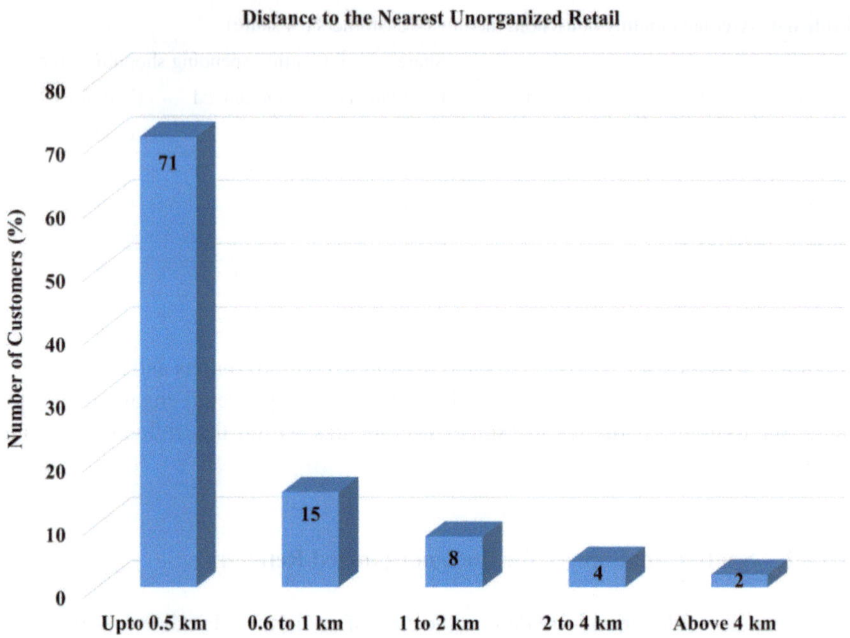

Fig. 6.3 Distance to the nearest unorganized retail (mean distance = 0.88 km)

6.3.2.3 Preference for Organized Versus Unorganized Retail Versus e-commerce

Customers patronizing organized retail articulated their preferences based on factors such as competitive pricing, superior product quality, streamlined one-stop shopping experiences, family-oriented shopping environments, a diverse assortment of brands and products, and consistent availability of inventory. Conversely, consumers engaging with unorganized pharmaceutical retail outlets attributed their patronage to considerations including residential proximity, credit availability, establishment reputation, bargaining opportunities, the availability of loose items, accommodating business hours, and the convenience of home delivery services. Customers in India opt for e-commerce due to the convenience of shopping from home, a diverse product range, competitive pricing, and the ease of comparing products. Additionally, factors such as doorstep delivery, digital payment incentives, customer reviews, and exclusive offerings further contribute to the growing preference for online shopping. The survey findings illuminate a noteworthy trend, indicating that consumers do not limit their shopping exclusively to either organized, unorganized retail, or e-commerce. Instead, they engage in shopping activities at all three types of establishments, with the allocation of expenditures differing across various product categories.

6.4 Conclusion

In this study, the process and pace at which the entry of organized chain pharmacy and e-pharmacy lead to the reduction in customer footprints and overall turnover of independent small unorganized pharmacies are investigated. For that purpose, a block in West Bengal, India, has been chosen as the study area. Unorganized retailers and customers are interviewed and the results are analyzed. The main assumption is that as organized retailers come into the market, the inventory stocking levels of these small, unorganized pharmaceutical retailers will slowly come down because of the erosion in demand as more and more customers switch to organized and e-pharmaceutical retailers. After a while, it will be unsustainable to stock so many medicines. So, it will have a reinforcing effect, such as the lesser the customer, the lesser the inventory, which will lead to more dissatisfied customers due to more stock-outs, which will eventually decrease even more customers coming to that particular shop. Small unorganized retailers usually have a profit percentage of around 18 to 25% from distributors [18]. Due to the predatory pricing of organized retailers and e-commerce, these independent pharmacies will either eventually degrade or may switch businesses. In those cases, rural households may not have a local pharmacy within a 5–10 km radius or more, making the rural healthcare system much worse than it is in the present scenario. In India's burgeoning retail market, this research will shed light on the relevance of the unorganized pharmaceutical retail sector.

References

1. Berenbrok, L.A., Tang, S., Gabriel, N., Guo, J., Sharareh, N., Patel, N., Hernandez, I.: Access to community pharmacies: a nationwide geographic information systems cross-sectional analysis. J. Am. Pharm. Assoc. **62**(6), 1816–1822 (2022)
2. Blanco, E.E., Fransoo, J.C.: Reaching 50 million Nanostores: Retail Distribution in Emerging Megacities, BETA Working Paper Series, Vol. 404. Technische Universiteit Eindhoven (2013)
3. Dugar, A., Chamola, P.: Retailers with traits of consumer: exploring the existence and antecedents of brand loyalty in small unorganized retailers. J. Retail. Consum. Serv. **62**, 1–9 (2021)
4. Zhang, K., Tang, C.S., Zhou, S.: Replenishment strategies for micro-retailers in developing countries. Prod. Oper. Manag. **26**(12), 2207–2225 (2017)
5. Escamilla, R., Fransoo, J.C., Tang, C.S.: Improving agility, adaptability, alignment, accessibility, and affordability in nanostore supply chains. Prod. Oper. Manag. **30**(3), 676–688 (2021)
6. Fransoo, J.C., Blanco, E.E., Argueta, C.M.: (2017) Reaching 50 million nanostores: retail distribution in emerging megacities. CreateSpace Independent Publishing Platform, Cambridge, MA
7. Child, P., Kilroy, T., Naylor, J.: Modern grocery and the emerging-market consumer: a complicated courtship. McKinsey & Company (2015). http://bit.ly/1Uf9aBY
8. Jerath, K., Sajeesh, S., John Zhang, Z.: A model of unorganized and organized retailing in emerging economies. Mark. Sci. **35**(5), 756–778 (2016)
9. Kohli, R., Bhagwati, J.: Organized retailing in India: issues and outlook. In: Indian Economic Policies in the Twenty-First Century. Cambridge University Press, Cambridge, UK (2012)

10. Ge, J.: Traditional retail distribution in megacities. Phd thesis 1, Research TU/e/Graduation TU/e, Industrial Engineering and Innovation Sciences. Technische Universiteit Eindhoven (2017)
11. India Brand Equity Foundation (IBEF), an initiative of the Ministry of Commerce & Industry, Government of India (2022). www.ibef.org
12. ReportLinker. Pharmacy Retail Market in India (2022). https://www.reportlinker.com/p05873822/Pharmacy-Retail-Market-in-India.html
13. Technopak: Impact of Indian Retail on Employment & Taxation. (September) Report (2020). https://www.technopak.com/
14. Miller, R., Goodman, C.: Do chain pharmacies perform better than independent pharmacies? Evidence from a standardised patient study of the management of childhood diarrhoea and suspected tuberculosis in urban India. BMJ Glob. Health **2**(3), 1–11 (2017)
15. Mossialos, E., Courtin, E., Naci, H., Benrimoj, S., Bouvy, M., Farris, K., Sketris, I.: From "retailers" to health care providers: transforming the role of community pharmacists in chronic disease management. Health Policy **119**(5), 628–639 (2015)
16. Joseph, M., Soundararajan, N., Gupta, M., Sahu, S.: Impact of organized retailing on the unorganized sector. In: Indian Council for Research on International Economic Relations, New Delhi, India, Working Paper, 222 (2008)
17. Pharmacy Council of India (2024). https://www.pci.nic.in
18. Panda, T.K., Sahadev, S.: Sales and Distribution Management. Oxford University Press, Oxford (2019)

Part II
Human-Centered Design and Ergonomics

Chapter 7
Eye-Tracking-Based Packaging Analysis for Toy Industry Using Bayesian Belief Networks and Evidential Reasoning Approach

Prachi Tulsyan, Amrita Ghosh, and Sobhan Sarkar

Abstract Toy packaging goes beyond protection; there has been limited study of the various factors affecting toy packaging and their influence on consumer choices. There is a lack of clarity on the relative importance of factors: brand reputation, product family, game features, and age appropriateness in influencing purchasing decisions. This research paper aims to study the eye-tracking metrics of consumers using the packaging data from two educational toy industries: DISET and EDUCA, through Bayesian belief networks (BBN) and the evidential reasoning approach (ERA) to explore the relationships between parental demographics and their engagement with packaging images. To streamline eye-tracking data analysis, we introduced the Interest-Saliency Index (ISI). The paper further discusses and ranks the factors influencing these purchases, shedding light on the relative importance of various packaging features and Areas of Interest (AOI), which advances theoretical understanding and furnishes actionable insights for professionals in this sector to curate improved strategies.

Keywords Bayesian belief networks (BBN) · Evidential reasoning approach (ERA) · Interest-saliency index (ISI) · Demographic factors · Area of interest (AOI)

Prachi Tulsyan, Amrita Ghosh are contributed equally to this work.

P. Tulsyan
Marketing and Strategy, Business Administration, IIM Ranchi, Ranchi, Jharkhand, India
e-mail: tulsyanprachi13@gmail.com

A. Ghosh (✉)
Mechanical Engineering, IIT Kharagpur, Kharagpur, West Bengal, India
e-mail: kgpamrita@gmail.com

S. Sarkar
Information Systems and Business Analytics, IIM Ranchi, Ranchi, Jharkhand, India
e-mail: sobhan.sarkar@iimranchi.ac.in

J. Maiti et al. (eds.), *Recent Advances in Industrial and Systems Engineering*, Lecture Notes on Multidisciplinary Industrial Engineering,
https://doi.org/10.1007/978-981-96-8323-9_7

7.1 Introduction

The toy industry is pivotal in shaping children's development and learning experiences. With an ever-expanding range of toys available on the market, it becomes increasingly crucial to analyze the toy industry from various perspectives. Understanding the factors influencing toy selection is essential for parents to make informed choices that positively impact their children's development and learning experiences in an industry that significantly shapes these outcomes. Toy customers are typically adults, while toy consumers are children [1]. Parents often seek to make informed choices when selecting toys for their children, yet there remains a lack of comprehensive research on the factors influencing toy selection. This study identifies various factors affecting parent toy purchase decisions, including purposes of use, emotional aspects, information availability, costs, children's demographics, and parental demographics. This will help bridge research gaps and enhance understanding of toy selection factors, thereby supporting children's holistic development and informed parental decision-making.

The relationship between toys and development is bidirectional. This means that more competent children tend to seek out a greater variety of appropriate play objects, and appropriate play objects assist development. Intrinsic motives of curiosity and mastery drive children to explore and manipulate interesting toys and objects. For example, children may learn how to share, take turns, and cooperate when playing with toys with others. These interactions can help shape and support social-cognitive development [2]. Effective packaging design in conveying toys' value and educational aspects ultimately influences consumer behavior. Effective packaging is more than just a way to contain and protect a product. As evidenced by the global toy packaging market's significant growth, with a CAGR of 3.8% during the historical period (2015–2021), manufacturers recognized the pivotal role of attractive packaging in influencing consumer behavior and driving impulse purchases. It can help to attract attention, convey information, and create a positive emotional response, creating a connection between the product and the consumer and making them more likely to buy it [3]. Moreover, the sensory aspects of products, such as their smell, sound, touch, taste, and look, can significantly impact how consumers perceive and interact with them. Some packaging may incorporate interactive elements to create a stronger connection with potential owners. This highlights the significance of creating unique and imaginative toy packaging that becomes an integral part of the toy experience, further enhancing the moment's excitement [4]. In short, the neurological effects of products and their presentation to consumers can be a powerful tool for companies to create a positive customer experience and build brand loyalty [5].

By using effective design, businesses can create packaging that protects and pre-serves their products and helps sell them [6]. The existing literature on the toy industry lacks comprehensive research on packaging analysis and consumer demographics. This has left unaddressed questions about which specific aspects of packaging design, information placement, or visual appeal have the most substantial impact on consumers based on purchasing decisions through emotional responses.

This research paper aims to shed light on the importance of multiple packaging factors when consumers with diverse demographic backgrounds make product selections. By analyzing the toy industry from their perspective, we can bridge the gap in existing research and industrial strategies. Thus, manufacturers and marketers can strategically position their offerings in the market.

7.1.1 Research Issues

Based on the aforementioned discussion, following research problems are identified:

(i) **Insufficient analysis of packaging design elements**: There is a dearth of comprehensive studies focusing on packaging elements like visual esthetics, information placement, structural design, iconography, and branding elements, that have the most substantial impact on consumers [7]. The lack of clarity in this area poses challenges for both designers and marketers in optimizing packaging strategies to captivate consumers' attention and influence purchasing decisions.

(ii) **Limited exploration of parental demographics and packaging engagement**: A noticeable gap exists in in-depth research examining the relationships between parental demographics like gender, marital status, and family composition and their engagement with toy packaging in terms of their influence on preferences [8].

(iii) **Limited study on emotional and neurological responses**: Exploration of the neurological aspects of toy selection remains largely untouched in existing literature. Prior research has not extensively addressed the integration of behavioral economics perspectives and psychophysiological measures [9] to understand emotional responses in the context of toy selection.

(iv) **Lack of definitive ranking for factors guiding toy purchases**: Despite the significance of factors influencing toy purchases, a clear and definitive ranking of their relative importance remains absent [10]. Key elements such as brand reputation, product family affiliation, features, and age suitability lack a systematic hierarchy. This gap in knowledge hinders the understanding of consumer preferences and the formulation of targeted marketing strategies [11].

7.1.2 Research Contributions

To address the aforementioned issues, our research work is directed to contribute the following:

(i) The research paper aims to fill a significant gap in the existing literature by exploring the interrelationships among various factors in toy product packaging design using Bayesian belief networks (BBN).

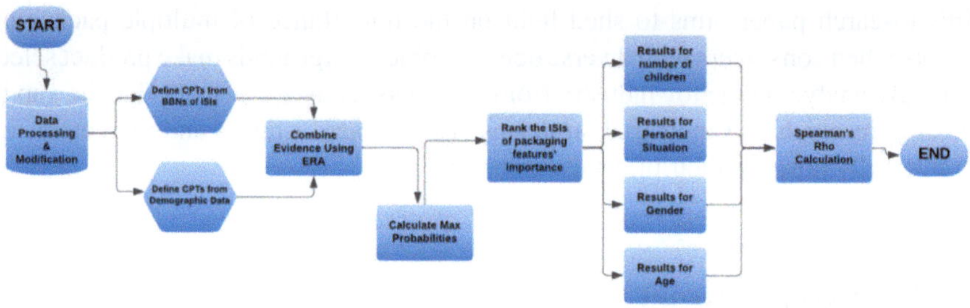

Fig. 7.1 Proposed research methodological flowchart

(ii) By employing an evidential reasoning approach (ERA), this paper aims to understand consumer behavior by identifying the key factors that influence toy purchases and ranking these attributes according to their relative importance.
(iii) By utilizing eye-tracking data, the study aims to analyze and interpret parental behavior and preferences when selecting toys, offering valuable insights from a neuromarketing perspective.

The rest of the article is structured as follows: In section 7.2 of the paper, the proposed methodology is presented, Section 7.3 covers results and discussions, Section 7.4 presents the conclusion, along with limitations and future scopes for the study.

7.2 Proposed Methodology

Figure 7.1 represents the overall proposed methodology. In this section, we have discussed about the methodology of our proposed approach: (1) Data collection, description, and modification; (2) Advantages of BBN as analytical framework; (3) Modeling of BBNs; (4) Implementation of ERA; (5) Performance metrics.

7.2.1 Data Collection, Description and Pre-processing

The data used in this study was collected from the Mendeley Data repository,[1] a case study conducted in the educational toy industry using eye-tracking technology. There are two primary data sources utilized in this study. The first data source is a file named *observations.csv*, which contains the eye-tracking data obtained with gazepoint analysis. This dataset consists of 350 records (samples) and 16 columns (variables), providing information on various aspects of the eye-tracking analysis. The second data source is a file named *users.csv*, which includes information about

[1] https://data.mendeley.com/datasets/pfc2zdr2g3/1.

the users who participated in the study. This dataset comprises 25 records (participants) and 8 columns, providing details about their social situations: gender, personal situation, number of children, and age.

The *observations.csv* and *users.csv* datasets were cross-referenced to enhance the analysis and interpretation of the eye-tracking data. As a result, a new dataset was derived, containing 13 columns and 350 rows that combined the relevant information from the eye-tracking observations with the corresponding user characteristics, providing a consolidated dataset for further analysis and interpretation. There are several Area of Interests (AOI) identified in the dataset as shown in Fig. 7.2—(i) AOI 1: Number of Questions and Topics; (ii) AOI 2: Draw of topics (EDUCA)/ Descriptive message (DISET); (iii) AOI 3: Brand; (iv) AOI 4: Recommended age; (v) AOI 5: Product family; (vi) AOI 6: Product name; and (vii) AOI 7: Game picture.

To improve the interpretation of the eye-tracking data, we performed a comprehensive dataset modification as follows: We merged the seven distinct AOIs present in the *observations.csv* dataset with their corresponding eye-tracking features. These features included *revisits*, *fixation*, *time to first visit*, and *time viewed*. This integration resulted in the creation of 28 columns, each representing a unique combination of AOI and eye-tracking features. We incorporated data from both DISET and EDUCA into the consolidated dataset, further enriching the information available for analysis by checking different issues, including imbalance issues [13]. The demographic data of the 25 participants, including gender, personal situation, number of children, and average children's age, were mapped to this modified dataset. This allowed

Fig. 7.2 Packaging of Educa (left) and Diset (right) showing the 7 AOIs [12]

us to establish connections between user characteristics and eye-tracking data. The dataset modification was crucial in enabling us to explore the relationships between user demographics and eye-tracking metrics. The resulting dataset is a synergy of eye-tracking technology, user demographics, and educational content assessment. It positions us to unravel valuable insights into how parents interact with educational toys, offering practical implications for the educational toy industry. The paper further utilizes the BBN method as one of the analysis techniques followed by an ERA to rank the attributes based on importance while packaging an educational toy.

7.2.2 Advantages of BBN as Analytical Framework

Several key factors guided our decision to select BBN as the primary analytical framework. BBNs offer transparent and interpretable models, allowing researchers to understand the probabilistic relationships between variables [14]. In contrast, deep learning models, such as neural networks, often function as "black boxes," making it challenging to interpret their decision-making processes [15], especially in complex domains like educational toy packaging analysis. The robustness of BBN in handling limited data makes them particularly well-suited for our scenario, where datasets may be small or incomplete. They enable the explicit incorporation of prior knowledge, essential in domains like the educational toy industry where expert insights play a pivotal role [16].

In these domains where data may be noisy or uncertain, BBN inherently captures the uncertainty and handles incomplete information through probabilistic inference. This capability is crucial. On the other hand, deep learning models typically struggle with uncertainty quantification and may yield overly confident predictions in such scenarios. Lastly, the flexibility and updateability of BBNs allow for iterative refinement as new data emerges or domain knowledge evolves, ensuring adaptability to changing contexts [17]. In contrast, others often require retraining from scratch when incorporating new information, making them less adaptable to evolving contexts. Considering these advantages, BBNs emerge as the most suitable choice for our research, providing a principled framework for modeling complex dependencies and uncertainties while effectively incorporating domain expertise from the industry.

7.2.3 Modeling of the BBNs

A BBN is a visual depiction that showcases the various probabilistic connections between different random variables within a given set. The BBN calculates probabilities through joint probability, specifically by considering the condition *P* (attribute/ parent). This refers to the probability of an attribute being true given the parent attribute. BBN or Bayesian Network or Belief Network is a Probabilistic Graphical

Model (PGM) that represents conditional dependencies between random variables through a Directed Acyclic Graph (DAG) [18].

The probabilities are calculated in the belief networks by the following Eq. (7.1):

$$P(X_1, X_2, X_3, \ldots\ldots, X_n) = \prod_{i=1}^{n} P\left(\frac{X_i}{P_a(X_i)}\right) \tag{7.1}$$

In Eq. (7.1), $P\ (X_i)$ represents the parent set of X_i in the BN model, where $i = 1$, 2, 3,..., n and n is the number of random variables. BBN is a probabilistic graphical model that represents variables and their dependencies using a DAG. It enables the modeling and analysis of complex systems by incorporating prior knowledge and updating beliefs based on observed evidence. The BBN model is constructed using the variables related to AOIs and user characteristics [19]. By utilizing BBN, the paper aims to infer insights and understand the relationships between different AOIs, user behavior, and design factors in the packaging of educational toys. These probabilistic graphical models were employed to capture and represent the complex interrelationships between demographic features and user engagement with packaging AOIs. Gender, a fundamental variable, encompasses two categorical states: "male" and "female." The number of children introduces states ranging from "1" to "3," revealing insights into family size's influence on preferences. The continuous state variable, average age of children, categorizes youngsters as "toddler," "preschooler," or "preteen," depicting the role of child age in shaping choices. Lastly, personal situation, defined by marital status, includes states such as "married," "divorced," and "couple," offering insights into family dynamics' impact on preferences.

In the preliminary stage, four BBNs were constructed, each correlating four demographic features (gender, family situation, number of children, average child age) as parent nodes with seven packaging features related to eye-tracking metrics as daughter nodes. For instance, the first BBN as shown in Fig. 7.3 linked gender, personal situation, number of children, and average child age to the time viewed on brand, time viewed on product name, time viewed on the number of questions, and so forth. Similar to Fig. 7.3 three other BBNs were created. This initial structure aimed to model the intricate relationships between demographics and user engagement metrics but resulted in analytical challenges. The large number of nodes in each BBN increased the complexity of conditional dependencies, hindering a straightforward interpretation of results. Interconnections between multiple eye-tracking metrics and packaging features led to intricate relationships that required more sophisticated approaches.

Need for ISI in Eye-Tracking Metrics

The extensive web of relationships between demographic features and eye-tracking metrics made it challenging to extract clear patterns and insights. To streamline the analysis and enhance interpretability, the ISI was introduced as shown in Eq. (7.2). The ISI is a unified score combining Fixation and Revisits (indicating sustained interest) with Time to 1st Visit and Time Viewed (providing insights into attention

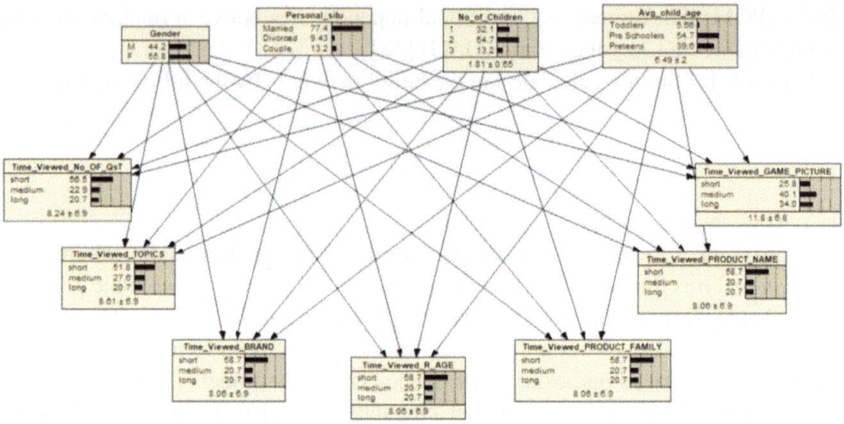

Fig. 7.3 BBN of demographic features of the users and the eye-tracking features of packaging AOIs

initiation and total engagement time). The ISI formula served as a comprehensive assessment tool, simplifying the complex relationships within the BBNs. Following the introduction of ISI, new BBNs were constructed with demographic features as parent nodes and ISI of packaging features as daughter nodes. For example, a BBN correlated aver- age child Age with ISI of brand, ISI of product name, ISI of number of questions, etc. This restructuring allowed for a more focused and streamlined analysis, enhancing the interpretability of results.

Development of the ISI

The rationale behind the ISI is to capture both user interest and the temporal dynamics of attention initiation. The ISI formula (7.2) is given as follows:

$$\mathrm{ISI} = \frac{\text{Fixation} + \text{Revisits}}{\text{Time to 1st visit} + \text{Time viewed}} \tag{7.2}$$

Equation (7.2) was derived to provide a comprehensive assessment of eye-tracking engagement with AOIs, combining fixation and revisits to signify sustained interest and engagement and simultaneously, considering Time to 1st visit and Time viewed, pro- viding insights into the speed of attention initiation and the total time invested in an AOI. Frequent fixations and revisits indicate sustained user interest and engagement with an AOI. "Time to 1st visit" addresses the temporal aspect of attention initiation. A shorter time to 1st visit indicates quicker engagement, highlighting AOI saliency. "Time Viewed" represents the total duration of AOI engagement. A higher time viewed suggests more extensive exploration and user interest. The ISI serves as a tool for interpreting user interaction with AOIs. A higher ISI score indicates both interest and quick attention initiation, signifying an AOI's appeal and saliency. Conversely, a lower ISI suggests slower attention initiation or reduced engagement with an AOI (Fig. 7.4).

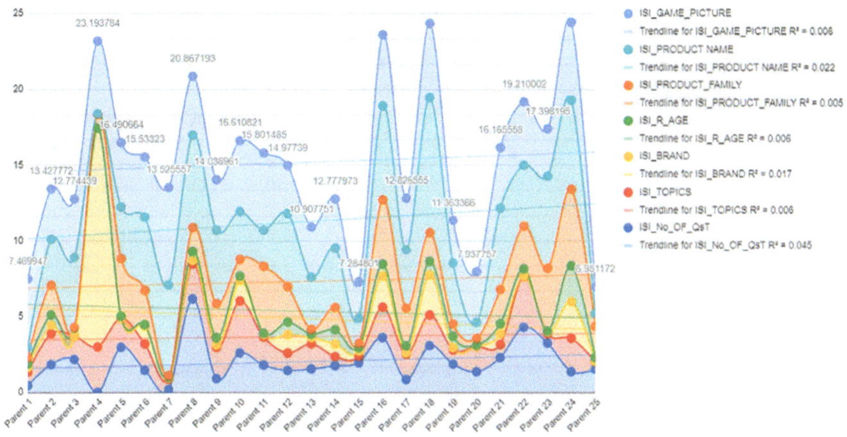

Fig. 7.4 Stacked area chart of ISI of eye-tracking values for the 7 AOIs by 25 users

After the introduction of ISI, we built four distinct BBNs, each designed to explore the relationship between a specific demographic feature and user engagement with packaging features. These demographic features served as the parent nodes in each BBN, while the daughter nodes represented the ISI of each packaging element. *BBN* 1: *Gender Analysis* (Fig. 7.6): We focused on gender as the parent node, and the daughter nodes comprised of the ISI of the seven packaging features, providing insights into the differential impact of gender on AOI engagement. *BBN* 2: *Analysis on Number of Children* (Fig. 7.7): The parent node was defined as the number of children, and the daughter nodes represented the ISIs of the seven packaging features. This analysis aimed to uncover patterns related to family size and user preferences in the educational toy context. *BBN* 3: *Average Children Age Analysis* (Fig. 7.5) The parent node, average children's age, was examined for its impact on user engagement. The daughter nodes comprised the ISIs of packaging features, shedding light on the role of average child age in shaping preferences for educational toys. *BBN* 4: *Personal Situation Analysis* (Fig. 7.8): Personal situation, as defined by the participants, was the focal point of BBN 4. This demographic attribute served as the parent node, and the daughter nodes encompassed the ISIs of packaging features. This analysis aimed to discern how personal circumstances influence user engagement in the Educational Toy Industry.

7.2.4 Implementation of ERA

The ERA is utilized following the application of a BBN. ERA is employed as a complementary method to enhance the analysis of design features in toys using eye-tracking data. The main function of ERA in this research paper is to address the challenges associated with uncertain and imprecise information in the analysis of design

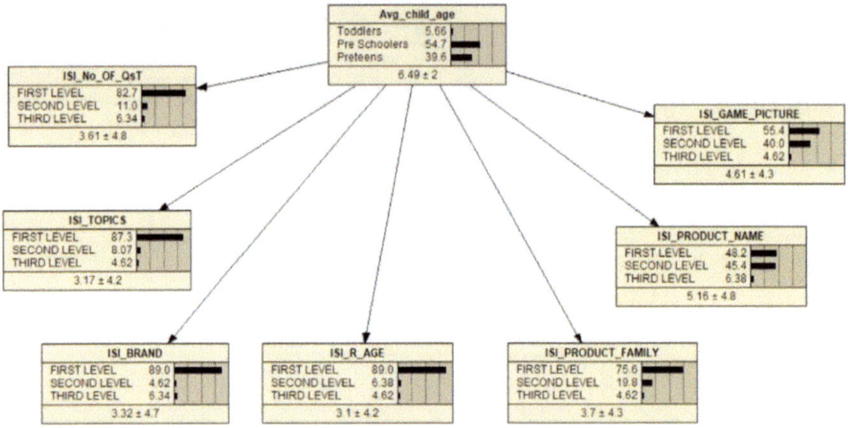

Fig. 7.5 BBN of average child age of the users and ISI of the packaging AOIs

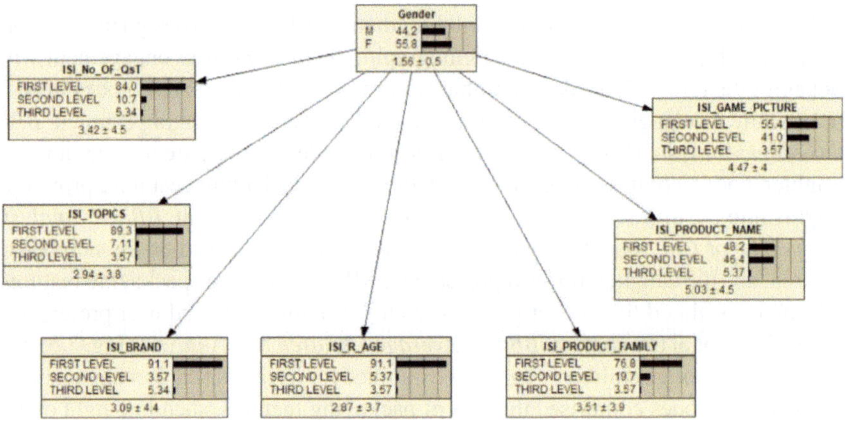

Fig. 7.6 BBN of gender of the users and ISI of the packaging AOIs

features in toys. By combining eye-tracking data with the BBN model, ERA allows for the fusion of multiple sources of evidence and the representation of uncertainties. This enables a more comprehensive and robust analysis, leading to better insights and decision-making in the field of toy design. The first aspect pertains to the arrangement and organization of evidence that is relevant to the specific problem at hand. The second aspect involves the selection of a framework for representing uncertainties within the evidence, along with the corresponding mathematical calculus employed to merge the evidence for decision-making [20].

Combined Belief Calculation Formula: Let—P (Demographic Feature $= x$) represents the custom probability associated with a specific demographic feature, where 'x' is the value of that feature. P (ISI $= y|$Demographic Feature $= x$) represents the conditional probability of the ISI being 'y' given a specific value 'x' of the

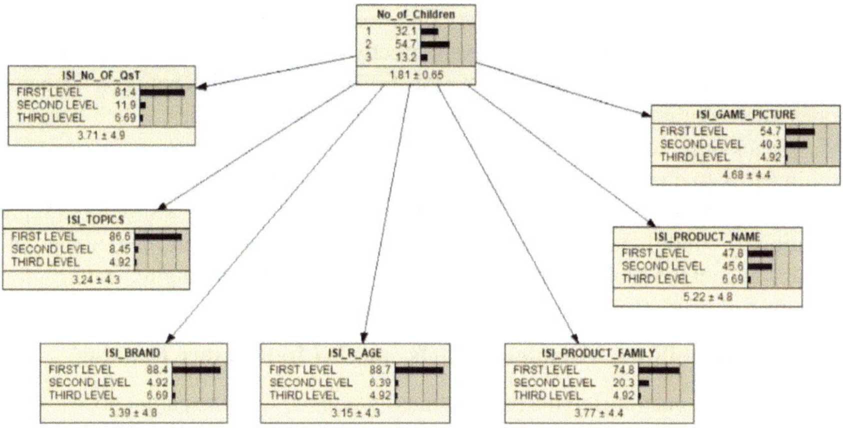

Fig. 7.7 BBN of number of children of the users and ISI of the packaging AOIs

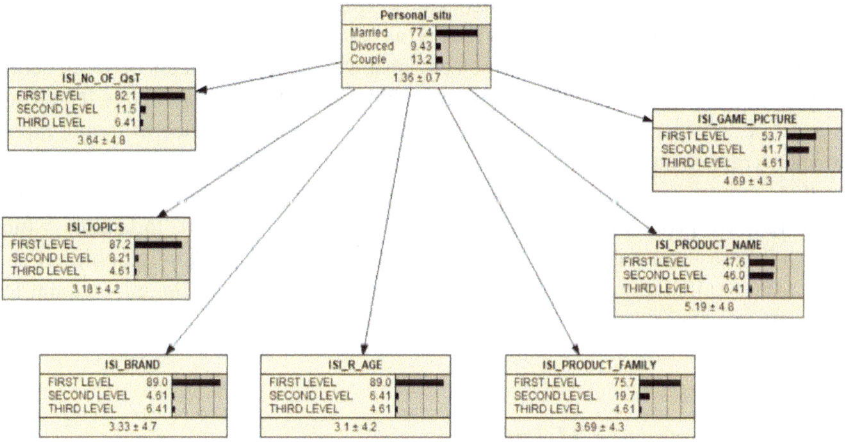

Fig. 7.8 BBN of personal situation of users and ISI of the packaging AOIs

demographic feature. The combined belief for the influence of the demographic feature on ISIs can be calculated through the Eq. (7.3).

$$\begin{aligned} &\text{Combined Belief (Demographic Feature} = x) \\ &= P(\text{Demographic Feature} = x) \cdot \\ &\sum \left[P(\text{ISI} = y | \text{Demographic Feature} = x) \cdot P(\text{ISI} = y) \right] \end{aligned} \tag{7.3}$$

In this formula, Σ represents the summation over all possible values of ISI, ‘y’ for a given demographic feature. P (ISI $= y$) represents the marginal probability of ISI being ‘y’ without considering the demographic feature.

At the core of the ER process lie the Conditional Probability Tables (CPTs). These tables serve as pivotal resources derived directly from the BBNs constructed earlier in our methodology. CPTs encode the probabilistic relationships between the parent nodes, representing demographic features, and the daughter nodes, which house the ISIs of packaging features within each BBN. The primary objective of utilizing CPTs within ER is to quantitatively assess the influence exerted by demographic features on user engagement with packaging features. These tables offer a structured representation of conditional probabilities, showing how variations in the demographic features impact the probabilities associated with the ISIs of packaging features, thus allowing for the systematic evaluation of the importance of each demographic factor.

Custom Probabilities, Combined Beliefs, and ISI Importance Ranking

In our ERA, we customized the probabilities associated with the parent nodes. These probabilities are derived from the dataset and serve as the foundation for quantifying the influence of demographic features on ISIs. The custom probabilities reflect the real-world dynamics of our study's context, ensuring that the analysis is both contextually relevant and robust. This involves the calculation of combined beliefs for each demographic feature's influence on ISIs. This calculation is achieved by multiplying the probabilities from the CPTs, which encapsulate the dependencies between demographic features and ISIs, by the custom probabilities of the parent nodes. The result is a set of combined beliefs that encapsulate the overall influence of demographic attributes on packaging feature engagement. Following this, the ranking of ISIs based on their importance within specific demographic categories was done. This importance ranking provides a clear hierarchy of packaging features, indicating which ones hold the most sway over user engagement for different demographic groups. This offers valuable guidance for targeted decision-making in the educational toy industry.

7.2.5 Performance Metrics

In our analysis, we employed performance metrics to evaluate the effectiveness of our methodology in ranking the ISI of packaging features for various demographic categories. The primary metric used for this evaluation was Spearman's Rank Correlation Coefficient, denoted as Spearman's Rho. It is a statistical measure used to assess the degree of association between two ranked variables. In our context, it serves as a vital indicator of the consistency between our approach's rankings of ISIs and the reference rankings. Spearman's Rho is particularly suited for ranking-based evaluations, as it does not require linear relationships and is robust to outliers. It is calculated using the following formula, Eq. (7.4):

$$\rho = 1 - \frac{6\sum d^2}{n(n^2 - 1)} \tag{7.4}$$

where ρ represents Spearman's Rank Correlation Coefficient, d represents the difference between the ranks of ISIs, and n is the number of ISIs being compared. This formula quantifies the degree of agreement between rankings, with a value of -1 indicating a perfect negative correlation, which aligns with our ranking convention (lower ranks imply higher importance). For each demographic category, we calculated Spearman's Rho to gauge the concordance between our approach's ISI rankings and the ground truth or reference rankings. These findings further support the practical applicability of our methodology for educational toy product design, marketing strategies, and audience targeting.

7.3 Results and Discussions

7.3.1 Construction of CPTs and Importance Scores of the AOIs

In this section, we present the results of our experiments that aimed to assess the influence of demographic features on user engagement with packaging features in the Educational Toy Industry. These CPTs are tailored to each ISI and specific to the demographic feature under consideration. They constitute a granular representation of the probabilistic dependencies, quantifying the conditional likelihood of each ISI state given the specific demographic context. Through these tables, we could establish a comprehensive foundation for dissecting the nuanced impact of demographic attributes on user engagement with distinct packaging elements (Figs. 7.9 and 7.10).

CPT of BBN with parent node: Gender			
Packaging feature	states	(1) Male	(2) Female
ISI_PRODUCT_NAME	FIRST LEVEL	0.52	0.451613
	SECOND LEVEL	0.44	0.483871
	THIRD LEVEL	0.04	0.0645161
ISI_BRAND	FIRST LEVEL	0.88	0.935484
	SECOND LEVEL	0.04	0.0322581
	THIRD LEVEL	0.08	0.0322581
ISI_R_AGE	FIRST LEVEL	0.92	0.903226
	SECOND LEVEL	0.04	0.0645161
	THIRD LEVEL	0.04	0.0322581
ISI_GAME_PIC	FIRST LEVEL	0.52	0.580645
	SECOND LEVEL	0.44	0.387097
	THIRD LEVEL	0.04	0.0322581
ISI_TOPICS	FIRST LEVEL	0.84	0.935484
	SECOND LEVEL	0.12	0.0322581
	THIRD LEVEL	0.04	0.0322581
ISI_PRODUCT_FAM	FIRST LEVEL	0.8	0.741935
	SECOND LEVEL	0.16	0.225806
	THIRD LEVEL	0.04	0.0322581
ISI_NO_OF_QsT	FIRST LEVEL	0.8	0.870968
	SECOND LEVEL	0.12	0.0967742
	THIRD LEVEL	0.08	0.0322581

CPT of BBN with parent node: Gender

CPT of BBN with parent node: Avg_child_age				
Packaging feature	states	(1) Toddler	(2) Pre-Schooler	(3) pre-teen
ISI_PRODUCT_NAME	FIRST LEVEL	0.2	0.419355	0.608696
	SECOND LEVEL	0.6	0.516129	0.347826
	THIRD LEVEL	0.2	0.0645161	0.0434783
ISI_BRAND	FIRST LEVEL	0.6	0.935484	0.869565
	SECOND LEVEL	0.2	0.0322581	0.0434783
	THIRD LEVEL	0.2	0.0322581	0.0869565
ISI_R_AGE	FIRST LEVEL	0.6	0.903226	0.913043
	SECOND LEVEL	0.2	0.0645161	0.0434783
	THIRD LEVEL	0.2	0.0322581	0.0434783
ISI_GAME_PIC	FIRST LEVEL	0.2	0.645161	0.478261
	SECOND LEVEL	0.6	0.322581	0.478261
	THIRD LEVEL	0.2	0.0322581	0.0434783
ISI_TOPICS	FIRST LEVEL	0.6	0.935484	0.826087
	SECOND LEVEL	0.2	0.0322581	0.130435
	THIRD LEVEL	0.2	0.0322581	0.0434783
ISI_PRODUCT_FAM	FIRST LEVEL	0.4	0.741935	0.826087
	SECOND LEVEL	0.4	0.225806	0.130435
	THIRD LEVEL	0.2	0.0322581	0.0434783
ISI_NO_OF_QsT	FIRST LEVEL	0.4	0.870968	0.826087
	SECOND LEVEL	0.4	0.0967742	0.0869565
	THIRD LEVEL	0.2	0.0322581	0.0869565

CPT of BBN with parent node: Avg Child Age

Fig. 7.9 CPT I and II

CPT of BBN with parent node: Personal_situ				
Packaging feature	states	(1) Married	(2) divorced	(3) couple
ISI_PRODUCT_NAME	FIRST LEVEL	0.488372	0.571429	0.333333
	SECOND LEVEL	0.465116	0.285714	0.555556
	THIRD LEVEL	0.0465116	0.142857	0.111111
ISI_BRAND	FIRST LEVEL	0.930233	0.714286	0.777778
	SECOND LEVEL	0.0232558	0.142857	0.111111
	THIRD LEVEL	0.0465116	0.142857	0.111111
ISI_R_AGE	FIRST LEVEL	0.930233	0.714286	0.777778
	SECOND LEVEL	0.0465116	0.142857	0.111111
	THIRD LEVEL	0.0232558	0.142857	0.111111
ISI_GAME_PIC	FIRST LEVEL	0.511628	0.714286	0.555556
	SECOND LEVEL	0.465116	0.142857	0.333333
	THIRD LEVEL	0.0232558	0.142857	0.111111
ISI_TOPICS	FIRST LEVEL	0.906977	0.714286	0.777778
	SECOND LEVEL	0.0697674	0.142857	0.111111
	THIRD LEVEL	0.0232558	0.142857	0.111111
ISI_PRODUCT_FAM	FIRST LEVEL	0.813953	0.571429	0.555556
	SECOND LEVEL	0.162791	0.285714	0.333333
	THIRD LEVEL	0.0232558	0.142857	0.111111
ISI_NO_OF_QsT	FIRST LEVEL	0.860465	0.714286	0.666667
	SECOND LEVEL	0.0930233	0.142857	0.222222
	THIRD LEVEL	0.0465116	0.142857	0.111111

CPT of BBN with parent node: Personal situation

CPT Table of BBN with parent node: No_of_children				
Packaging feature	states	(1) 1	(2) 2	(3) 3
ISI_PRODUCT_NAME	FIRST LEVEL	0.526316	0.483871	0.333333
	SECOND LEVEL	0.421053	0.451613	0.555556
	THIRD LEVEL	0.0526316	0.0645161	0.111111
ISI_BRAND	FIRST LEVEL	0.894737	0.903226	0.777778
	SECOND LEVEL	0.0526316	0.0322581	0.111111
	THIRD LEVEL	0.0526316	0.0645161	0.111111
ISI_R_AGE	FIRST LEVEL	0.894737	0.935484	0.666667
	SECOND LEVEL	0.0526316	0.0322581	0.222222
	THIRD LEVEL	0.0526316	0.0322581	0.111111
ISI_GAME_PIC	FIRST LEVEL	0.789474	0.483871	0.222222
	SECOND LEVEL	0.157895	0.483871	0.666667
	THIRD LEVEL	0.0526316	0.0322581	0.111111
ISI_TOPICS	FIRST LEVEL	0.894737	0.870968	0.777778
	SECOND LEVEL	0.0526316	0.0967742	0.111111
	THIRD LEVEL	0.0526316	0.0322581	0.111111
ISI_PRODUCT_FAM	FIRST LEVEL	0.736842	0.774194	0.666667
	SECOND LEVEL	0.210526	0.193548	0.222222
	THIRD LEVEL	0.0526316	0.0322581	0.111111
ISI_NO_OF_QsT	FIRST LEVEL	0.842105	0.806452	0.777778
	SECOND LEVEL	0.105263	0.129032	0.111111
	THIRD LEVEL	0.0526316	0.0645161	0.111111

CPT of BBN with parent node: No. of children

Fig. 7.10 CPT III and IV

These CPTs capture the probabilistic relationships between the parent demographic feature and the daughter nodes representing ISIs. The importance scores of the 7 AOIs corresponding to the parental demographics is shown in the line plot in Fig. 7.11. The rankings and importance values for each demographic feature is shown in Fig. 7.12.

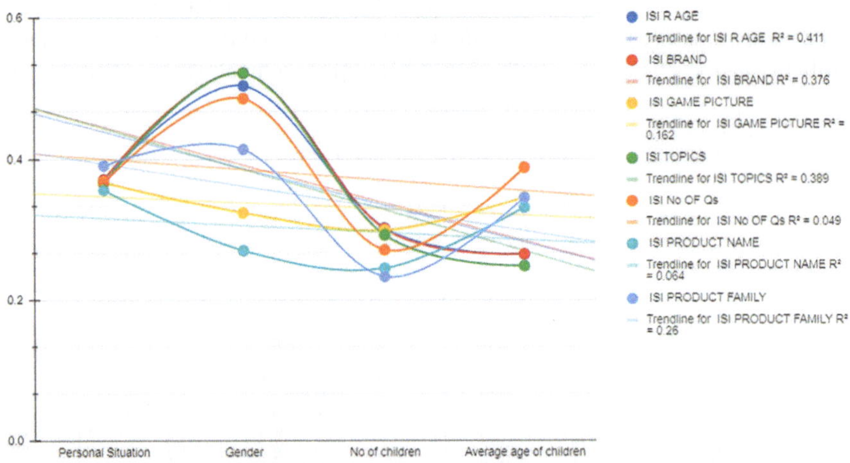

Fig. 7.11 Smooth line plot showing importance scores of various AOIs

Demographics:	No. of children		Gender		Personal Situation		Avg. Child Age	
Spearman's RCC:	ρ = 1.245		ρ = 1.1347		ρ = 1.1517		ρ = 1.2239	
Packaging Element	Importance	Rank	Importance	Rank	Importance	Rank	Importance	Rank
Number of Questions & Topics	0.271	5	0.486	4	0.366	5	0.388	1
Draw of topics / Descriptive message	0.292	4	0.522	2	0.365	6	0.248	7
Brand	0.301	2	0.522	1	0.372	3	0.265	5
Recommended Age	0.302	1	0.504	3	0.371	2	0.264	6
Product Family	0.233	7	0.414	5	0.391	1	0.345	3
Product Name	0.245	6	0.27	7	0.356	7	0.331	4
Game picture	0.299	3	0.324	6	0.368	4	0.345	2

Fig. 7.12 Importance scores and rankings of the packaging elements for each demographic feature along with their Spearman's Rho coefficients

7.3.2 Feature Rankings and Its Interpretation

We obtained feature rankings that highlighted the packaging features' relative importance within specific demographic contexts. These rankings offered a clear and actionable hierarchy of features that can guide stakeholders in the Educational Toy Industry in their decision-making processes. We also calculated and documented the maximum probabilities associated with each packaging feature within the context of demographic influence. These maximum probabilities provided a numerical representation of the peak importance of each feature, further enhancing our ability to prioritize packaging elements effectively. To enhance the interpretability of our results, we sorted the packaging features based on their maximum probabilities in descending order. This sorting process arranged the features from the most influential to the least influential within specific demographic categories, aiding stakeholders in focusing their efforts and resources on the most critical aspects of packaging design and marketing.

We utilized Spearman's rank correlation coefficient to provide an additional layer of insight into the relationships between the ISI rankings and demographic features. It is noteworthy that Spearman's rank correlation coefficient consistently exhibits a negative value across different demographic features. This negative correlation implies that as the ranking of ISIs increases (indicating higher importance), the rank of Spearman's correlation coefficient decreases. However, only the absolute value is mentioned in the paper.

7.3.3 Demographic Influence on ISIs

Our investigation into the impact of the number of children on packaging preferences unveils a fascinating interplay. The alignment of packaging with the recommended age of children emerges as the paramount influencer. This finding emphasizes the critical importance of tailoring packaging to specific age groups within the educational toy industry. Further, Brand, Game Picture, and Topics, in descending order, demonstrate the enduring significance of brand identity, captivating visuals, and thematic content. The number of children significantly shapes user engagement by driving preferences for age-appropriate and visually compelling packaging (Fig. 7.13).

Our analysis of personal situations, product family emerges as the dominant influencer. The high rankings of recommended age and brand further underscore the importance of aligning packaging with family-oriented scenarios, such as age—appropriate content and brand resonance. Thus, marketing strategies and packaging designs should resonate with diverse family contexts. By acknowledging the significance of personal situations, industry stakeholders can connect with consumers on a deeper level, enhancing the appeal of educational toys within various family dynamics. Our gender-based analysis transcends stereotypes to reveal a nuanced picture of packaging feature preferences. Brand and Topics jointly claim the top rank in importance, signaling the universal appeal of brand identity and thematic content across gender lines. The recommended age is slightly low in rank. These insights challenge gender norms by highlighting the potential for gender-neutral packaging strategies that focus on universal appeal rather than stereotypes. The industry can embrace inclusivity and broaden its audience, fostering a diverse and engaged consumer base. Average child age as a demographic factor reveals intriguing insights into packaging preferences. Number of Qs emerges as the most influential, signaling the pivotal

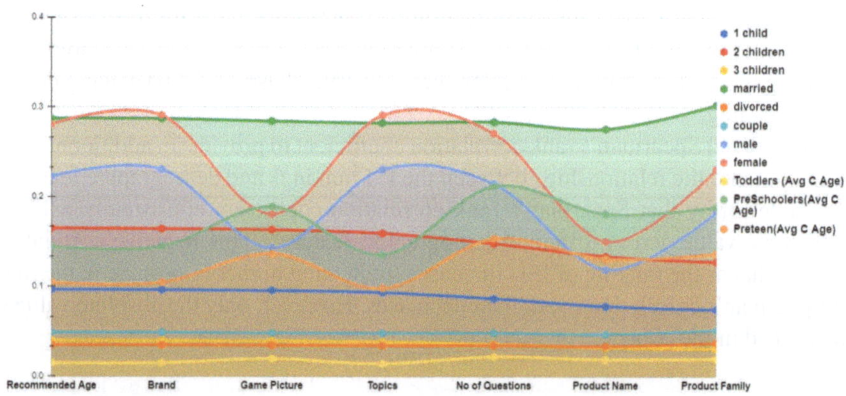

Fig. 7.13 Area chart showing the importance score of each demographic state for the packaging elements

role of the number of questions on packaging in engaging users. This finding underscores the educational value of packaging content, reflecting users' desire for interactive and intellectually stimulating experiences. Game picture and Product family also exhibit substantial importance, highlighting the influence of visual content and product cohesion, and prioritizing interactive packaging elements.

7.4 Conclusions

Our findings reveal the interplay between demographic attributes and packaging features, empowering stakeholders in the toy industry to make informed decisions that enhance user engagement. Tailoring packaging to specific age groups, family contexts, and gender preferences has the potential to elevate educational toys and drive consumer satisfaction and revenue growth. Emphasizing age-appropriate packaging, brand identity, and thematic content suggests avenues for refining marketing strategies and product development.

Theoretical Contribution: Our study's theoretical contributions offer solutions to packaging problems, transcending traditional approaches. The introduction of the ISI provides a comprehensive tool for handling eye-tracking metrics in packaging feature preferences, enabling precise understanding of consumer behavior. In conjunction with BBNs, our methodology quantifies the impact of demographic factors on packaging engagement, offering clarity to consumer preferences. Complementing BBNs, ERA systematically navigates uncertainty, paving the way for more robust analyses. Spearman's rank correlation coefficient enhances this toolbox, offering a quantitative means to evaluate consumer behavior analyses. These contributions provide methodologies to untangle packaging dilemmas, offering novel solutions in this field.

Industrial Application: Understanding the relative importance of packaging features and their alignment with various demographic contexts empowers industry stakeholders to allocate resources effectively. Age and brand consistently rank high across demographics, underlining the importance of clear brand logos and safety icons. This empowers companies to prioritize designs that effectively communicate brand identity and essential safety information. For families, bundled packaging catering to multiple children demonstrates the highest preference. Additionally, implementing a color-coding system based on age groups can enhance user experience and aid in selection. While personal situations highlight the dominance of specific product lines, catering to diverse gender preferences through varied packaging designs can strengthen consumer connections. Interestingly, age-appropriate and inclusive branding strategies, evident in the top-ranking brand and topics, demonstrate universal appeal. Aligning packaging elements with insights enhances product appeal, deepens consumer connection, and informs competitive decisions, benefiting both consumers and stakeholders. Our research underscores the multifaceted nature of user engagement with toy packaging, paving the way for data-driven decisions.

7.4.1 Limitations and Future Scopes

While our study provides valuable insights, it faces limitations and suggests future research avenues. Focusing on a specific dataset may not universally apply to diverse toy categories and industries, and the reliance on data from limited participants may not fully capture the complete nature of consumer preferences in the expansive toy market. Moreover, limitations of eye-tracking technology in comprehensively capturing consumer behavior and preferences need acknowledgement, along with the failure to account for evolving consumer preferences over time. The saliency index suggested by us needs to be validated by external validation techniques, which was not feasible in this research. External validation techniques, such as eye-tracking heatmaps and subjective user feedback, could further bolster the reliability of our methodology, thus presenting an avenue for future research. Additionally, conflicting design goals for users who are not purchasers, such as toys and baby products, pose challenges in creating attractive designs for both target groups [21].

To address these challenges, future research could encompass more extensive and diverse participant demographics, conduct longitudinal studies, combine eye-tracking data with diverse data sources, compare different packaging designs and strategies, employ machine learning models for more in-depth analysis, gather direct consumer feedback through surveys, explore international markets and cultural variations, and consider eco-friendly packaging preferences. Advanced clustering techniques with optimization strategy can be used on image data to obtain better classification accuracy. Rough set theory or Z-number can be used for handling the uncertainty in image classification. A decision support system can be built for smooth operations in decision- making. These future research directions can address intricate packaging challenges across various industries and provide pragmatic solutions.

References

1. Ni, J.Z., Flynn, B.B., Jacobs, R.: The effect of a toy industry product recall announcement on shareholder wealth. Int. J. Prod. Res. **54**, 5404–5415 (2016)
2. Bradley, R.H.: Social-cognitive development and toys. Top. Early Child. Spec. Educ. **5**(3), 11–29 (1985)
3. Alhamdi, F.: Role of packaging in consumer buying behavior. Manag. Sci. Lett. **10**(6), 1191–1196 (2020)
4. Ju´arez-Var´on, D., Tur-Vin˜es, V., Rabasa-Dolado, A., Polotskaya, K.: An adaptive machine learning methodology applied to neuromarketing analysis: prediction of consumer behaviour regarding the key elements of the packaging design of an educational toy. Soc. Sci. **9**(9), 162 (2020)
5. Krishna, A., Cian, L., Aydınoğlu, N.Z.: Sensory aspects of package design. J. Retail. **93**(1), 43–54 (2017)
6. Chen, J.C.-C.: The impact of nostalgic emotions on consumer satisfaction with packaging design. J. Bus. Retail. Manag. Res. **8**(2) (2014)
7. Al-Samarraie, H., Eldenfria, A., Dodoo, J.E., Alzahrani, A.I., Alalwan, N.: Packaging design elements and consumers' decision to buy from the web: a cause and effect decision-making model. Color. Res. Appl. **44**(6), 993–1005 (2019)

8. Lindsey, E.W., Mize, J.: Contextual differences in parent–child play: implications for children's gender role development. Sex Roles **44**, 155–176 (2001)
9. Lydia, V.P., Jenkins, E., Moore, A., Perrine, C., Pritchard, H.: The importance of play: a literature review (2014)
10. Ulfa, M., Djamaludin, M.D.: The influence of parent's perception and involvement in purchasing decision of toys for children. J. Consum. Sci. **1**(1), 59–71 (2016)
11. Al Kurdi, B.: Investigating the factors influencing parent toy purchase decisions: reasoning and consequences. Int. Bus. Res. **10**(4), 104–116 (2017)
12. Varón, D.J.: Neuromarketing analysis of the educational toy industry. In: Mendeley Data (2020). https://data.mendeley.com/datasets/pfc2zdr2g3/1
13. Sarkar, S., Khatedi, N., Pramanik, A., Maiti, J.: An ensemble learning-based undersampling technique for handling class-imbalance problem. In: Proceedings of ICETIT 2019: Emerging Trends in Information Technology, pp. 586–595. Springer International Publishing (2020)
14. Geffner, H., Dechter, R., Halpern, J.Y.: Probabilistic and causal inference: the works of Judea pearl. ACM (2022)
15. Ghahramani, Z.: Probabilistic machine learning and artificial intelligence. Nature (7553), 452–459 (2015)
16. Jiang, Z., Zheng, T., Liu, Y., Carlson, D.: Incorporating prior knowledge into neural networks through an implicit composite kernel. arXiv preprint arXiv:2205.07384 (2022)
17. Yu, H.: Incremental learning of Bayesian networks from concept-drift data, pp. 701–704 (2019) https://doi.org/10.1109/ICCCBDA.2019.8725689
18. Guney, A.: Introduction to Bayesian belief networks. Towards Data Science (2019). https://towardsdatascience.com/introduction-to-bayesian-belief-networks-c012e3f59f1b
19. Rao, S.: Basic understanding of Bayesian belief networks. GeeksforGeeks (2020).https://www.geeksforgeeks.org/basic-understanding-of-bayesian-belief-networks/
20. Srivastava, R.P.: An introduction to evidential reasoning for decision making under uncertainty: Bayesian and belief function perspectives. Int. J. Account. Inf. Syst. **12**(2), 126–135 (2011)
21. Tiwasing, W., Sahachaisaeree, N.: Distinctive design perception: a case of toy packaging design determining children and parents' purchasing decision. Procedia—Social Behav. Sci. **42**, 391–398 (2012) https://doi.org/10.1016/j.sbspro.2012.04.203; ASEAN Conference on Environment Behaviour Studies (AcE-Bs), Riverside Majestic Hotel, 7–8 July 2010, Kuching, Sarawak, Malaysia

Chapter 8
Subjective Health Complaints and Associated Risk Factors Among Academicians

Julliet Pradhan and Mohammed Rajik Khan

Abstract Musculoskeletal disorders (MSDs) represent a prevalent occupational health concern among academicians, adversely affecting their quality of life and posing significant economic burdens. The present study aims to investigate the prevalence of musculoskeletal pains and the related ergonomic risk factors among university professors during classroom teaching. Here, 50 institute professors (42 men and 8 women) were evaluated for pain using the Nordic questionnaire, and postural assessment was carried out using the observational assessment method. Chi-square (χ^2) test was performed to identify the association of individual and work-related factors to MSDs. The results of pain prevalence among professors for the past 12 months implies that the pain is most commonly felt in the neck (58%), right shoulder (64%), left shoulder (56%), upper back (62%), lower back (54%) and right wrist (52%). The right upper limb assessment (RULA) also identifies the ergonomic risk factors associated with different teaching postures. The work-related factors like experience, age, classes per week, and body postures deviated from neutral positions shows a significant relation with the prevalence of MSDs ($p < 0.05$). High risk was identified for teaching activities performed above eye level. Research focusing on targeted ergonomic interventions and modifications in teaching practices or educational programs on proper posture, could help mitigate the negative effects identified in this study.

Keywords Musculoskeletal pains · Academicians · Nordic questionnaire · RULA · Risk factors

---

J. Pradhan · M. R. Khan (✉)
Department of Industrial Design, National Institute of Technology Rourkela, Rourkela, India
e-mail: khanmr@nitrkl.ac.in

J. Pradhan
e-mail: pradhanjulliet@nitrkl.ac.in

J. Maiti et al. (eds.), *Recent Advances in Industrial and Systems Engineering*, Lecture Notes on Multidisciplinary Industrial Engineering,
https://doi.org/10.1007/978-981-96-8323-9_8

8.1 Introduction

Inflammatory or degenerative illnesses affecting the joints, muscles, bones, ligaments, tendons, peripheral nerves, or supporting blood vessels are referred to as musculoskeletal disorders (MSDs). They gradually worsen and show up as musculoskeletal aches, pains, or discomfort, which subsequently impairs function [1, 2]. The majority of work-related MSDs are brought on by or made worse by certain job tasks or by the affects of the surrounding workplace. Numerous individual, occupational, and psychosocial aspects have been found to have an impact on the intricate process of work-related musculoskeletal disorders (WRMSDs).

In recent years, the teaching profession has garnered attention for its association with WRMSDs, marking a significant departure from its prior neglect in discussions surrounding occupational health [3, 4]. Characterized by discomfort, pain, or dysfunction in muscles, tendons, joints, and other soft tissues, WRMSDs pose a substantial risk to the well-being and effectiveness of academicians worldwide. Despite their pivotal role in shaping future generations, teachers face unique occupational hazards that predispose them to these debilitating conditions. Epidemiological studies have shed light on the alarming prevalence of WRMSDs among school teachers, painting a concerning picture of their occupational health landscape [5, 6]. Compared to other professional groups, teachers consistently report disproportionately high rates of WRMSDs across various body regions. Shockingly, prevalence rates of self-reported WRMSDs among educators have been documented to range between 39% and a staggering 95% [5]. This prevalence not only underscores the severity of the issue but also underscores the urgent need for comprehensive understanding, prevention, and intervention strategies within the teaching profession.

The nature of teaching work, which involves not only classroom instruction but also planning, assessment, and involvement in extracurricular activities, contributes to both emotional and physical stress among teachers [7]. The impact of WMSDs extends beyond physical discomfort, affecting work attendance, disability rates, and financial costs for both employees and employers. People whose work contains highly repetitive tasks and awkward postures are more prone to shoulder and neck pains. The degree of risk depends on factors such as the duration of exposure, the frequency of exposure and the extent of exposure. It has been demonstrated that female teachers had a higher prevalence of MSDs (95.3%) than do their male colleagues (81.4%) [8].

A few studies were available on the incidence of WMSD and risk assessment in general teaching practice. A detailed study on the occurrence of WMSD and associated risk factors among school educators in Botswana was conducted where factors like gender, past injuries, smoking habits, years of employment, sleep disorders, age, and prolonged standing or sitting in unnatural positions was also taken into account [9]. However, psychological factors such as high work pressure, low job satisfaction, repetitive work, and low job control also play a role in the prevalence of WMSDs. Among elementary and secondary school teachers, WMSD risk variables were shown to be higher than those of other diseases in terms of worker disability and absence from work [10]. Due to WMSDs, there are increases in worker absenteeism, operating expenses, employee damage compensation, and premium payments. An Irish study of school teacher disability retirement found that WMSD was the leading cause of disability retirement among school teachers, accounting for 10% of teachers who retired due to disability. Most WMSD cases are reported to occur among school teachers, with 61% in the United States, 59.2% in China, and only 17.7% in Japan. However, critical review shows that WMSD can be reduced or prevented through high job satisfaction, regular exercise, and a healthy work environment [3].

Research on the prevalence of WMSDs among teachers and the assessment of associated risks is growing but remains somewhat limited compared to other occupational sectors. While more research is needed to fully understand the prevalence of WMSDs among teachers and the effectiveness of interventions, addressing ergonomic concerns in educational settings is essential for promoting the health and well-being of educators and ensuring the quality of teaching and learning experiences. Hence, the present study aims to identify the ergonomic risk factors, prevalence of WMSDs, postural evaluation and other health complaints among university professors in classroom teaching. The pain prevalence on institute professors was assessed through a subjective evaluation survey using the Nordic questionnaire. Observational exposure assessment involves the analysis of videos from an E-learning website, NPTEL, to identify ergonomic risk factors in various activities related to classroom teaching. The rapid upper limb assessment (RULA) tool is employed to report factors with higher risks, and ergonomic interventions are further proposed based on the findings. In summary, the study emphasizes the importance of addressing WMSDs among institute professors, recognizing the multifaceted nature of their work and the need for ergonomic interventions to create a healthier work environment. The findings could potentially inform strategies to mitigate the risks associated with WMSDs in the teaching profession.

8.2 Methodology

8.2.1 Study Sample

Two separate studies were conducted among institute professors to identify ergonomic risk factors in the classroom teaching process. The first is a subjective assessment using a modified Nordic questionnaire, where professors from National Institute of Technology (NIT) Rourkela, India were randomly selected to avoid the bias in this cross-sectional study. In the second study, observational evaluation is used to investigate different teaching postures using 50 randomly selected E-learning videos from the National Program for Technology Enhanced Learning (NPTEL) web. In the first study, the participants were chosen based on two criteria: (a) they could not have had a prior accident that resulted in pain, and (b) they had to have been teachers for at least a year. For this study, prior agreement has been obtained from 50 academics (42 men and 8 women) who have agreed to participate.

8.2.2 Questionnaire

The motivation behind the study was made clear to the selected candidates and they were provided with a self-administered questionnaire in English to complete in a relaxed atmosphere (no educational activity for at least 3 h prior to completing the responses). Two sections make up the questionnaire. The demographic information on the professor's age, height, weight, and marital status makes up the first section. The work-related issues, such as the number of teaching hours per week, the length of each day's break, and years of experience, are covered in the second section. Professors were also asked to score the degree of pain or discomfort they felt in different body parts while carrying out various tasks for their teaching activities. A score of 1 indicated no pain, whereas a score of 2–5 indicated moderate to severe discomfort. Data were kept about how long and how often specific body parts were held in abnormal positions. Additionally, the incidence of pain frequency in various body regions for the past 12 months was dichotomized as no, sometimes, often, and regular.

8.2.3 Pain Assessment

Pain ratings collected through the questionnaire survey were analyzed using IBM SPSS version 22 statistical tools to determine pain frequency. To assess pain during teaching activity, a pain rating of 1 was dichotomized as "no MSD" as no pain occurred, and pain ratings 2–5 were considered as "MSD." Similarly, pain data for the past 12 months were collected, and responses of "no" were counted as "no MSD" and other responses were counted as "MSD."

8.2.4 Posture Analysis

In the observational exposure assessment, postures adopted during teaching activities were classified as follows: (a) standing and teaching, (b) writing on the board, (c) pointing and explaining at the board, and (d) cleaning board using duster. From the selected videos by NPTEL, the postures of the subjects under each activity were recorded at a rate of 30 frames per second (fps) in order to ascertain the angles of the neck, upper and lower arms, wrists, trunk, and legs from a neutral standing position. Additionally, all recorded postures were categorized as eye level, below eye level, and above eye level. At each eye level, ergonomic risk factors for each of the four teaching activities were determined using rapid upper limb assessment (RULA) [11, 12].

8.2.5 *Data Analysis*

Chi-square tests of independence (χ^2) were performed on data collected through self-administered questionnaires. A significance level of $p < 0.05$ was adopted to assess the association between demographic and other work-related variables and musculoskeletal pain. To determine the odds ratio (OR) and confidence interval (95% CI) of potential risk factors, binary logistic regression analysis was performed. The assumed significance level was $p \leq 0.1$. The independent variables were age, teaching experience, number of classes per week, and duration of postural deviations of different parts of the body. The dependent variables were pain scores in different regions of the upper extremities (proximal: neck, upper back, and shoulders; distal: elbows and wrists) and lower extremities (proximal: lower back and hips).

8.3 Results

8.3.1 *Population Study*

The general information of the participants collected from the self-administered questionnaire is shown in Table 8.1. The average age was 41.78 years. 84% of participants were male and 94% were married. 68% had more than 5 years of teaching experience. 38% have 10 or more classes per week, and only 20% have less than 5 classes per week. 76% of participants took less than 2 h of rest per day, 80% worked at a table for more than 3 h per day, and 94% are teaching exclusively.

8.3.2 *Pain Rating*

Professors reported experiencing the most pain in the right shoulder (64%), upper back (62%), neck (58%), left shoulder (56%), lower back (54%), and right wrist (52%) during teaching activities. Table 8.2 displays the prevalence of MSDs over the previous 12 months in the various body regions. According to the study, MSDs are most common in the neck (58%), upper back (50%) lower back (64%), right shoulder (54%), and left shoulder (44%) (Fig. 8.1).

8.3.3 *Correlation of Risk Factors with Musculoskeletal Pains*

Table 8.3 depicts the results of the χ^2 test of independence which associates individual and work-related factors with MSDs. Factors such as gender, marital status, and BMI did not show a significant relation with MSDs ($p > 0.05$). However, work-related

Table 8.1 General information about the study participants (N-50)

Variable	Mean (SD)	Range
Age (Years)	41.78 (7.03)	29–60
Height (cm)	167.87 (7.59)	150–182
Weight (kg)	71.38 (9.87)	50–100
BMI	25.3 (3.06)	16.9–33.7
Variable	**Number**	**Percentage (%)**
Gender		
Male	42	84
Female	8	16
Experience (years)		
< 1	2	4
1–5	14	28
5–10	19	38
> 10	15	30
Classes per week		
< 5	10	20
5–10	21	42
10–15	15	30
> 15	4	8
Break (in hrs)		
< 1	15	30
1–2	23	46
2–3	9	18
> 3	3	6
Other jobs		
Yes	47	6
No	3	94
Dominant side		
Right	48	96
Left	3	4
Marital status		
Married	47	94
Single	3	6

factors including experience, age, classes per week, and body postures deviated from neutral positions demonstrated a significant relation with the prevalence of WMSDs ($p < 0.05$).

The majority of results obtained during binary logistic regression analysis were found to be non-significant. The results are raw and unadjusted. Among the individual

Table 8.2 Prevalence of MSDs among professors ($N = 50$) in different body parts for the past 12 months

Parts	Frequency	Percentage (%)
Neck	29	58
Upper back	25	50
Lower back	32	64
Left shoulder	22	44
Right shoulder	27	54
Left elbow	12	24
Right elbow	17	34
Left wrist	15	30
Right wrist	19	38
Left hip	17	34
Right hip	16	32
Left knee	18	36
Right knee	13	26
Left ankle	15	30
Right ankle	13	26

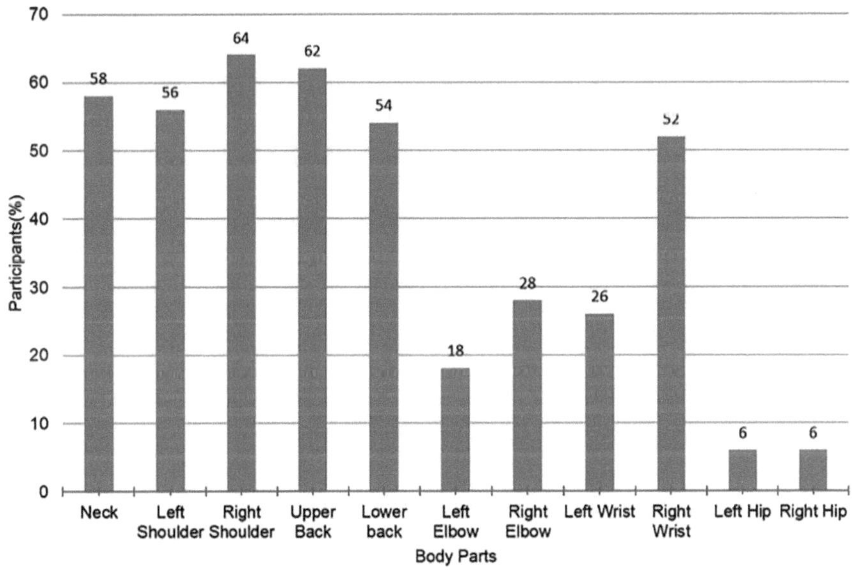

Fig. 8.1 Prevalence of MSDs among professors ($N = 50$) in different body parts during teaching

Table 8.3 Association of individual and work-related factors to MSDs ($N = 50$)

Factors	MSDs		X^2	p
	Yes	No		
Age (years)				
< 30 (1)	0	1	28.29	0.01
30–40 (21)	14	7		
> 40 (28)	28	0		
Experience (years)				
< 1 (2)	0	2	19.035	0.01
1–5 (14)	9	5		
5–10 (19)	18	1		
> 10 (15)	15	0		
Gender				
Male (42)	35	7	0.087	0.771
Female (8)	7	1		
BMI				
< 18.5 (1)	1	0	50	0.98
18.5–24.9 (21)	18	3		
25–29.9 (25)	20	5		
> 30 (3)	3	0		
Marital status				
Married (47)	40	7	0.713	0.403
Single (3)	2	1		
Classes per week				
< 5 (10)	6	4	6.066	0.025
5–10 (21)	18	3		
10–15 (15)	14	1		
> 15 (4)	4	0		
Neck posture (min)				
< 15 (13)	8	5	7.094	0.012
15–25 (29)	26	3		
> 25 (8)	8	0		
Upper arm posture (min)				
< 15 (8)	0	8	50	0.00
15–25 (42)	42	0		
> 25 (0)	0	0		
Lower arm posture (min)				
< 15 (16)	8	8	20.23	0.00
15–25 (34)	34	0		

(continued)

Table 8.3 (continued)

Factors	MSDs		X^2	p
	Yes	No		
> 25 (0)	0	0		
Wrist posture (min)				
< 15 (6)	0	6	35.795	0.00
15–25 (44)	42	2		
> 25 (0)	0	0		
Trunk posture (min)				
< 15 (30)	22	8	6.349	0.016
15–25 (19)	19	0		
> 25 (1)	1	0		
Shoulder posture (min)				
< 15 (22)	14	8	12.121	0.01
15–25 (27)	27	0		
> 25 (1)	1	0		

risk factors, experience was substantially correlated with the occurrence of MSDs in different body parts. For the subjects in the experience group 5–10 years, significant association was found with musculoskeletal symptoms in the neck, left shoulder, right shoulder, upper back, and lower back, respectively, as OR 0.19; 95% CI [0.036, 0.976], OR = 0.273; 95% CI [0.057, 1.296], OR = 0.655; 95% CI [0.134, 3.186], OR = 0.2; 95% CI [0.042, 0.983], and lower back OR = 0.25; 95% CI [0.048, 1.292]. Among the factors related to work, upper arm posture was significantly correlated with the pervasiveness of MSDs in various body regions. For the subjects in the upper arm posture group 15–25 min, significant correlation was found with musculoskeletal symptoms in the neck, left shoulder, right shoulder, upper back, and lower back, respectively, as OR = 0.071; 95% CI [0.008, 0.639], OR = 0.143; 95% CI [0.016, 1.265], OR = 0.227; 95% CI [0.041, 1.259], OR = 0.107; 95% CI [0.012, 0.95], and OR = 0.133; 95% CI [0.024, 0.756].

8.3.4 Posture Analysis

Table 8.4 represents risk levels identified when professors' postures were analyzed during different teaching activities. At above eye level, it is observed that the activities "Write on the board," "Point to the board while explaining," and "Use duster" each achieve the highest RULA score of 7 designating them as high-risk level. Similarly, at below eye level, a highest RULA score of 5 was seen for the activities like "Write on the board" and "Using duster." Accordingly, teaching activities at EL has least risk among the three levels.

Table 8.4 Work posture analysis of professors using RULA

Activity	Above eye level (AEL)			Eye level (EL)			Below eye level (BEL)		
	Score	Risk level	Action	Score	Risk level	Action	Score	Risk level	Action
Standing and teaching	3	Low	Change: least required	2	Negligible	Change: not required	4	Low	Change: least required
Writing on board	7	High	Change: highly required	4	Low	Change: least required	5	Medium	Change: required
Pointing board and explaining	7	High	Change: highly required	3	Low	Change: least required	4	Low	Change: least required
Using duster	7	High	Change: highly required	3	Low	Change: least required	5	Medium	Change: required

8.3.5 Health Problems

Questionnaire surveys regarding current health issues and challenges faced by professors during their teaching were also evaluated and is shown in Fig. 8.2.

Figure 8.2 shows that 70% of the professors are in good health. 20% of them suffer from gastrointestinal disorders, 8% have diabetes and the remaining 2% have respiratory disorders and none have cardiovascular disorders. As seen in Fig. 8.2b, 80% of the professors expressed dissatisfaction with the internal layout of the classroom, extreme temperatures plagued 54% of them, with lack of fresh air and noise coming in second and third, respectively, as the other two main issues.

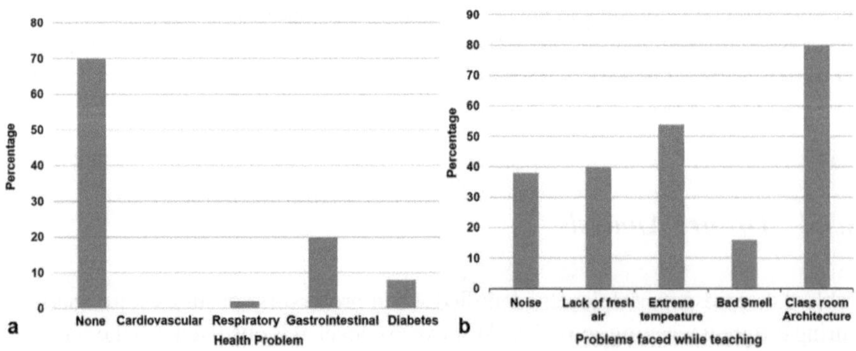

Fig. 8.2 According to questionnaire study professors ($N = 50$) suffering from **a** Health problems **b** Problems faced during classroom teaching

8.4 Discussion

Professors have been observed to feel pain during teaching, especially in their right shoulder, upper back, neck, left shoulder, lower back, and right wrist (Fig. 8.1). The professor's right hand makes up 96% of their dominant hand, hence, their right wrist and shoulder are where most of their pain is felt. Additionally, the RULA scores obtained from different educational activities at above and below eye level are higher than those obtained at eye level. As a result, when engaging in those activities, the lower back, left shoulder, upper back, and neck are mostly out of their neutral positions, which might be painful. Table 8.3 demonstrates the linear relationship between MSDs and postures created by various body regions, including the neck, shoulders, upper and lower arms, and back, with a significance level of $p < 0.05$ indicating the intensity of pain experienced. This makes sense because teaching activities and prolonged standing both raise the risk of musculoskeletal pain, particularly in the lower and upper back. Similar to this, assuming awkward postures while writing on the board and giving an explanation causes pain in specific body parts.

Given the correlation between individual characteristics and work-related factors and MSDs (Table 8.3), professors are more susceptible to MSDs as they age, suggesting a linear relationship between exposure-dose-response to MSDs. Professors with greater expertise have been found to have fewer MSDs than those with less experience. As one's teaching experience grows, they become accustomed to the style of teaching and get used to reducing stress and painful feelings. The limited sample size of this study, however, limits the ability to clearly illustrate the reaction. Again, professors who teach more classes per week tend to have more pain and are more likely to suffer from musculoskeletal disorders. This is because longer workdays cause them to feel more pain from their unfavorable postures throughout various teaching-related tasks. The regression analysis results were found to be inconclusive. The sample size of our study is not enough to obtain decisive statistical data and it puts a limitation of this study in concern to binary regression analysis.

RULA's analysis of the postures of university professors has shown that the postures adopted are dangerous and that corrective measures should be taken immediately. RULA score for all the activities at eye level is the lowest, reaching a maximum value of 4 (Table 8.4). This is because it is a neutral/comfortable position for different parts of the body. Additionally, Table 8.4 shows that, among the four activities, writing on the board has the highest RULA score (7), with the majority of body parts—including the neck, upper and lower arms, trunk, upper and lower back, and wrist—engaging and deviating from neutral postures. RULA scores greater than 4 are at high risk and immediate ergonomic changes are essential.

Although this study offers valuable insights into the problem of MSDs among academicians, there are still certain drawbacks. A self-administered questionnaire, which may have some reporting bias, served as the basis for the survey. The survey data gathered is of professors from a single institute. Random video samples may introduce bias into observational studies since the measured angles might not match those obtained from real-time measurements. These elements could mislead one to estimate the result too high or too low.

8.4.1 Ergonomic Interventions

Considering the postures and problems faced by the professors while teaching in a classroom, some of the ergonomic changes are suggested below in order to reduce the risk levels:

(a) The position of mid-line of the board should be at the eye level so that the neck and back are in neutral positions while standing. Provision to slide vertically could accompany professor's height.
(b) The width of the board with respect to eye level should be in a range such that the maximum RULA score obtained for the deviated postures should be less than or equal to 4.
(c) The classroom seating architecture should be well designed in steps at a gradient level. The step height and classroom length should have flexible design so that risk levels of professor's postures are low.
(d) Proper ventilation should be provided.
(e) Temperature should be maintained considering the outside temperature.
(f) Stretching exercises should be performed by the professors in the morning and in between of continuous classes.

8.5 Conclusion

In the present study, prevalence of MSDs and ergonomic risk factors among institute professors during various classroom teaching activities have been determined through questionnaire study and posture evaluation. Occupational and postural factors such as classes per week, deviation of neck, upper arm lower arm, wrist, trunk and shoulder postures from respective neutral postures for long duration increases the odds of MSDs among teachers. The findings highlighted the need for intervention models underscores the practical application of our research findings. The conclusions drawn from the results suggest a need for targeted ergonomic interventions and modifications in teaching practices to address the identified risk factors and alleviate musculoskeletal discomfort among university professors. Overall, a clear direction is suggested for advancing the scientific literature and addressing a significant occupational health concern.

References

1. Khan, M.R., Singh, N.K.: Prevalence of musculoskeletal disorders among Indian railway sahayaks. Int. J. Occup. Environ. Health **24**(1–2), 27–37 (2018)
2. Khan, M.R., Singh, N.K., Shinde, D.: An ergonomic study: bicycle repairer in rural India. In: Research into Design for a Connected World: Proceedings of ICoRD 2019 Volume 2, pp. 509–517. Springer Singapore (2019)
3. Erick, P., Smith, D.: Musculoskeletal disorder risk factors in the teaching profession: a critical review. OA Musculoskelet Med **1**(3), 29 (2013)
4. Fahmy, V.F., Momen, M.A.M.T., Mostafa, N.S., Elawady, M.Y.: Prevalence, risk factors and quality of life impact of work-related musculoskeletal disorders among school teachers in Cairo. Egypt. BMC Public Health **22**(1), 2257 (2022)
5. Erick, P.N., Smith, D.R.: A systematic review of musculoskeletal disorders among school teachers. BMC Musculoskelet. Disord. **12**, 1–11 (2011)
6. Yue, P., Liu, F., Li, L.: Neck/shoulder pain and low back pain among school teachers in China, prevalence and risk factors. BMC Public Health **12**, 1–8 (2012)
7. Chong, E.Y., Chan, A.H.: Subjective health complaints of teachers from primary and secondary schools in Hong Kong. Int. J. Occup. Saf. Ergon. **16**(1), 23–39 (2010)
8. Althomali, O.W., Amin, J., Alghamdi, W., Shaik, D.H.: Prevalence and factors associated with musculoskeletal disorders among secondary schoolteachers in Hail, Saudi Arabia: a cross-sectional survey. Int. J. Environ. Res. Public Health **18**(12), 6632 (2021)
9. Erick, P.N., Smith, D.R.: The prevalence and risk factors for musculoskeletal disorders among school teachers in Botswana. Occup Med Health Aff **2**(4), 178 (2014)
10. Ng, Y.M., Ibrahim, N., Maakip, I.: Prevalence and risk factors of musculoskeletal disorders (MSDs) among primary and secondary school teachers: a narrative review. Austin J. Musculoskelet. Disord. **4**(2), 1046 (2017)
11. Naik, G., Khan, M.R.: Prevalence of MSDs and postural risk assessment in floor mopping activity through subjective and objective measures. Saf. Health Work **11**(1), 80–87 (2020)
12. Khan, M.R., Naik, G.: An experimental investigation on postural risks in floor mopping. In: Design for Tomorrow—Volume 1: Proceedings of ICoRD 2021, pp. 39–50. Springer Singapore (2021)

Chapter 9
Whole-Body Vibration Analysis Using FEM in Public Transportation

Akshar J. Patel, Virajkumar G. Patel, and D. H. Pandya

Abstract Low-frequency vibrations are the most harmful forces produced in public transport, which have many adverse effects on the human body, like imbalances, muscle pain, nausea, vomiting, and dizziness, and they often resemble motion sickness. These vibrations primarily arise from various sources, such as road irregularities and moving mechanical components in vehicles. This article focuses on buses, which make up a significant portion of public transport in India. A 3D CAD multi-degree of freedom vibratory model weighing 65 kg of an Indian male is made according to a survey by NIN (Channel in National institute of nutrition changes average weight, height measurements for Indian men and women, 2020 [1]) in three different standing postures. Modal and harmonic analyses are carried out to identify the damping behavior and modal frequency of the model. Also, FEM random vibration analysis with frequency domain excitation is carried out to obtain frequency responses at the head. For the identification and comparison of the posture, feet-to-head transmissibility and vertical deformation at the head are considered. From this study, it is obscrvcd that thc most harmful position whilc travcling on a bus is standing normally without any support, while the most comfortable and less vulnerable compared to the other two postures is standing with an extended hand holding the handrail.

Keywords Low-frequency vibration · FEM analysis · Whole-body vibration · FFT · Vibrations in bus · Damping

Nomenclature

FFT Fast Fourier Transform
CAD Computer-Aided Design

A. J. Patel (✉) · V. G. Patel · D. H. Pandya
KSV University, Gandhinagar, India
e-mail: aksharpatel2647@gmail.com

V. G. Patel
e-mail: virajp016@gmail.com

J. Maiti et al. (eds.), *Recent Advances in Industrial and Systems Engineering*, Lecture Notes on Multidisciplinary Industrial Engineering,
https://doi.org/10.1007/978-981-96-8323-9_9

FEM Finite Element Method
WBV Whole-Body Vibration
DOF Degree of Freedom
NIN National Institute of Nutrition
PSD Power Spectral Density

9.1 Introduction

Exposure to whole-body vibration (WBV) for a healthy human can be an alarming factor influencing agility. According to [2] (ISO 2631-1:1997), exposure to a WBV for more than 8 h can cause serious health problems like brain damage, muscle pain, nausea, dizziness, bone joint pain, and many more as buses play a vital role in daily as well as occasional transportation in India. Irregularity and inappropriate practices in the maintenance of buses and roads generate vibrations with a wide range of frequencies and amplitudes to which the human body is exposed daily. Takahashi et al. [3] found that vibration with a frequency of less than 50 Hz can cause great damage to the human body by their experimental evaluation. Dutra et al. [4] conducted the experiment to assess the change in smartphone typing performance of passengers due to whole-body vibration in public transportation. By utilizing the experimental approach, they exposed passengers to three frequencies, 5, 10, and 15 Hz, with two magnitudes, 0.6 and 1.2 m/s^2, while carrying out subjective assessment to rate comfort and objective assessment by identifying the number of errors and typing speed. Their results show that subjects made more typing errors when exposed to lower frequencies compared to higher ones while, at lower magnitudes, fewer typing errors were observed. Also, by using both assessment methods, it was clear that a direct proportionality exists between frequency or magnitude and discomfort; on the other hand, the number of errors showed an inverse proportionality with the typing speed.

Bhuiyan et al. [5] undertook a comprehensive analysis of the intricate relationships between the numerous elements of whole-body vibration and the physiological and cognitive factors that contribute to driver fatigue. The study assessed the impact of vibration parameters such as frequency, amplitude, waveform, and duration, excluding hand-arm vibration, shocks, and acute or transient vibrations. The researchers discovered that the frequency, amplitude, waveform, and duration of the vibration work together to influence drowsiness. The optimal conditions can induce significant drowsiness within 30 min, and low-frequency whole-body vibrations, especially those in the range of 4–10 Hz, are the most effective in causing drowsiness.

To study and identify the possible solution for this exposure to WBV and reduce human discomfort, many studies have been carried out by various methods like FEM analysis, experimental methods, mathematical approach, lumped-mass system, and more.

Kumar et al. [6] conducted an experiment on six healthy male subjects. He exposed them to random whole-body vibration having a magnitude of 0.5 m/s^2 and 1 m/s^2 r.m.s vibration magnitude of frequency range 1–20 Hz and found that the first peak was observed at 4 to 7 Hz for floor-to-head transmissibility. In contrast, higher transmissibility was observed at the knee than at the head; also, holding the handle has lesser transmissibility than holding the handrail. Thuong and Griffin [7] investigated the discomfort experienced by individuals in fore-and-aft, lateral, and vertical positions when exposed to sinusoidal vibrations. Utilizing the magnitude estimation method, 12 subjects were asked to rate the discomfort caused by 4-Hz sinusoidal vibrations at ten different intensities. Their results revealed that subjects were less sensitive to lateral vibrations than to fore-and-aft vibrations ($K_y/K_x = 0.71$) and more sensitive to vertical vibrations than to horizontal vibrations ($K_z/K_x = 1.95$; $K_z/K_y = 2.77$). Previous research on the frequency-dependent discomfort of standing individuals was referenced to determine the frequency weightings for each direction, which differed from those appropriate for seated individuals and differed from the weightings in current standards that were mainly derived from understanding the discomfort of seated individuals. Singh et al. [8] conducted an experiment in two different postures of hand and arms in straight and 90° bend in standing position. He was exposed at an acceleration magnitude of 2.65–5.25 m/s^2 with a frequency range of 2.5–2500 Hz. Moreover, it has been found that grip force and position of the wrist and arms mainly affect transmissibility from the vibration section through the body. The transmissibility was greater in the vertical direction when the subject hand was on the handle rather than holding the handrail. Furthermore, it has been found that g-forces increase with an increase in frequency.

Kumar et al. [9] adopted finite element modeling (FEM) for the study of the biodynamic response of the human body and conducted modal analysis for two different Indian male subjects of masses 54 and 76 kg in standing posture, exposing them to undamped free vibrations. Calculations have been done using anthropometric data and formulae mentioned for the making of 3D CAD models of human subjects. In conclusion, the effect of vibration transmitted from the head to the lower arms and hands is maximum with the increase in natural frequency in both scenarios. In contrast, its effect in the upper segments (head, neck, hands) was found to be more in comparison with the lower segments. Moreover, it was concluded that, with an increase in natural frequency, the vibration transmits from the head to the lower arm to the hands. It was observed that the effect on the upper segment (head, neck, hands) is comparatively higher than that on the lower segments. AlShabi et al. [10] made a mathematical model of the seated human body and simulated it using Simulink/MATLAB. They also included the development of models for individual body segments, utilizing Simulink/MATLAB. Model analysis and state-space methods were employed to validate the results obtained from the simulations. The effects of mechanical vibrations on the head and lumbar regions were examined explicitly by them, as these areas are typically the most affected. Kinematic states of the head and lumbar segments were considered, and the characteristics of the vibration response in these areas were analyzed. Two types of forces are used to test the whole seated human body under low-frequency conditions: a sinusoidal wave signal based on literature and an impulse function. Their results show that the head and lumber were the most affected organs by vibration in accordance with force. Also, they concluded that there may be some fatigue, which causes strain, leading to pain in the rest of the organs according to the magnitude of the vibration.

Shibata et al. [11] investigated the differences in the experience of whole-body vibration between men and women when standing. Twenty-four individuals, comprised of twelve males and twelve females, rated their level of discomfort during the examination, which included three types of vibration (fore-and-aft, lateral, and vertical) at three different intensity levels (0.2, 0.4, and 0.8 m/s^2 r.m.s., unweighted) with a consistent power spectrum density at frequencies ranging from 1.0 to 20 Hz. The order of the test stimuli was randomized, and each stimulus was repeated three times. The participants utilized the category judgment method to rate the discomfort caused by whole-body vibration exposure in each direction, and then, they calculated the mean Stevens' power law exponent for both genders at each vibration direction. Their results showed that women experienced greater discomfort than men under fore-and-aft and lateral whole-body vibration exposure, as their power exponents were higher in these directions.

Tatsuno and Maeda [12] conducted a study to investigate the effects of whole-body vibration (WBV) exposure on standing balance function in the context of slip, trip, and fall accidents in the land transportation industry. It is hypothesized that these accidents are related to WBV exposure during vehicle travel. Ten participants underwent 60 min of virtual driving in a driving simulator (DS) to induce WBV exposure. Standing balance measurements were taken before and after exposure, as well as 2 and 4 min after exposure. Four conditions were tested by combining two WBV magnitudes and driver/passenger conditions. The study analyzed two standing balance indexes—total length and enveloped area—and calculated the rate of change relative to the pre-vibration value. The rate of change remained constant at 1.0 in the control condition without vibration exposure but varied under vibration exposure conditions. Interestingly, the rate of change at 2 min after exposure was high in the driver condition but decreased to almost 1.0 in the passenger condition. Since there was no difference in vibration acceleration between the driver and passenger conditions, it was assumed that the difference was due to fatigue caused by accelerator-pedal operation. Based on the percentage of standing balance that returned to 1.0 after 4 min in most conditions, the study recommends allowing a rest period of several minutes after exposure to WBV at 0.5 m/s^2 rms for 60 min in the DS. Singh et al. [13] presented a nine-DOF lumped parameter model for free vibration of a 54 kg Indian male in standing and sitting posture. They established the decoupling of vertical and fore-and-aft oscillation, which had been a typical behavior observed in human subjects with a maximum frequency of 39.8 Hz. Compared with others, the results obtained in this paper have been near the upper limit, as damping has not been included in the free vibration analysis.

After all the literature review, it has been observed that very less work using FEM analysis has been done. In this work, we would like to compare the three different standing postures in public transportation and identify the best posture among them. The identification and comparison are made by evaluating the feet-to-head transmissibility and vertical deformation at the head.

9.2 Modeling

For this study, [1] the NIN survey says the average weight of an Indian male is 65 kg and a height of 175.1 cm. A 3D CAD model using Siemens Solid Edge is developed, resembling the weight and height shaping of the human body structure. The material magnesium alloy is considered as it resembles the bone density of humans [14] in terms of density.

9.2.1 *Mass Distribution*

As there is no uniform distribution of mass in the body, each body part needs to be designed according to its weight distribution. The stiff parts of the body are considered solid cylinders like arms and legs, or else other parts are designed according to the weight required. All the joints in the body that provide multiple degrees of freedom are designed as the assembly of male and female ball joints. The considered ball joints have frictionless surface contacts, providing a free relative motion between connected parts.

Figure 9.1 shows the nomenclature used while CAD designed the model, and Table 9.1 gives the weight of individual components and their total weight in the body.

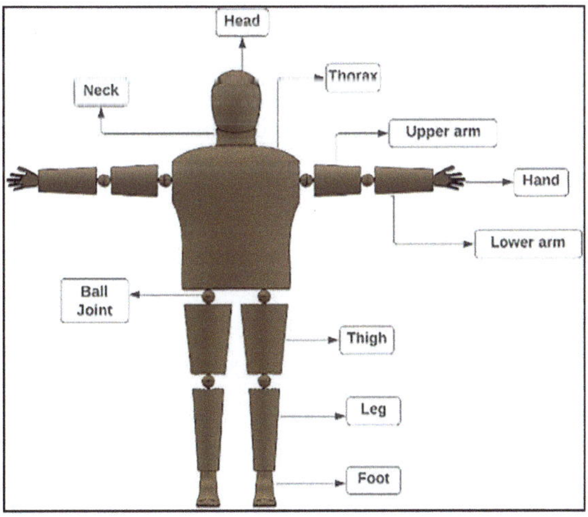

Fig. 9.1 Human model parts nomenclature

Table 9.1 Weight distribution of considered human model

Body Part	Mass (Kg)	NOS	Total mass (Kg)
Male ball joint	0.132	8	1.056
Female ball joint	0.029	8	0.232
Head	3.668	1	3.668
Thorax	38.052	1	38.052
Upper arm	2.079	2	4.158
Lower arm	1.775	2	3.55
Hand	0.173	2	0.346
Thigh	3.669	2	7.338
Leg	2.76	2	5.52
Foot	0.812	2	1.624
Total weight of body model			65.544

9.2.2 Posture

In buses, most people travel in standing positions and holding handrails. To cover major of the population who are involved in bus mobility, three postures are considered: one posture of standing, usually without holding or taking support of anything, which will create a base condition for comparison, and two postures holding handrails in two different styles.

Figure 9.2 resembles the human standing on a bus without any hand support, just on the legs; the hands of the human model are hanging downwards and are accessible from any supporting contact, and the foot of the model is in direct contact with the base excitation plate. Hence, vibration exposure to the body will result from the leg only, as the vibration enters the body from its bottommost part and is transmitted up to the top-most part of the body; nevertheless, it does not exit the body as there is the absence of any extra contact on the upper body.

Figure 9.3 shows the CAD model of a human standing in a bus and holding the handrail with the raised hand, which provides additional support to the body, due

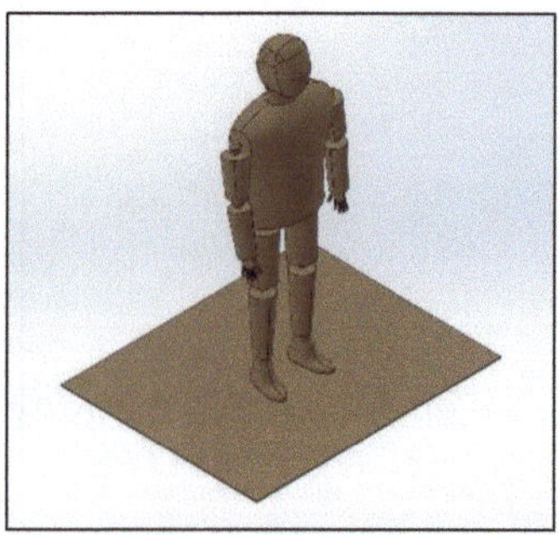

Fig. 9.2 CAD model of standing normal posture

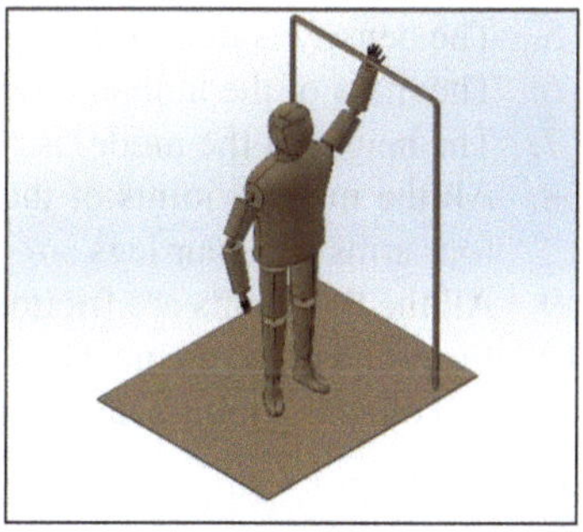

Fig. 9.3 CAD model of standing with raised hand holding handrail posture

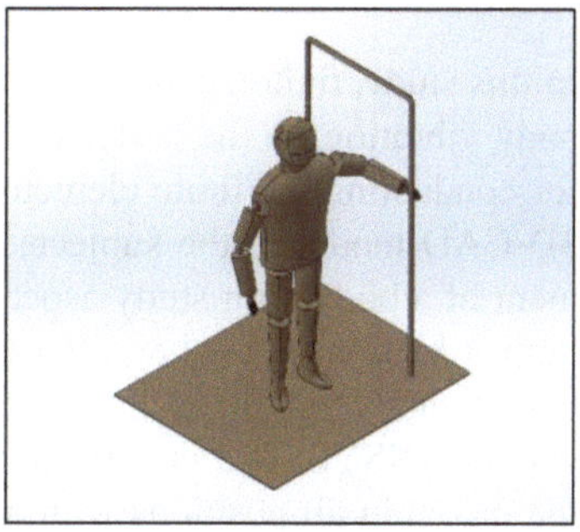

Fig. 9.4 CAD model of standing with extended hand holding handrail posture

to which the center of mass of the human shifts upwards from its previous position when compared to standing normally.

Similarly, in Fig. 9.4., the human model is holding a handrail with an extended hand below the head as support; here, the position of the point of center of mass shifts upwards when compared with standing normally, but not as up as the raised hand and somewhere in the middle of points obtained in previous two models, vibration transmission will be from leg as well as extended hand.

9.2.3 Boundary Condition

While conducting this study, some boundary conditions and assumptions are taken, which are mentioned:

1. The base excitation data were collected by FFT spectrum analyzer in unit time (s) vs. displacement (mm) and converted to frequency (Hz) vs. PSD amplitude (mm^2/Hz).
2. Coco 80-X FFT spectrum analyzer is used to record data.
3. Feet are fixed with the plate, and three directional excitations are provided.
4. There is a surface contact between the foot and plate and between the hand and handrail.

5. The density is the same over the entire model, which is 1800 kg/m^3.
6. The mass of the human model is 65.544 kg.
7. The height of the model is 175.1 cm.
8. All the moving joints of the arm and leg are considered to be ball joints, i.e., four arms and four legs.
9. All the ball joints are frictionless, and all connections are bonded.
10. Flesh and blood are not considered for the design of the model; only the skeleton and dynamic joint behavior of the body are considered.

9.3 Methodology

In this study, finite element method (FEM) is used to evaluate the effect of whole-body vibration on the body while traveling in a bus in a standing posture. Hence, for conducting the finite element analysis, the most mandatory part is creating the 3D CAD model of the subjects whose motion behavior best replicates the movement of a body with study aspects and possesses a multi-degree of freedom as the human body has. Further, after achieving the appropriate model with the required height and weight to identify the modal frequencies of the model, the modal analysis in ANSYS 23.2 is carried out by providing a fixed support under the feet of the model in all positions with a tetrahedral meshing of maximum element size of 10 mm; here, six modal frequencies are obtained. Further, with the same model and meshing, harmonic response analysis is conducted with a gravitational acceleration in a downward direction acting on the body. By this analysis, the frequency response plot of normal stress is plotted, and using the half-power bandwidth method, the value of the damping ratio (ζ) is obtained. Now, by using modal frequencies and damping ratio, the mass coefficient (α) and stiffness coefficient (β) are calculated. Then, with the same model and meshing, random vibration FEA is carried out, the frequency domain excitation is applied at fixed support, and the damping nature of the model is considered by inserting the values of damping ratio (ζ), mass coefficient (α), and stiffness coefficient (β) manually. By performing this whole process, the feet-to-head transmissibility and vertical deformation are calculated by comparing parameters to identify the potential effect of vibration on the head in the vertical direction. The whole process of FEM analysis is shown in a flowchart in Fig. 9.5.

9.3.1 Modal Analysis

The human body exhibits a highly complex and nonlinear dynamic response that significantly varies when subjected to vibration conditions. A tetrahedral meshing approach with a maximum element size of 10 mm is employed to address this. This method effectively covers all surface contours and provides higher accuracy during

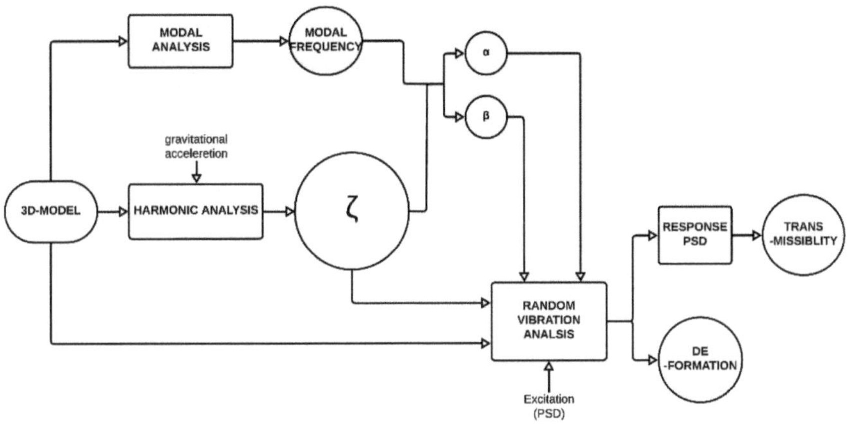

Fig. 9.5 Flowchart of methodology of analysis

Table 9.2 Number of elements and nodes for each posture

Posture	No. of elements	No. of nodes
Standing normal	145,144	367,271
Standing with raised hand holding handrail	151,607	405,000
Standing with extended hand holding handrail	165,463	410,705

computer calculations. Table 9.2 gives the number of nodes and elements for each model obtained by meshing.

Modal analysis is performed for all postures with a fixed support at the feet for standing in a normal posture as well as for holding a handrail posture. The fixed support is applied at both the feet and bottom surfaces of the handrails to ensure an accurate resemblance of the condition. Then, six modal frequencies for each model are derived, which will be taken into consideration for harmonic analysis and random vibration analysis and also be used for the calculation of mass coefficient (α) and stiffness coefficient (β). Table 9.3 gives the list of modal frequencies obtained from this analysis.

9.3.2 Harmonic Analysis

The model is added in the harmonic response module, with the exact meshing as in modal analysis. Further gravitational acceleration in the downward direction is applied to the model with a magnitude of 9.81 m/s^2, keeping the fixed support as it is. As the modal frequencies are a maximum of 17.906 Hz among all three models to cover all the aspects, the harmonic frequency range is set from 0 to 50 Hz with an interval of 0.5 Hz, which is more than 1.5 times the obtained modal frequency.

Table 9.3 Modal frequency in Hz for three standing postures

Mode number	Standing normal	Standing with raised hand holding handrail	Standing with extended hand holding handrail
1	2.0569	2.9498	2.166
2	6.3841	6.2384	3.231
3	7.0279	7.2725	3.4777
4	11.334	9.7534	4.0943
5	12.049	16.033	5.2591
6	12.574	17.906	6.0474

To identify the response at the head due to this harmonic acceleration applied on the body, the frequency response probe is inserted at the point on the head where the study is to be conducted. The frequency response in terms of normal stress (MPa) is considered because its results give the best identification of the peak frequency. The plotted response of frequency (Hz) vs. normal stress amplitude (MPa) is shown in Fig. 9.6 for standing normal posture with a peak at 11.5 Hz and normal stress amplitude of 0.000477 MPa. Similarly, by conducting the same procedure on the standing with raised hand holding handrail model, the response graph is shown in Fig. 9.7., where the peak frequency is 13 Hz and normal stress amplitude is 0.091563 MPa. Figure 9.8 shows the frequency response of standing with an extended hand holding handrail with a peak of normal stress amplitude of 0.004012 MPa at a frequency of 16 Hz. From these numbers, the half-power bandwidth method is used for the approximation of the damping ratio (ζ) [15].

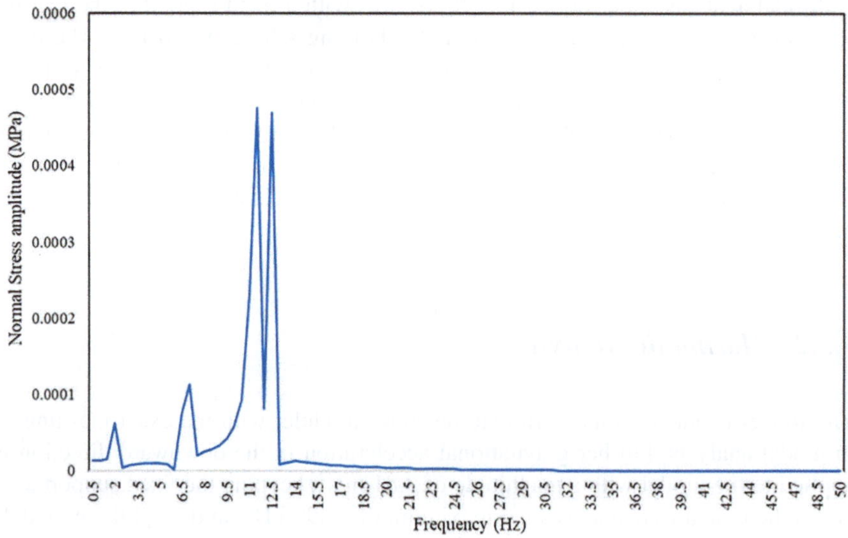

Fig. 9.6 Frequency response for standing normal

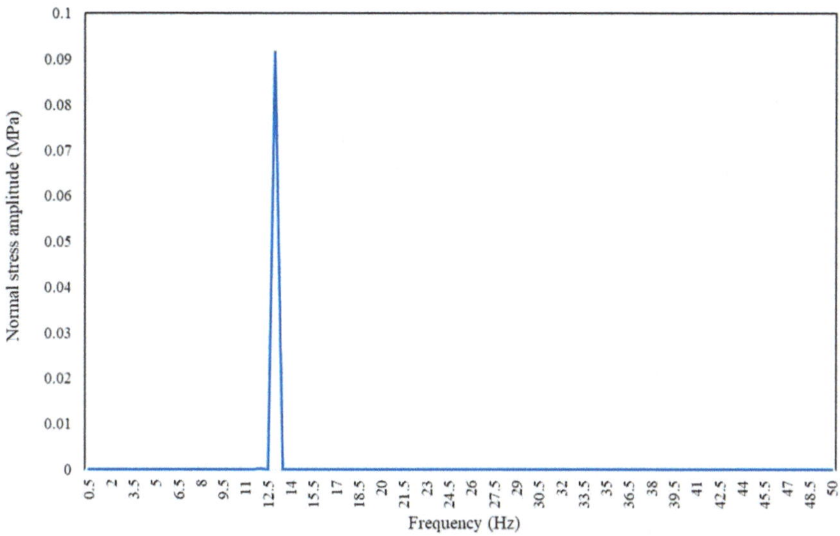

Fig. 9.7 Frequency response for standing with raised hand holding handrail

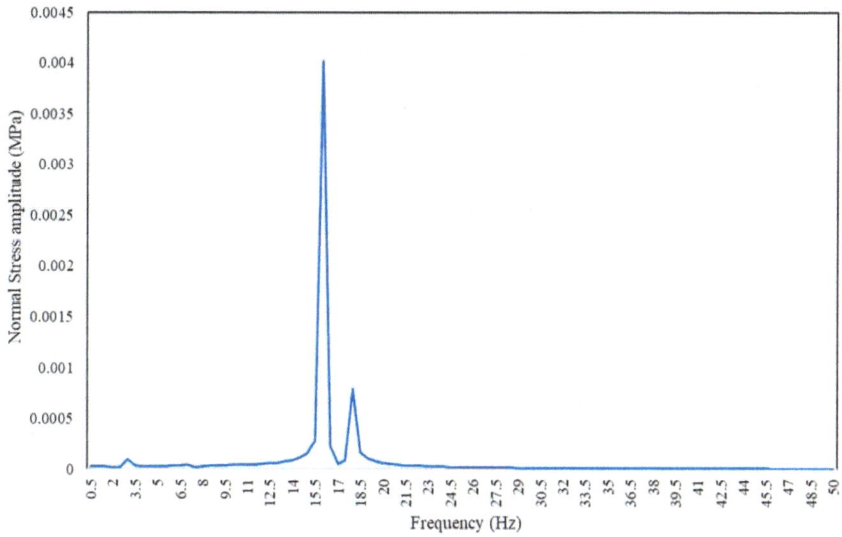

Fig. 9.8 Frequency response for standing with extended hand holding handrail

The damping behavior of a model is considered by three parameters: damping ratio (ζ), mass coefficient (α), and stiffness coefficient (β). By using the value of the damping ratio (ζ) obtained and the modal frequencies obtained in modal analysis, the values of mass coefficient (α) and stiffness coefficient (β) are calculated, which are mentioned in Table 9.4.

Table 9.4 ζ, α, and β for three different postures

Posture	ζ	α	β
Standing normal	0.0217	0.9138	0.000514
Standing with raised hand holding handrail	0.0173	0.2817	0.001023
Standing with extended hand holding handrail	0.0125	0.3145	0.000433

9.3.3 FEM Random Vibration Analysis

The vibrational study of humans in public transport and bus analysis lags behind the method of evaluation using FEM, and if done before, major of the studies used the sinusoidal acceleration excitation of the model to obtain results. As there are no symmetric road irregularities, the vibration produced in the bus is of a random nature. Hence, to simulate nearly the same conditions occurring in the bus, the vibration data of the bus floor are recorded using an FFT spectrum analyzer. On the bus floor, in terms of time (s) vs. acceleration (mm/s^2), the vibration response using the accelerometer probe was recorded for 5 s. To understand the distribution of magnitude along the floor, the data were recorded at the front, middle, and rear parts of the buses. From the recorded data, it was concluded that the maximum amplitudes are observed at the rear part of the bus, so that was used for excitation. Among all the three directional data recorded at the rear of the bus floor, it was observed that the maximum amplitude occurs in the vertical direction followed by fore and aft. In contrast, the least vibrational acceleration amplitudes were observed in the lateral direction of the bus [7].

The same model used for modal and harmonic analysis with similar meshing is then used to conduct FEM random vibration analysis with the base excitation recorded in the bus. The variables representing the damping behavior of the model are damping ratio (ζ), mass coefficient (α), and stiffness coefficient (β), which were calculated in harmonic analysis. These values are inserted manually for FEM analysis to consider to simulate the damping nature of the body.

The study focuses on the impact of postures on the head of the human body. The PSD response in terms of amplitude (mm^2/Hz) is required to evaluate the impact. The random vibration data recorded from the bus that are to be used for the excitation of the model as per the FEM requirements are converted from time (s) vs. acceleration (mm/s^2) to frequency (Hz) vs. PSD acceleration (mm^2/Hz). The converted data are used as excitation in terms of PSD response in the solver and are exerted on fixed support in all three directions. Here, in ANSYS 23.2, the input X, Y, and Z resemble lateral, vertical, and fore-and-aft directions, respectively.

To compare the impact on the head in all three models, feet-to-head transmissibility, which means the amount of vibration amplitude that reaches the head, and vertical direction deformation at the head are to be obtained. In solutions, to obtain the vertical direction deformation, the total vertical deformation of the body with a scale factor of 1 sigma is evaluated, and by using the probe tool, the vertical deformation of the head at a point on the head is obtained. Further, a PSD probe is inserted at the head of the model for every posture at the same point as of deformation, and PSD acceleration response in the Y direction in a frequency range of 0–10 Hz is obtained from the software. Then, the feet-to-head transmissibility is calculated by comparing those output responses with input responses at the feet with a similar frequency.

9.4 Result and Discussion

The modeling of the human model was the first step of the study in which certain considerations like body weight distribution and the overall height of the model should be near the average height of the Indian male subject, and frictionless male and female ball joints are taken. After that, three different body postures of standing conditions, like standing normally and two postures holding handrails in two different styles, are created according to the need. Further, those models are inserted in ANSYS 23.2 for the calculations of modal frequencies, harmonic analysis, and random vibrations on the body with a maximum mesh size of 10 mm. As a result of modal analysis, we get to obtain six natural frequencies for each posture. The result of harmonic analysis is the value of zeta(ζ), which is the value of the damping ratio for random vibration; also, the values of mass coefficient (α) and stiffness coefficient (β) are calculated using the half-power bandwidth frequency method by using the first- and second-mode frequencies. Using an FFT spectrum analyzer, raw three directional vibration data, that is, fore-and-aft, lateral, and vertical directions in the bus, are recorded in the time domain and then converted to the frequency domain. In random vibration analysis, as maximum influence is observed at the rear part of the bus, those data are input as base excitation in the frequency domain, along with the damping behavior of the model, which is considered by inputting the value of damping ratio (ζ), mass coefficient (α), and stiffness coefficient (β). By conducting random vibration analysis, the frequency domain response at the head and vertical direction deformation are obtained; further, using this, the comparing parameters feet-to-head transmissibility and vertical deformation at the head are calculated. The models are compared by comparing the numbers of these two parameters, and results have been obtained (Figs. 9.9, 9.10, and 9.11).

By carrying out all the FEM analysis steps on all three standing posture models, the obtained values of feet-to-head transmissibility and vertical deformation at the head are mentioned in Table 9.5. The feet-to-head transmissibility and vertical deformation at the head in a standing normal position are 0.0037 and 0.0544 mm, respectively. Among all the three considered postures, these are the maximum values. As there is no support other than the legs where the excitation is provided, the overall stability of the body is less compared to other postures. Due to the absence of support, as the vertical vibration energy enters the body, there is no way for it to exit, and hence, the traveler experiences the maximum exposure to vertical vibration and undergoes major deformation of the head. The head carries highly critical and fragile organs like the brain, which are highly vulnerable under minor displacements and contact with the skull. The eyesight of a person can also get damaged as constant exposure

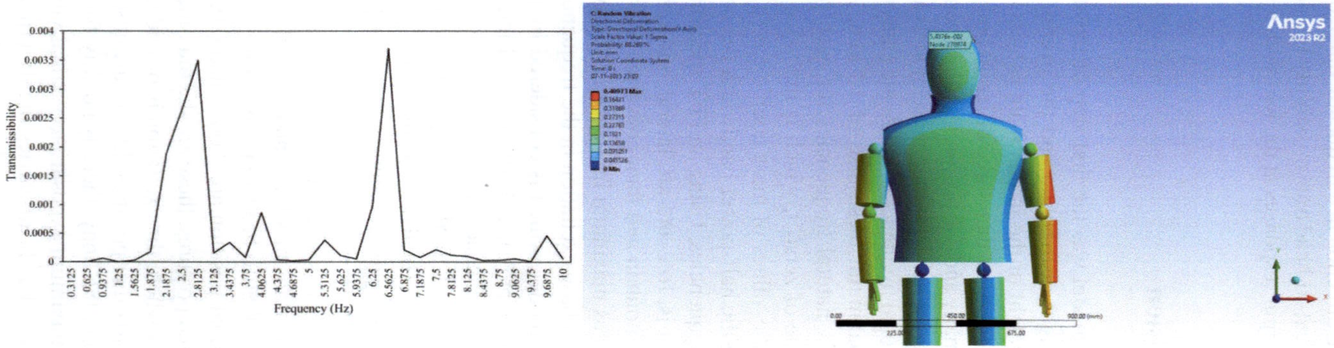

Fig. 9.9 Feet-to-head transmissibility plot and vertical deformation of the head for standing normal posture

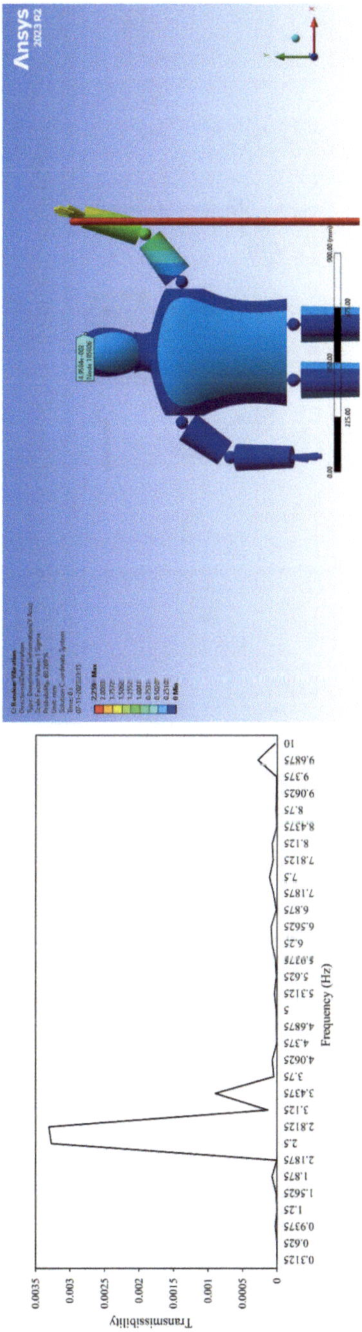

Fig. 9.10 Feet-to-head transmissibility plot and vertical deformation of the head for standing with raised hand holding handrail posture

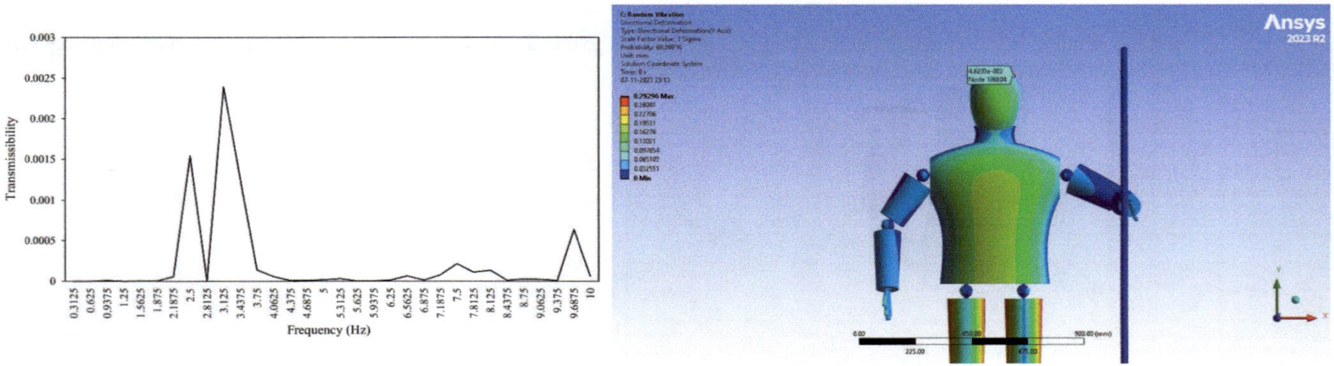

Fig. 9.11 Feet-to-head transmissibility plot and vertical deformation of the head for standing with extended hand holding handrail posture

to vibration can lead to disturbance in the cornea and microblood veins present in the eyes, resulting in eyesight harm. The body itself damps out the major proportion of the vibration to which it is exposed before reaching the head, but still, the resulting values are alarming. This posture gives the maximum numbers compared to others; hence, it is considered as the base reference condition for comparison of the other models.

Looking at the next posture of standing with a raised hand holding a handrail, the value of feet-to-head transmissibility is 0.0033, and the value of vertical deformation at the head is 0.0496 mm. In this posture, the human model is holding the handrail above the head with the left hand; this adds support to the body, and this additional contact at the upper body results in increased stability of the body compared to standing in a normal position. This reduction of 10.81% in feet-to-head transmissibility and 8.82% in vertical deformation at the head is observed because the addition of support to the body by holding the handrail reduced the tendency of the body to vibrate as it acts like an additional support at the end rather than a cantilever, as the free end of cantilever beam undergoes more deformation compared to two-end supported beam under same loading condition. So, from this phenomenon, by holding the handrail, an additional point of support is added, making the model more stable and reducing the effect of vibration exposure. In this posture, there may be induced fatigue at the joints of the shoulder and elbow. However, these organs are less vulnerable compared to the brain and other organs at the head, and the overall strength of the organ is high due to the existence of denser bones. Hence, this posture is recommended to be utilized compared to a normal standing posture.

Further, the hand support moves below the head for the next position of standing with an extended hand holding the handrail. It shows a value of 0.0024 for feet-to-head transmissibility and 0.0463 mm of vertical deformation at the head. In this posture, the center of mass of the body shifts toward the ground by a minor value compared to the posture of standing with a raised hand holding the handrail. However, it contributes to further enhancing the stability of the body and reduces the tendency of the body to vibrate more than holding a handrail with a raised hand. Hence, in this position, a 35.15% reduction in feet-to-head transmissibility and 14.89% of vertical deformation at the head is obtained, which is the lowest among all the considered

Table 9.5 Feet-to-head transmissibility and deformation at the head for all models and its percentage reduction as compared to standing normal posture

Posture	Transmissibility	% Reduction in transmissibility	Deformation (mm)	% Reduction in deformation
Standing normal	0.0037	–	0.0544	–
Standing with raised hand holding handrail	0.0033	10.81	0.0496	8.82
Standing with extended hand holding handrail	0.0024	35.13	0.0463	14.89

standing positions. Similar to standing with a raised hand holding handrail posture, the increase in support of the body by the addition of an extra support point on the body over the legs plays a vital role in reducing the exposure of whole-body vibration at the head.

Moreover, as the center of mass shifted below, and the arm went below the head, this added to a reduction of amplitudes, hence a reduction in the number of transmissibility. In this posture, the arm will not fatigue as in standing with a raised hand position as the magnitude of the force vector acting in the downward direction reduces as well as the distance from the ground reduced, resulting in a reduction in potential energy. Hence, the body needs to burn less carbohydrates to hold the hand in this position. This will further enhance the comfort of the traveler and create less joint pain in the arms. So, from all the considered three postures, it is highly recommended to travel by standing with an extended hand holding handrail as it will reduce the effect of whole-body vibration on critical organs in the head and will generate less fatigue on arm joints, increasing the travel comfort compared to other two postures.

The comparison of the percentage reduction of feet-to-head transmissibility and vertical deformation at the head is shown in Fig. 9.12. This clearly shows that by just changing the position of holding the handrail from raised hand to extended hand, a significant drop in the value of feet-to-head transmissibility is observed; however, an overall reduction in vertical deformation at head is also seen following the same trend among both postures but the drop is not as significant as of feet-to-head transmissibility. The results of both the feet-to-head transmissibility and vertical deformation show a similar pattern; this justifies the results and behavior of the body in these postures.

So, from this, it can be said that, while traveling in a bus standing at the rear part of the bus, the highest discomfort will occur when the traveler is standing in a normal position as there will be no additional contact other than legs so it will act like a cantilever and will experience highest exposure to whole-body vibration in the vertical direction. The effect of these vibrations will reduce when the person stands with hand support by holding a handrail; now, if the holding style is with an extended hand below the head, then the person will experience maximum comfort compared

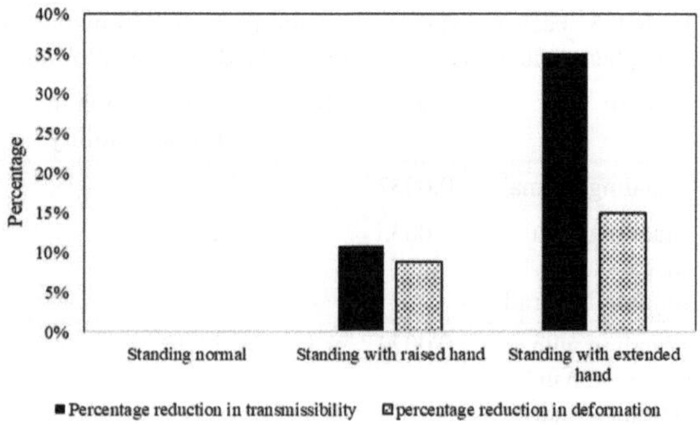

Fig. 9.12 Comparison of percentage change in feet-to-head transmissibility and vertical deformation at the head

to holding by raising the hand above the head. This will not eliminate but will reduce the effect on the critical organs of the head due to whole-body vibration exposure and will help increase comfort while traveling in a standing posture.

9.5 Conclusion

Low-frequency vibration exposure in the human body results in major as well as minor health problems to vulnerable organs of the human body like the brain, eyesight, and more. This study focused on the effect of this WBV on the human body and identified the change in vibration transmission due to different posture and support conditions to maximize comfort and minimize its adverse effects. The effect of WBV on the head while standing for three postures is evaluated using the FEM analysis method. To identify the damping of the body, modal and harmonic analyses are carried out, and by using the half-power bandwidth method, values of damping ratio (ζ) are calculated, and further using modal frequency, values of mass coefficient (α) and stiffness coefficient (β) are derived. The recorded data in the bus using the FFT spectrum analyzer are converted to the frequency domain from the time domain and used as excitation for the random vibration analysis. The frequency domain response at the head ranging between 0 and 10 Hz is obtained along with the vertical direction deformation at the head.

From all the results and discussion, it can be concluded that:

- This study will be helpful for the design and development of devices that provide support to the head or to critical organs of the body of a traveler in public transportation.
- In standing normal conditions, the maximum feet-to-head transmissibility of 0.0037 is observed. In contrast, a reduction of 10.81% in standing with a raised hand holding handrail posture is observed, and the least value of feet-to-head transmissibility is noticed in standing with an extended hand holding handrail of 0.0024, which is a reduction of 35.13%. Hence, the least amount of vibration reaches the head in the extended hand holding handrail position.
- A similar trend is observed in vertical deformation at the head; the minimum head deformation is observed in standing with an extended hand holding a handrail position and maximum in standing normally without holding a handrail, which is a reduction of 14.89% compared to standing in normal posture.
- Compared to the standing normal posture, the reduction in vibration exposure at the head is observed in postures holding the handrail due to the addition of an extra support point by hand.
- While the vulnerability of the head is not entirely eliminated, by holding the handrail, the effect of vibration happens on the joints of arms, which possess higher strength compared to organs in the head.

Hence, these results are justified as feet-to-head transmissibility and vertical deformation show similar trajectories. So, among all three evaluated standing postures, the most comfortable and less harmful position is standing with an extended hand holding the handrail.

References

1. Channel, W.: National institute of nutrition changes average weight, height measurements for Indian men and women (2020)
2. ISO 2631-1:1997. https://www.iso.org/standard/7612.html
3. Takahashi, Y., Yonekawa, Y., Kanada, K., Maeda, S.: A pilot study on the human body vibration induced by low frequency noise. Ind. Health/Ind. Health **37**, 28 (1999)
4. Dutra, R.M.A., Duarte, M.L.M., De Melo, G.C., Neves, J.A.B.: The effect of whole-body vibration on public transportation passenger performance while typing on smartphones. Ergonomics **65**, 1202 (2022)
5. Bhuiyan, M.H.U., Fard, M., Robinson, S.R.: Effects of whole-body vibration on driver drowsiness: a review. J. Safety Res. **81**, 175 (2022)
6. Kumar, V., Saran, V.H., Pawar, R.: Biodynamic response to random whole body vibration in standing posture. In: Proceedings of the 1st International and 16th National conference on machines and mechanisms (iNaCoMM2013), IIT Roorkee, India (2013)
7. Thuong, O., Griffin, M.J.: The vibration discomfort of standing people: relative importance of fore-and-aft. Lateral Vert. Vib. Appl. Ergon. Appl. Ergon. **43**, 902 (2012)
8. Singh, A., Singh, I., Kalsi, S.: FEM analysis of human CAD model in standing posture while travelling. ResearchGate (2019)
9. Kumar, R., Singh, I., Kalsi, S., Sehgal, S.S.: Modal analysis of standing human body using FEM. J. Sci. Ind. Res. (JSIR) **77**, 639 (2018)
10. AlShabi, M., Araydah, W., ElShatarat, H., Othman, M., Younis, M.B., Gadsden, S.A.: Effect of mechanical vibrations on human body. World J. Mech. **06**, 273 (2016)
11. Shibata, N., Ishimatsu, K., Maeda, S.: Gender difference in subjective response to whole-body vibration under standing posture. Int. Arch. Occup. Environ. Health **85**, 171 (2011)
12. Tatsuno, J., Maeda, S.: Effect of whole-body vibration exposure in vehicles on static standing balance after riding. Vibration **6**, 343 (2023)
13. Singh, I., Nigam, S.P., Saran, V.H.: Modal analysis of human body vibration model for indian subjects under sitting posture. Ergonomics **58**, 1117 (2014)
14. Density » IT'IS Foundation. https://itis.swiss/virtual-population/tissue-properties/database/density/
15. Wang, J.-T., Jin, F., Zhang, C.-H.: Estimation error of the half-power bandwidth method in identifying damping for multi-DOF systems. Soil Dyn. Earthq. Eng. **39**, 138 (2012)

Chapter 10
Ergonomic Evaluation of Workers in a Pump Manufacturing Industry

Chidambaram Vigneswaran, Gopalsamy Madhan Mohan, M. G. Vikram, S. Nitish, M. Naveenkumar, and R. Obulinarashiman

Abstract The pump manufacturing industry is one of the oldest and steadily expanding industries. The workers are involved in repetitive tasks and also are standing throughout entire operation, which causes musculoskeletal disorders (MSDs) to the workers. Frequency of MSDs in the pump manufacturing industry is documented by a relatively small number of researches. In manufacturing industries, the common cause for MSDs is the poor work posture and improper workstation design. The objective of the study is to investigate and identify the risk factor related to MSDs in the pump manufacturing industry. Thirty workers form different sections like winding, assembly, material handling, painting, and machine operators were selected for the study. Data collected through a questionnaire include the job-related factor, personal details, demographic data, and Cornell Musculoskeletal Disorder Questionnaire (CMDQ). Employees' working postures are evaluated using Rapid Upper Limb Assessment (RULA). The result shows that industrial workers have postural problems. The pump assembly section had the high RULA score of 6.4 which indicates immediate change in work environment. It can be concluded that due to poor ergonomic work environment, the workers in this sector exposed to various occupational health and safety issues. MSDs of the worker in the pump manufacturing sector can be reduced by altering the work environment.

Keywords Industrial ergonomics · Musculoskeletal disorders (MSDs) · CMDQ · RULA

C. Vigneswaran (✉) · G. M. Mohan · M. G. Vikram · S. Nitish · M. Naveenkumar · R. Obulinarashiman
Ergonomics Laboratory, Department of Production Engineering, PSG College of Technology, Coimbatore, India
e-mail: cvn.prod@psgtech.ac.in

J. Maiti et al. (eds.), *Recent Advances in Industrial and Systems Engineering*, Lecture Notes on Multidisciplinary Industrial Engineering,
https://doi.org/10.1007/978-981-96-8323-9_10

10.1 Introduction

When problem occurs in workplace environment in terms of less productivity and less efficiency, there might be a problem with workers. Therefore, some improvements should be done in workplace environment in favor of workers, necessitating ergonomics. Ergonomics is the study of workers efficiency in their workplace environment [1]. OSH stands for occupational safety and health which mainly deals with workers health in workplace environment. OSH contains set of rules and regulations to enhance the workers' health conditions and to avoid unwanted injuries in workplace [2]. By doing same job work for prolonged period of time, there occurs a fatigue in workers muscles which causes MSD. Having same work posture for long period of time is the main root cause for MSD. ABC Pumps and Motors have been chosen for this journal work. In this industry, various manufacturing operations like turning, winding assembling, and painting are done. A visit to the rotating machinery division in ABC Pump and Motor was made, and RMD is specialized in manufacturing the pump. The worker postures were analyzed in every station including lathe, coil winding, pump assembly, motor assembly, and painting. By using the RULA worksheet, specified score for every workstation has been calculated. Among these workstations, in the pump assembly station, workers' postures are poor with high RULA score, posing a high risk such as fatigue and MSDs [3]. The working postures of the machine operator for all workstations are shown below (Fig. 10.1).

The pump manufacturing industry is pivotal across various sectors, boasting longevity and sustained growth. However, the inherent nature of work within this industry often involves repetitive tasks, necessitating prolonged periods of standing, thereby contributing to the prevalence of MSDs [4]. Despite its significance, the frequency of MSDs within the pump manufacturing sector remains inadequately addressed, with limited studies probing into this issue. As highlighted in the abstract, poor work postures and inadequate workstation designs emerge as primary culprits behind MSDs in manufacturing industries. In response, our study endeavors to delve

Fig. 10.1 Overview of different workstations in the ABC pump manufacturing industry

into these concerns by identifying specific risk factors related to MSDs in the pump manufacturing industry.

The aim of the study is to gauge the level of discomfort experienced by workers through the CMDQ [5], to pinpoint ergonomic risks linked to both workplace design and working postures, with a specific focus on the pump assembly unit, and to assess the occurrence of MSDs among workers in the pump manufacturing unit. Finally, workers' work postures will be captured and manually analyzed using RULA. Ultimately, our overarching goal is to mitigate work-related musculoskeletal disorders (WMSD) in the pump manufacturing industry by implementing interventions aimed at optimizing both worker posture and workplace environment [6]. Through these efforts, a healthier and safer working environment for all employees involved in the pump manufacturing processes is created.

10.2 Methods

10.2.1 Task Description

The study was carried out at the ABC Pump and Motors facility in Coimbatore. A random 30 workers were selected for the study. This unit specializes in manufacturing a range of submersible and monoblock pumps. The study uses a five-stage approach, i.e., as follows:

- Workplace selection.
- Observation and interview with the worker for discomfort.
- Questionnaire survey and photographs of posture are taken.
- Postural analysis is done using RULA [7].
- Suggestion and modification are proposed.

10.2.2 Study Design and Method

The study was carried out for a month in ABC pumps and motor manufacturing industry located in Coimbatore, Tamil Nadu, India. Workplace observation is conducted to identify the nature of issues that arise from workers' discomfort. The initial observation includes the understanding the working methods, receiving complaints regarding discomfort or suffering from incorrect postures and overall work environment. Around 50 workers are working in the industry in a shift for 8–12 h which includes break of one to one and half-hour. Samples of 30 operators were selected randomly for the survey. The participant was not paid for the contribution of the survey. Then, the data collection was done using CMDQ survey which was answered by the operator.

Table 10.1 Criteria for selection and summary of the organization

Organization name	Criteria for selection	Summary
ABC pump manufacturing	1. Longevity and stability in the pump manufacturing industry 2. Availability to participate in the study 3. Diverse operations covering various aspects of pump manufacturing 4. Willingness to collaborate on research endeavors	ABC pump manufacturing is a well-established entity within the pump manufacturing sector, known for its longevity and stability. The organization encompasses a range of operations, including winding, assembly, material handling, painting, and machine operations, providing a comprehensive scope for research With a commitment to promoting occupational health and safety, ABC Pump Manufacturing demonstrated eagerness to collaborate on this study, making it an ideal partner for our research endeavors

By adhering to these criteria in Table 10.1, ABC Pump Manufacturing emerged as a suitable candidate for conducting our study, ensuring both relevance and feasibility in our research objectives [8].

10.2.3 Data Collection

10.2.3.1 Cmdq

The CMDQ comprises of 54 items which includes body map diagram and personal questions. The questions focus on musculoskeletal pain and discomfort across 20 body regions. The questionnaire will also include scores related to the frequency of pain, the severity of aches, and the level of discomfort experienced over the past week. Table 10.2 provides the ratings for frequency, discomfort level, and interference scales.

The questionnaire is sub-divided into two parts:

Part 1: This section includes personal and work-related profile, i.e., age, gender, work experience height, and weight.

Table 10.2 Rating for the frequency, discomfort severity, and interference scales

Frequency	Level of discomfort	Interference
0 = Never	1 = slightly uncomfortable	1 = not at all
1.5 = 1–2 times/week	2 = moderately uncomfortable	2 = slightly interfered
3.5 = 3–4 times/week	3 = very uncomfortable	3 = substantially interfered
5 = once in every day		
10 = several times every day		

Part 2: This section includes the pain or ache in different body parts such as neck, arm, shoulder, trunk, back, and lower limb.

10.2.3.2 Rula

RULA was created to assess each worker's exposure to ergonomic risk factors related to upper extremities of MSD [9]. This assessment tool uses structured process to asses upper extremities for biomechanical and MSD risk [10]. The assessment considers load requirement for the tasks done on the upper extremities. The operator working posture is videotaped and snipped to form stick diagram. The positions are taken with the frame reference from the video, and scores are given according to the RULA [11]. The entire process is considered during the work, and the final score is obtained.

10.2.4 Data Analysis

The obtained raw data are extracted in Microsoft Excel, and the frequency and severity of discomfort is tabulated. The overall frequency and severity score was calculated for each process in the industry. The RULA scores obtained for every posture are analyzed, and the observation is discussed with the supervisor for further changes. The assessment is done for the posture which contributes to discomfort, repeated task, and muscle loading activity. The grand score is given for every posture. The score is sub-divided into two groups: Group A and Group B. The final score is derived from the table C in the RULA sheet whose value ranges from 1 to 7 score. The obtained RULA score is compared with the action level table, and accordingly, the action is taken.

10.3 Results and Discussion

10.3.1 Preliminary Findings

The investigation was conducted to find the ergonomic problem and frequency of MSD among the workers of pump manufacturing industry. The study is based on the information given by the worker, initial observations, and the CMDQ form. The postural stresses are reported to be high in the industry and cause MSD among the workers.

Table 10.3 Demographic variables of pump manufacturing industry

Parameters	Mean	SD	Minimum	Maximum
Age (Years)	36.46	2.018	19	45
Weight (Kg)	162.38	5.625	48	76
Height (cm)	61.84	6.788	150	174
Working years	18.69	6.869	1	32

10.3.1.1 Demographic Data

The study was carried out in the pump manufacturing industry. A random sample of 30 workers participated in the survey. Table 10.3 represents the demographic and occupational factors of pump manufacturing sectors. The workers' age ranges from 19 to 45 years (mean = 36.46, SD = 2.018). Total of 30 workers participated; all were male, and the participant height ranges from 150 to 174 cm (mean = 61.84, SD = 6.788) and had working experience of 1–32 years (mean = 18.69, SD = 6.869).

10.3.1.2 Body Discomfort Survey

The report of the physical discomfort assessment and CMDQ survey is shown in Fig. 10.2. This can be inferred that the major of workers has significant proportion of discomfort in right upper arm, neck, shoulder, and lower back (80%, 76%, 67%, and 63%, respectively). Moderate pain was reported in upper back, right forearm, hip, right knee, and left knee. The major contribution of the discomfort is caused due to repetitive task and improper posture.

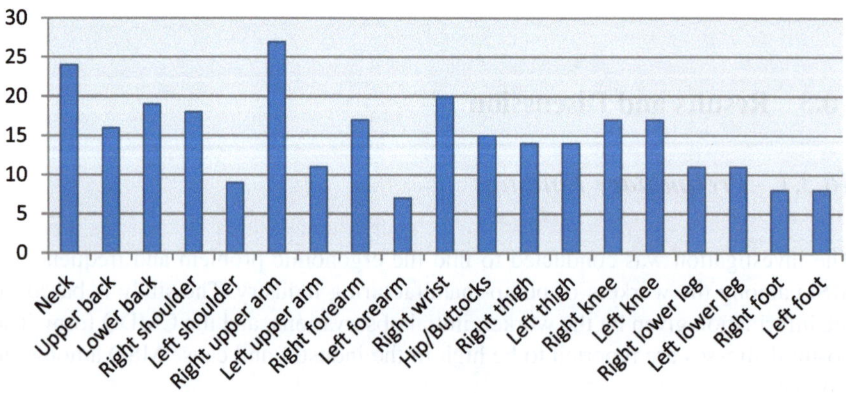

Fig. 10.2 Reported frequency of MSDs symptoms among workers

Fig. 10.3 Actual working posture with body angles of pump assembly operator in ABC industry

10.3.2 Assessment

As previously mentioned, the risks related to working postures were assessed using the RULA technique. Numerous positions from various sections—such as winding, assembly, painting, inspection, and machine operation—were examined. Pictures were captured and examined. Stick diagrams were created to mark and measure the angles of body parts like the lower arm, wrist, upper arm, elbow, neck, and trunk after these photographs were loaded into paint program. The risk scores and required remedial measures were then determined using the RULA and scoring sheet (Fig. 10.3).

The manual RULA scores for pump assembly station, coil winding station, motor assembly, painting station, and lathe operators are obtained. The average of all the critical posture grand scores is given in Table 10.4. The pump assembly station has high RULA score of 6 followed by the lathe operator of score 5. The motor assembly station has score 4, winding score as 3, and the painting score as 2.

Table 10.4 RULA score for the entire workstations

Workstation	RULA score*
Lathe	5
Winding	4
Motor assembly	3
Painting	2
Pump assembly	6

* The RULA scores presented are the averages of the critical postures identified for each worker. For ease of interpretation, decimals have been rounded to the nearest whole number

Table 10.5 Distribution of RULA score in the pump industry [12]

RULA score	Risk level	Action	Total Number of workers	Percentage of workers
1–2	Negligible	Acceptable	7	24
3–4	Low	Investigate and change may be needed	13	43
5–6	Medium	Investigate, change soon	4	13
7	High	Investigate, change immediately	6	20
		Total	30	100

Table 10.5 presents the distribution of workers in the pump manufacturing industry based on their RULA scores, categorized by risk levels and corresponding recommended actions. The majority of workers (24%) exhibit negligible risk levels with RULA scores ranging from 1 to 2, indicating acceptable working postures requiring no immediate intervention. A significant portion of workers (43%) fall into the low-risk category with RULA scores of 3 to 4. While their working postures are generally acceptable, some investigation and potential changes may be warranted to mitigate emerging ergonomic risks. A smaller proportion of workers (13%) are classified as having medium risk, with RULA scores of 5 to 6, suggesting the need for closer scrutiny and expedited changes in working conditions to address identified ergonomic concerns. Approximately 20% of operators are categorized as high risk, with a RULA score of 7, necessitating immediate investigation and intervention to rectify critical ergonomic issues and prevent potential MSD.

It is evident that the RULA score for pump assembly operators in assembly section has high risk level (20%) and about 13% have medium risk. The lower risk level is neglected. The high risk levels indicate that inadequately designed work environments and workstation setups result in uncomfortable and potentially awkward postures for workers, leading to muscular strain and discomfort.

10.4 Conclusion

The current study shows ergonomic problems, occupational discomfort, unfavorable workplace environment, environmental issues, and safety concerns in the pump manufacturing industry. Workers in the sector exhibit signs of stress, tiredness, and fatigue due to poorly designed workstations that force them to adapt to the environment, leading to awkward postures. These postures significantly contribute to MSDs, and sustained awkward positions result in fatigue. The assessment reveals that 20% of the operators are exposed to extreme risk levels. Study investigates that there is high risk in the pump assembly operation and needs immediate change in the workstation design. Overall, by applying ergonomic principles, it can make various improvements to reduce the risks associated with workers adopting awkward posture, ultimately promoting a healthier workforce.

The study's outcomes hold major impact for both the pump manufacturing industry and broader occupational health and safety initiatives. By identifying specific risk factors associated with MSDs and poor ergonomic conditions within the pump manufacturing sector, this research offers valuable insights into areas requiring immediate attention. The categorization of workers based on RULA scores provides a practical framework to introduce tailored interventions intended to reduce ergonomic risks and promoting worker well-being.

Although this study offers valuable insights, it has some limitations. One major limitation is the relatively small sample size, which could influence the generalizability of the findings to a wider population of pump manufacturing industry. Additionally, the use of self-reported data through questionnaires introduces the possibility of response bias and inaccuracies. Moreover, the absence of a formal pilot survey may have impacted the refinement of data collection instruments and procedures. Furthermore, the cross-sectional research approach restricts the ability to determine causality or evaluate long-term trends in ergonomic concerns and the frequency of MSDs. Therefore, future research endeavors could adopt a larger-scale, longitudinal approach to gather more robust data on MSDs and ergonomic risk factors within the pump manufacturing industry. Incorporating objective measures, such as direct observation or wearable technology, could enhance the accuracy and reliability of data collection. Furthermore, extending research efforts to encompass other related industries or geographic regions would facilitate comparative analyses and broaden the applicability of findings. Ultimately, by building upon the foundation laid by this study, futurc rcscarch holds thc possibility to support scicntifically informcd actions and regulations directed at promoting safer and healthier workplaces within the pump manufacturing sector and beyond.

References

1. El Saadany, A.M.A., Jaber, M.Y., Bonney, M.: Environmental performance measures for supply chains. Manage. Res. Rev. **34**(11), 1202–1221 (2011). https://doi.org/10.1108/01409171111178756
2. Fallentin, N.: Regulatory actions to prevent work-related musculoskeletal disorders-the use of research-based exposure limits. Scand. J. Work Environ. Health **29**(4), 247–250 (2003). https://doi.org/10.5271/sjweh.728
3. Chidambaram, V., Gopalsamy, M.M., Kanchan, B.K., Mouleeswaran, S.: A holistic methodology for mitigating awkward postural risks: evidence from South Indian small-scale industries. Work 1–15 (2023). https://doi.org/10.3233/WOR-230210
4. Hossain, M.D., et al.: Prevalence of work related musculoskeletal disorders (WMSDs) and ergonomic risk assessment among readymade garment workers of Bangladesh: a cross sectional study. PLoS ONE **13**(7), e0200122 (2018). https://doi.org/10.1371/journal.pone.0200122

5. Shariat, A., Tamrin, S.B.M., Arumugam, M., Ramasamy, R.: The Bahasa Melayu version of Cornell musculoskeletal discomfort questionnaire (CMDQ): reliability and validity study in Malaysia. Work **54**(1), 171–178 (2016). https://doi.org/10.3233/wor-162269
6. Soares, C.O., Pereira, B.F., Gomes, M.V.P., Marcondes, L.P., de Campos Gomes, F., de Melo-Neto, J.S.: Preventive factors against work-related musculoskeletal disorders: narrative review. Revista Brasileira de Medicina do Trabalho **17**(3), 415 (2019). https://doi.org/10.5327/Z1679443520190360
7. Kulkarni, V.S., Devalkar, R.V.: Postural analysis of building construction workers using ergonomics. Int. J. Constr. Manag. **19**(6), 464–471 (2018). https://doi.org/10.1080/15623599.2018.1452096
8. Massaccesi, M., Pagnotta, A., Soccetti, A., Masali, M., Masiero, C., Greco, F.: Investigation of work-related disorders in truck drivers using RULA method. Appl. Ergon. **34**(4), 303–307 (2003). https://doi.org/10.1016/s0003-6870(03)00052-8
9. Namwongsa, S., Puntumetakul, R., Neubert, M.S., Chaiklieng, S., Boucaut, R.: Ergonomic risk assessment of smartphone users using the rapid upper limb assessment (RULA) tool. PLoS ONE **13**(8), e0203394 (2018). https://doi.org/10.1371/journal.pone.0203394
10. Nishanth, R., Muthukumar, M.V., Arivanantham, A.: Ergonomic workplace evaluation for assessing occupational risks in multistage pump assembly. Int. J. Comput. Appl. **113**(9), 9–13 (2015). https://doi.org/10.5120/19852-1764
11. Peppoloni, L., Filippeschi, A., Ruffaldi, E.: Assessment of Task Ergonomics With an Upper Limb Wearable Device (2014). https://doi.org/10.1109/med.2014.6961394.
12. McAtamney, L., Nigel Corlett, E.: RULA: a survey method for the investigation of work-related upper limb disorders. Appl. Ergon. **24**(2), 91–99 (1993). https://doi.org/10.1016/0003-6870(93)90080-s

Chapter 11
Effects of Schoolbag Design on Biomechanics: Considerations for Backpack Straps, Dimensions, and Carrying Methods: A Review-Based Research

Sanjay Kumawat, Narayan Lal Jain, and Yashpal

Abstract Carrying schoolbags is a common daily activity for students, but the design of the bag and the way it is carried can have a significant influence on the biomechanics of the body. This review paper examines the effects of schoolbag design on biomechanics, with a particular focus on backpack straps, dimensions, and carrying methods. The paper provides a comprehensive overview of previous research on the topic, covering studies that have investigated the effects of various schoolbag design factors on the body. The results of these studies suggest that the design and development of the backpack straps is particularly important, as poorly designed straps can lead to increased pressure on the shoulders and neck, as well as changes in spinal curvature. In addition to straps, the dimensions of the schoolbag can also play a role in the biomechanical effects of carrying a bag. Research has shown that backpacks that are too heavy or too large can lead to changes in posture and gait and may increase the risk of injury. Finally, the paper explores different carrying methods for schoolbags, such as over one shoulder, across the chest, or using a rolling backpack. Each of these methods comes with its own set of advantages and disadvantages, and the paper offers a comprehensive analysis of the biomechanical effects associated with each approach. Overall, this review paper highlights the importance of considering the biomechanical effects of schoolbag design when designing and selecting backpacks for students. By taking into account factors such as strap design, bag dimensions,

S. Kumawat (✉) · Yashpal
Department of Mechanical Engineering, Poornima University, Jaipur, India
e-mail: sanjay.kumawat@poornima.org

Yashpal
e-mail: yashpal.kaushik@poornima.edu.in

N. L. Jain
Department of Mechanical Engineering, Poornima College of Engineering, Jaipur, India
e-mail: narayan.jain@poornima.org

J. Maiti et al. (eds.), *Recent Advances in Industrial and Systems Engineering*, Lecture Notes on Multidisciplinary Industrial Engineering,
https://doi.org/10.1007/978-981-96-8323-9_11

and carrying method, it may be possible to mitigate the risk of injury and discomfort associated with carrying schoolbags and promote healthier habits among students.

Keywords Musculoskeletal health · Ergonomics · Student health · Load carriage · Biomechanical stress

11.1 Introduction

Carrying schoolbags is a daily activity for millions of students around the world, but it can have a significant impact on their musculoskeletal health. Researchers have investigated various factors that affect the biomechanics of carrying schoolbags, including the design of the backpack, the weight and dimensions of the bag, and the way it is carried. This review paper seeks to offer a comprehensive summary of the existing knowledge in this field, incorporating insights from various preceding research studies.

Studies have found that the design of backpack straps can have a significant impact on the biomechanics of carrying a schoolbag. For instance, Abdelraouf and colleagues [1] discovered that altering the lengths of shoulder straps resulted in modifications to cervical posture and the upper trapezius pressure pain threshold. In another study, Abaraogu and colleagues [2] observed that the load of the backpack and the choice of straps could impact the pulmonary function of children. Wide padded straps have been found to reduce pressure on the shoulders and neck compared to non-padded straps [3, 4].

Another critical factor in schoolbag design is the dimensions and weight of the bag. Studies have shown that carrying a backpack that is too heavy or too large can lead to changes in spinal curvature and gait patterns, as well as an increased risk of musculoskeletal injury [5–7]. Moreover, the way the backpack is worn can also affect its biomechanical effects on the body. For instance, carrying a backpack over one shoulder has been found to lead to changes in spine curvature and increased muscle activation in the trapezius muscles compared to carrying it across the chest or using a rolling backpack [6, 8, 9].

Furthermore, several studies have investigated the prevalence of musculoskeletal pain among schoolchildren and the risk factors associated with such pain. For example, Angarita-Fonseca and colleagues [10] found that non-specific low back pain was prevalent among schoolchildren in Colombia. Similarly, Alami and colleagues [11] investigated ergonomic factors of school bags and their adaptation to the weight of students.

In summary, this review paper aims to offer a thorough overview of the impact of schoolbag design on the biomechanics associated with carrying schoolbags. By synthesizing the findings of previous research studies, this review will provide insight into the most significant factors that affect the musculoskeletal health of students and inform the development of guidelines for the design and use of schoolbags.

Carrying a schoolbag is an essential activity for students, as they transport books, laptops, and other materials to and from school. However, the increased load on their backs can lead to musculoskeletal problems such as pain, discomfort, and postural changes [12]. Studies have shown that schoolchildren in various countries around the world are at risk of developing musculoskeletal problems due to the weight of their backpacks [13–15]. Moreover, the design, load, and carrying methods of schoolbags can also affect their musculoskeletal health [3, 16, 17].

Numerous studies have delved into the impact of schoolbag design on the musculoskeletal health of students. For instance, Abaraogu and colleagues [2] investigated the influence of variable backpack load and strap options on the pulmonary function of children, using a simulation of treadmill walking. Another study by Abdelraouf et al. [1] explored how the length of backpack shoulder straps affects cervical posture and the pressure pain threshold in the upper trapezius. Furthermore, [11] delved into the ergonomic factors of school bags and their adaptation to the weight of students.

Moreover, studies have also looked at the effects of schoolbag weight and design on the posture and gait of schoolchildren. A randomized controlled experimental study investigated the adolescent standing postural response to backpack loads [18]. Another study examined the influence of schoolbag carrying on gait kinetics [19]. Furthermore, researchers have delved into the impact of school bag design and load on the spinal posture of children during stair use [4].

In spite of the numerous studies on the design of schoolbags and their influence on the musculoskeletal health of students, there remains a necessity for additional research on the subject. This review strives to consolidate the present evidence regarding the impact of schoolbag design on biomechanics, taking into account factors like backpack straps, dimensions, and carrying methods. Through a comprehensive analysis of existing literature, the review intends to shed light on the determinants affecting the musculoskeletal health of schoolchildren. Ultimately, it aims to offer recommendations for schoolbag design to mitigate the risk of musculoskeletal problems.

11.2 Studies Selected for Systematic Review

A systematic review of the literature concerning the effects of schoolbag design on biomechanics indicates a growing concern regarding the impact of backpacks on children's health. The studies included in this review indicate that improper backpack use can lead to musculoskeletal pain, postural changes, and impaired pulmonary function, among other issues [20]. For instance, studies have shown that carrying heavy backpacks can cause back and shoulder pain and that the weight of the backpack can affect spinal curvature and gait pattern. Moreover, studies have found that different carrying methods, such as backpacks with single or double straps, can have different effects on the musculoskeletal system, with some methods causing more strain than others.

The dimensions and weight of the backpack are also important factors in determining the risk of injury. Studies have suggested that the weight of the backpack should not exceed 10–15% of a child's body weight and that the weight distribution within the backpack should be even to avoid excessive pressure on certain parts of the body. Furthermore, the design of the backpack straps can also affect the risk of injury, with wider straps and adjustable padding being associated with lower levels of discomfort and strain.

Despite the growing body of evidence on this topic, many questions still remain unanswered. For instance, it is not clear whether certain types of backpacks or carrying methods are more suitable for children of different ages and sizes, or whether there are specific interventions that can help prevent or alleviate musculoskeletal pain. Further research is needed to address these issues and provide more guidance for parents, educators, and healthcare professionals on how to promote healthy backpack use in children.

The quality assessment of the selected studies was conducted by a single author (JL) using standardized tools, including the Cochrane Risk of Bias 2 (RoB2) tool and the Strengthening the Reporting of Observational Studies in Epidemiology version 4 (STROBE(V4)) tool. The Cochrane RoB2 tool assesses seven bias categories for randomized trial studies. These categories include the randomization of sequence generation, concealment and bias of selection and allocation, reporting bias through selective reporting, blinding of participants and personnel, blinding of outcome assessment, incomplete outcome data that may lead to attrition bias, and potential bias not addressed by the aforementioned categories. The study underwent classification as "Good," "Fair," or "Poor," determined by the criteria met through the Cochrane RoB2 tool. Conversely, the STROBE(V4) statement is structured to assess the quality of observational studies using a 22-item tool. A quality assessment standard was established by referencing other systematic reviews, with a higher score indicative of a higher study quality. Three score levels were determined: 26–33 classified as good quality, 15–25 as intermediate quality, and 0–14 as poor quality for the study. By using these standardized tools, the study's quality could be assessed objectively, and any biases could be identified and addressed to ensure the validity of the study's findings.

The study included a total of 36 studies, which were identified through a systematic search process and included in the review. The high ICC value for the rater involved in article selection suggests a robust and dependable study selection process. After screening, some irrelevant articles were excluded, which did not meet the study's parameters or criteria. In conclusion, a total of 36 studies were incorporated into the analysis. This set comprised 30 experimental studies, encompassing both randomized and control trials, along with three observational studies, all of which were classified as intermediate quality. The assessment of the included studies' quality was conducted using two tools: the Cochrane tool for evaluating the risk of bias in randomized trials (RoB2) and the Strengthening the Reporting of Observational Studies in Epidemiology (STROBE) version 4. The included studies were synthesized and categorized into four groups: biomechanics consideration, strap, dimension, and carrying method. The review findings offered valuable insights into recommendations for school backpack design. These insights were summarized, and comparisons among various types of bags were illustrated. The results of the 30 reviewed randomized trial/RCT studies were presented in Table 11.1, and the details of their quality assessment were included in Appendix. Overall, this systematic review provides a comprehensive and rigorous assessment of the available evidence on school backpack design and can guide future research and development efforts in this area.

11.3 Selection Criteria for Review Papers

A thorough and methodical methodology was used to choose the 36 papers that made up this review. The following criteria were applied:

3.1. Database and Keyword Search: We performed a comprehensive search across many databases, such as PubMed, Google Scholar, and Scopus, by employing topic-specific keywords and phrases including "schoolbag design," "biomechanical effects," "children's posture," "gait analysis," and "musculoskeletal health."

3.2. Inclusion Criteria: Population: Research concentrating on kids and teenagers between the ages of 5 and 18. Study Type: Only original, peer-reviewed research articles that included both experimental and observational investigations were considered. Research that looked specifically at how schoolbag design such as weight, kind of strap, and carrying technique affects biomechanical outcomes like posture, walking, and musculoskeletal pain.

3.3. Screening and Selection Process: A total of sixty articles were found in the initial database search. After duplicates were removed, 48 publications were screened using titles and abstracts. A total of 45 articles were assessed for eligibility. This process involves a comprehensive review by two independent researchers who assessed the validity and quality of each study. In cases of disagreement regarding included studies, a third researcher was consulted to reach consensus.

3.4. Final Selection: After applying these criteria, 36 studies were found to be suitable for inclusion in this review. These studies were selected based on methodological rigor and the importance of their results to the purpose of the review.

11.4 Rationale for Selecting the Six Areas

The six topics discussed in this article were chosen because they are important for understanding how school bags affect children's biomechanics. The following reasons guided the selection process:

(i) Strap Design: The strap design plays a major role in affecting body and comfort by distributing the weight of the bag across your shoulders and back. Many studies

Table 11.1 Summary of the references papers

References	Study objective	Study participants	Key findings
[2]	To investigate the effects on children's pulmonary function of various backpack loads and strap choices	20 healthy Nigerian children aged 6–11 years	Backpack load and strap option have a significant effect on pulmonary function parameters in children
[1]	To investigate how the length of shoulder straps on backpacks affects both cervical posture and the pressure pain threshold in the upper trapezius	30 male university students	The cervical posture and upper trapezius pressure pain threshold are greatly impacted by the length of backpack shoulder straps
[11]	To investigate the ergonomic elements of school bags and how well they accommodate the weight carried by students	548 Iranian school children aged 10–15 years	School bags' weight and design are associated with musculoskeletal problems
[21]	To investigate the prevalence of low back pain among students pursuing health sciences	456 Saudi Arabian health sciences students	Low back pain is a common issue among students studying health sciences
[10]	To pinpoint the factors linked to non-specific low back pain in children aged 10–12	518 Colombian children aged 10–12 years	Non-specific low back pain is correlated with factors such as carrying a heavy schoolbag, maintaining poor posture, and having a family history of back pain
Ben [22]	To examine the prevalence, contributing variables, and outcomes of low back, shoulder, and neck pain in kids enrolled in secondary schools	2000 Tunisian secondary-school children	Neck, shoulder, and low back pain are prevalent issues among secondary-school children, with high schoolbag weight and poor posture identified as significant risk factors
[23]	To examine the impact of frontpacks, shoulder bags, and handheld bags on the three-dimensional back shape and posture in young university students	60 British university students	Various bag types exert a notable influence on the shape and posture of the back
[24]	To investigate the health effects on offspring of pregnant women who follow the Mediterranean diet	Five systematic reviews and meta-analyses	Adherence to the Mediterranean diet during pregnancy is associated with better health outcomes in children

(continued)

Table 11.1 (continued)

References	Study objective	Study participants	Key findings
[25]	To collect initial subjective load carriage injury data using interviews and questionnaires	160 British soldiers and marines	Load carriage injuries are prevalent and have a significant impact on soldiers' and marines' physical performance
[5]	To explore the weight of schoolbags among early school-age pupils and its impact on body posture	205 Polish children aged 7–10 years	Heavy schoolbags are associated with changes in body posture, particularly in the head, neck, and trunk
[26]	To determine the factors that pose a risk for low back pain during childhood and adolescence	21 studies	The primary risk factors for low back pain in childhood and adolescence encompass gender, age, physical activity, and psychosocial factors
[27]	Examine the effects of backpack weight and placement on physical strains experienced by male students while they walk	20 Chinese male schoolchildren aged 10–12 years	Backpack load and position have a significant effect on body strains in schoolchildren

have shown the difference between straps, padding, and adjustability in terms of stress on the shoulders and back.

(ii) Bag Weight: The main factor affecting children's musculoskeletal health is the weight of the school bag. It is quite beneficial with the possibility of developing body-related problems. Many studies have shown a relationship between the weight of the school bag and the occurrence of back pain and discomfort in young people.

(iii) Carrying Method: The body's posture and mobility are dependent on the way in which a bag is carried, such as briefcase or two bags or stroller bag. In fact, several researchers have carried out studies that define biomechanical variations between these methods and are capable of justifying the advantages and disadvantages of each approach.

(iv) Bag Size and Capacity: Depending on the capacity and size of a bag, its total load together with how it is distributed can be carried by a child's weight. The study investigated the impact of bag size on delivery of cargo, as well as children's biomechanical results.

(v) Posture and Gait Changes: Knowing how carrying a backpack affects a child's posture and mobility is essential in establishing long-term musculoskeletal potential. According to studies, shoulder, walking, and spine muscles react differently when carrying different styles of backpacks.

(vi) Ergonomic Features: Ergonomic features meant to enhance comfort and relieve pressure include such things as padded back panels and lumbar supports. The purpose of this study was to determine the effectiveness of these features in mitigating the ill effects of overloading.

These topics were chosen because they cover most of the important variables that affect the biomechanics of bag use. Each question is supported by extensive research, providing a better understanding of how various structures and behaviors affect children's health. In addition, these areas, which are the main purpose of this study, provide important information for creating a safe and comfortable school environment.

Focusing on these six areas, the study explores the most important aspects of school bag design and their impacts and offers recommendations for all stakeholders, including teachers, parents and designers.

11.5 Research Gaps

Finding study gaps in each of the six areas covered in the paper can assist direct future investigations and advance knowledge of how schoolbag design affects biomechanics. The possible research gaps in each area are broken down as follows:

5.1 Strap Design: The main areas of research have been strap width, cushioning, and adaptability.

Research Gaps: Long-term research studies evaluating the impacts of different strap designs on kids' musculoskeletal health over time are scarce. There has not been much research done on how different strap designs influence kids of different ages, shapes, and genders. Research on novel and inventive strap designs like those that automatically adapt to the user's activity or body shape is required.

5.2 Bag Weight: There is a documented correlation between bag weight and musculoskeletal pain.

Research Gaps: To determine acceptable weight thresholds in relation to children's body weight and physical development phases, more research is required. There are not many studies examining how psychological elements (such peer pressure and self-esteem) affect how much weight kids are willing to carry. Interventions that seek to lessen carrying weight and evaluate the effect on health outcomes are required.

5.3 Bag Size and Capacity: Children's Overall burden and load distribution are influenced by the size and volume of the bag.

Research Gaps: The best bag size and capacity that strikes a balance between use and health concerns is not well established. There is a dearth of research on bags that can dynamically change their capacity in response to load or user requirements. To learn more about how parents' and kids' preferences for bag size affect their decisions and the ensuing health effects, more research is required.

5.4 Carrying Method: The way you carry "one strap," "two straps," or a "trolley" affects your posture and walk a lot.

Research Gaps: More thorough comparisons between various carrying strategies in a range of demographics are required. There are not many biomechanical models available to forecast the long-term effects of different carrying techniques. There is a dearth of research on carrying strategies that adjust to various activities and locations (such as walking vs. cycling).

5.5 Posture and Gait Changes: Walking patterns, shoulder posture, and spinal alignment are all impacted by carrying schoolbags.

Research Gaps: Few research have looked at how different age groups' posture and gaits are affected by schoolbag use. There is little research that really addresses how gender differs in posture and how using a schoolbag can alter stride. For a thorough examination of changes in posture and gait, wearable sensors and motion capture are two examples of cutting-edge technology that are still underutilized.

5.6 Ergonomic Features: The goal of ergonomic features is to lessen strain and increase comfort.

Research Gaps: Further research is required to determine how well some ergonomic elements, like lumbar supports, might lessen discomfort and the chance of damage. Little study has been done on ergonomic components that can be altered to satisfy the needs of various users. There are few studies assessing whether adding ergonomic elements to schoolbags is cost-effective when taking into account both the economic and health benefits.

By filling in these research gaps, the field can advance its knowledge and produce more innovative recommendations by gaining a deeper understanding of the intricate relationships between schoolbag design and children's health.

11.6 Positioning of Backpack and Posture

Multiple studies have sought to identify the optimal position of the backpack's center of gravity (CG) with the aim of minimizing spinal postural changes and discomfort. However, the results of these studies have been inconclusive, possibly due to variations in experimental protocols and outcome factors. Variations were noted, for example, in the types of backpacks used in the experiments, in the participant age and ethnic groups, and in the reference points used to position the backpacks. Currently, conducting a meta-analysis is challenging, and it is recommended that future investigations adhere to a comprehensive gold standard of protocols to achieve more convincing and conclusive results. Grimmer et al. [18] provided a higher level of evidence in their study, incorporating a randomized control trial (RCT) with a larger sample size and a lower risk of biased results. Nevertheless, additional studies, preferably utilizing meta-analysis and RCT approaches, are necessary, with adjustments for factors such as ethnicity, age group, and types of backpacks. Meanwhile, it is critical to understand that there must be a trade-off because there is not a single backpack position that can guarantee ideal postural change in every spinal site. While users can select the optimal position based on their individual pain site, designers should take into account the frequency of various pain sites within their target market when determining the loading position.

Ergonomic design features, including the shape of the backpack interface, adjustable straps, and the positioning of hip straps and chest, should be taken into consideration. Additionally, designers could implement sensory-based nudges to encourage users to adjust the backpack to their optimal level [28]. Even though single-strap packs are a common carrying method, there is a noticeable lack of research on the vertical loading position of these packs and the health consequences that are associated with them. Further research is recommended in this area [6, 18, 27].

11.7 Foot Weight Distribution and Positioning of Backpack

Positioning of foot weight and backpack distribution are two important factors that can affect the posture and gait of children when carrying heavy loads [29]. Previous studies have suggested that the position of the backpack's center of gravity (CG) can affect spinal posture and discomfort [6, 18, 27], but the effect of foot weight distribution has not been extensively studied. A recent review by Kim et al. [30]; Chow et al. [6] aimed to investigate the effect of foot weight and backpack position distribution on posture and gait parameters in school-aged children. The participants' posture and gait were found to be significantly impacted by their foot weight distribution and backpack position.

In terms of backpack position, the study found that when the backpack was positioned higher on the back, there was a significant increase in the forward head angle and thoracic kyphosis angle, indicating a more forward head posture and a more rounded upper back. When the backpack was positioned lower on the back, there was a significant increase in the lumbar lordosis angle, indicating an increased curvature of the lower back. These findings are consistent with previous studies that have shown different effects on spinal posture depending on the backpack position [6, 18, 27].

Regarding foot weight distribution, the study found that when more weight was distributed on the rear foot, there was a significant increase in the forward head angle and thoracic kyphosis angle, indicating a more forward head posture and a more rounded upper back. When more weight was distributed on the forefoot, there was a significant increase in the lumbar lordosis angle, indicating an increased curvature of the lower back. These findings suggest that foot weight distribution can also affect spinal posture and should be considered when designing backpacks or prescribing exercises for children carrying heavy loads.

In conclusion, the positioning of backpack and foot weight distribution are both important factors that can definitely affect the posture and gait of school-aged children when carrying heavy loads. Designers should consider these factors when designing backpacks, and healthcare professionals should take them into account when prescribing exercises or interventions for children with back pain or postural problems. To validate these results and determine the ideal foot weight distribution and backpack positions for various age and ethnic groupings, more research is required.

11.8 Effects of Backpack Positioning: Perception, Lung Function, and Metabolic Cost

The positioning of backpack can have significant effects on the body's metabolic cost, perception, and lung function. A study by Ramadan & Al-Tayyar [31] investigated the impact of backpack position on the metabolic cost of walking and found that a lower position of the backpack resulted in reduced metabolic cost. Similarly, a study by Sahli et al. [32] explored the impact of backpack positioning on lung function and found that a higher backpack position had a greater negative impact on lung function. Perception of load is also influenced by backpack positioning, as shown in a study by Ramadan & Al-Tayyar [31], which found that backpacks positioned higher on the back were perceived as heavier than those positioned lower. It is clear that the positioning of backpacks has significant implications for the body, and designers should consider the impact of different positions on metabolic cost, lung function, and perception when creating new backpack designs.

In addition to the effects on perception, metabolic cost, and lung function, the positioning of a backpack can also have an impact on other aspects of physical health. For example, a study by Lim et al. (2020) found that a higher backpack position resulted in increased muscle activity and spinal loading during gait, which could lead to greater fatigue and discomfort over time. On the other hand, a lower backpack position may distribute the load more evenly across the back and hips, reducing the risk of musculoskeletal injury. Another study by Abdelraouf et al. [1] investigated the impact of backpack position on trunk muscle activity and found that a lower position resulted in reduced activity of the erector spinae and rectus abdominis muscles, indicating a more relaxed posture. This finding is consistent with previous research showing that excessive muscle activity can contribute to muscle fatigue and discomfort Kim et al. [30]. Finally, the positioning of a backpack can also affect the stability and balance of the wearer, particularly during dynamic activities such as running or jumping. A study by Sahli et al. [32] found that a higher backpack position resulted in reduced balance control during single-leg stance, which could increase the risk of falls and other injuries. Overall, these findings suggest that the positioning of a backpack can have significant implications for physical health and performance and should be carefully considered in the design and use of backpacks.

11.9 Positioning of Backpack and Pressure of Shoulder Straps

The arrangement of the backpack and the pressure exerted by the shoulder straps are pivotal factors that can significantly influence biomechanical responses and the user's comfort level. Previous research indicates that the vertical positioning of the backpack can impact the amount of pressure experienced on the shoulder straps. For example, Hadid et al. [33] discovered that backpacks situated higher on the back can induce more pressure on the shoulders, leading to discomfort and fatigue. Similarly, a study conducted by Chen and Mu [27] revealed that backpacks positioned at the T12 level resulted in less shoulder pressure compared to higher positions. Ensuring proper adjustment of the shoulder straps is crucial to evenly distribute the weight of the backpack across the shoulders and back. An uneven distribution of weight may cause asymmetrical loading, potentially resulting in musculoskeletal disorders like neck pain, back pain, and shoulder pain. Adequate adjustment of the shoulder straps not only helps mitigate these issues but also reduces pressure on the shoulders, contributing to an overall improvement in user comfort. In conjunction with the shoulder straps, the design and padding of the back panel are equally crucial in ensuring the even distribution of the backpack's weight and minimizing pressure on the shoulders. Correct positioning of the backpack and the appropriate adjustment of shoulder straps contribute not only to enhanced comfort but also to a decreased risk of musculoskeletal disorders associated with the use of backpacks.

11.10 Positioning of Backpack and Muscle Activities

The placement of backpacks also influences muscle activities. Numerous studies have delved into the impact of different backpack positions on the muscle activities in the back and shoulder regions. For example, Chen & Mu [27] conducted a study examining the impact of backpack position on the muscle activity of the upper trapezius and erector spinae muscles during gait in school-aged children. The findings revealed that the muscle activity of the upper trapezius significantly increased when the backpack was positioned at the T7 level compared to the T12 and L3 levels. Conversely, the erector spinae muscle activity was significantly lower when the backpack was positioned at the L3 level compared to the T7 and T12 levels.

Another study by A Chen & Mu [27] investigated the influence of backpack position on the muscle activity of the trapezius, deltoid, and erector spinae muscles during a lifting task. The results indicated that the muscle activity of the trapezius and deltoid muscles was significantly higher when the backpack was positioned at the T7 level compared to the L3 level. Conversely, the erector spinae muscle activity was significantly lower when the backpack was positioned at the L3 level compared to the T7 level.

In addition to the backpack position, the pressure of shoulder straps can also affect muscle activities. In a study conducted by Golriz et al. [3], the effects of varying shoulder strap pressures on the muscle activity of the trapezius and erector spinae muscles were investigated during standing and walking tasks. The findings revealed a significant increase in trapezius muscle activity with higher shoulder strap pressure in both standing and walking tasks, while no significant effect was observed on erector spinae muscle activity. Another study by Kim et al. [34] delved into the impact of shoulder strap pressure on the muscle activity of the trapezius, erector spinae, and serratus anterior muscles during a lifting task. The results indicated a significant increase in the muscle activity of the trapezius and erector spinae muscles with elevated shoulder strap pressure, whereas no significant effect was observed on the serratus anterior muscle activity.

In general, shoulder strap pressure and backpack placement have a big influence on how much muscle is used. These findings suggest that designers should consider the impact of these factors when developing school backpacks to reduce the risk of musculoskeletal disorders among children and adolescents. Furthermore, users should be educated on the optimal positioning and adjustment of backpacks to reduce the risk of muscle strain and fatigue.

11.11 Shoulder Strap Width

The width of the shoulder straps is another important factor in backpack design that affects the comfort and health of the wearer. The shoulder straps distribute the weight of the backpack on the shoulders, and if they are too narrow, they can cut into the skin and cause discomfort or even pain. On the other hand, if they are too wide, they may interfere with the arm movements and cause discomfort or strain on the upper body muscles. Several studies have investigated the effects of shoulder strap width on the musculoskeletal system and overall comfort of the wearer. One study found that increasing the shoulder strap width from 25 to 50 mm resulted in a decrease in the pressure on the shoulders and a reduction in the perceived discomfort of the backpack [3]. Another study investigated the effect of the shoulder strap width on the muscle activity of the trapezius and found that increasing the width from 15 to 40 mm resulted in a significant decrease in muscle activity [35]. However, other studies have reported no significant difference in comfort or muscle activity with different shoulder strap widths [30, 36]. Thus, the optimal width of the shoulder straps may depend on various factors such as the weight of the backpack, the body type of the wearer, and the type of activity performed while wearing the backpack. Overall, designers should consider the width of the shoulder straps as an important factor in backpack design and aim to provide a comfortable and functional solution that distributes the weight of the backpack evenly and avoids unnecessary strain on the musculoskeletal system.

11.12 Future Research Potentials and Recommendations

This article provides a thorough overview of the ways in which school bag designs affect student health. It is titled Biomechanical Impact of School Bag Design: Considerations for Backpack Straps Size and Carrying Methods. Numerous variations such as backpack types sizes and carrying techniques were noted by the research. According to this article, wearing shoulder straps incorrectly can cause back neck and shoulder pain in addition to altering your spinal curvature. A bulky or overweight backpack can make you look unattractive and can lead to injuries due to physical issues. According to reviews, these risks could be reduced by using an ergonomic backpack that is the right size and has wide padded straps. The investigation also examined various methods of lifting the backpack including rolling it putting it across the chest and only using one side.

Studies have shown how considering biomechanical impacts when designing and selecting school bags can encourage children to lead a healthy life. In future, the main aim of research should be developing better methods of experimentation. Research needs to look at the influence of one backpack and foot weight distribution among different age groups and different ethnicities. Additionally, more studies are warranted on how backpack placement affects cognition, lung function, and metabolic rates. The integration of ergonomic design features such as detachable shoulder straps and suggestions that promote optimal backpack usage would also be beneficial in further improving student health. Based on all said above, this review explains how backpack designs affect student biomechanics and provides recommendations for alleviating musculoskeletal problems related to carrying school bags.

11.13 Conclusion

In conclusion, this systematic review aimed to consolidate the existing literature on the design parameters of school backpacks and their impact on the musculoskeletal system of school-age children. The review encompassed 50 studies exploring various aspects of backpack design, including backpack positioning, shoulder strap width and padding, weight distribution, backpack size and shape, and carrying methods. The study findings were categorized into four groups: biomechanics consideration, strap, dimension, and carrying method. The findings imply that variables like age, ethnicity, and the kind of backpack under investigation may have an impact on where the center of gravity of the backpack should be positioned. Additionally, a balance must be struck between postural changes and discomfort in different anatomical sites. Designers should account for the prevalence of pain in different areas among their target audience when determining the loading position. Furthermore, the width and padding of shoulder straps, along with their angle and tightness, play a vital role in reducing muscle activities and perceived discomfort in school children. The distribution of backpack weight, ideally within 10–15% of the child's body weight, also impacts the musculoskeletal system and the metabolic cost of carrying the backpack.

Overall, this review underscores the significance of taking various design parameters into account when creating school backpacks to mitigate adverse effects on the musculoskeletal system of school-age children. It also advocates for more high-quality studies to be conducted to identify the most effective design parameters for school backpacks. Designers are encouraged to provide adjustable features and incorporate sensory-based nudges to motivate children to optimize their backpack positioning. Lastly, it emphasizes the importance of informing parents and educators about the selection of appropriate backpacks and educating children on how to wear and adjust their backpacks properly to minimize the negative impact on their musculoskeletal health.

References

1. Abdelraouf, O.R., Hamada, H.A., Selim, A., Shendy, W., Zakaria, H.: Effect of backpack shoulder straps length on cervical posture and upper trapezius pressure pain threshold. J. Phys. Ther. Sci. **28**, 2437–2440 (2016)
2. Abaraogu, U.O., Kizito, E.B., Okafor, U.A., Okoye, G.C.: Effect of variable backpack load and strap option on the pulmonary function of children: a simulation using treadmill walking. Work **55**(3), 525–530 (2016)
3. Golriz, S., Hebert, J.J., Bo Foreman, K., Walker, B.F.: The effect of shoulder strap width and load placement on shoulder-backpack interface pressure. Work **58**(4), 455–461 (2017). https://doi.org/10.3233/WOR-172651
4. Hong, Y., Fong, D.T., Li, J.X.: The effect of school bag design and load on spinal posture during stair use by children. Ergonomics **54**(12), 1207–1213 (2011) https://doi.org/10.1080/00140139.2011.615415
5. Brzęk, A., Dworrak, T., Strauss, M., Sanchis-Gomar, F., Sabbah, I., Dworrak, B., Leischik, R.: The weight of pupils' schoolbags in early school age and its influence on body posture. BMC Muscoskel. Disord. 18 (1), 117 (2017). https://doi.org/10.1186/s12891-017-1462-z
6. Chow, D.H.K., Ou, Z.Y., Wang, X.G., Lai, A.: Short-term effects of backpack load placement on spine deformation and repositioning error in schoolchildren. Ergonomics **53**(1), 56–64 (2010)
7. Korovessis, P., Koureas, G., Papazisis, Z.: Correlation between backpack weight and way of carrying, sagittal and frontal spinal curvatures, athletic activity, and dorsal and low back pain in schoolchildren and adolescents. J. Spinal Disord. Tech. **17**(1), 33–40 (2004)
8. Chow, D.H.K., Li, M.F., Lai, A., Pope, M.H.: Effect of load carriage on spinal compression. Int. J. Ind. Ergon. **41**(3), 219–223 (2011)
9. Dahl, K.D., Wang, H., Popp, J.K., Dickin, D.C.: Load distribution and postural changes in young adults when wearing a traditional backpack versus the BackTpack. Gait Posture **45**(90–96), 2016 (2016)
10. Angarita-Fonseca, A., Boneth-Collantea, M., Ariza-Garcia, C.L., Parra, J., CorredorVargas, J.D., Villamizar-Nino, A.P.: Factors associated with non-specific low back pain in children aged 10–12 from Bucaramanga, Colombia: a cross-sectional study. J. Back Musculoskelet. Rehabil. **32**(1–2), 1–9 (2019)
11. Alami, A., Tehrani, H., Lael-Monfared, E., Moghaddam, F.S., Boghsani, G.T., Jafari, A.: Ergonomic factors of school bags and their adaptation to the weight of students. Work **65**(4), 809–820 (2020). https://doi.org/10.3233/WOR-203133

12. Goodgold, S., Corcoran, M., Gamache, D., Gillis, J., Guerin, J., Coyle, J.Q.: Backpack use in children. Pediatr. Phys. Ther. **14**(3), 122–131 (2002)
13. Delele, M., Janakiraman, B., Abebe, A.B., Tafese, A., van de Water, A.T.M.: Musculoskeletal pain and associated factors among Ethiopian elementary school children. BMC Muscoskel. Disord. **19**(1), 276 (2018). https://doi.org/10.1186/s12891-018-2192-6
14. Forjuoh, S.N.: School backpack weights: a survey of students in Ghana, Guatemala and the USA. Inj. Contr. Saf. Promot. **1**(4), 287–289 (2004)
15. Kędra, A., Kolwicz-Ganko, A., Sitarski, D., Kędra, P., Czaprowski, D.: Prevalence of back pain and the knowledge of preventive measures in a cohort of 11619 Polish school-age children and youth—an epidemiological study. Medicine (Baltim.) **98**(22), e15729 (2019)
16. Golriz, S., Walker, B.: Backpacks. Several factors likely to influence design and usage: a systematic literature review. Work **42**(4), 519–531 (2012). https://doi.org/10.3233/WOR-2012-1383.
17. Hong, Y., Li, J.X.: Influence of load and carrying methods on gait phase and ground reactions in children's stair walking. Gait Posture **22**(1), 63–68 (2005)
18. Grimmer, K., Dansie, B., Milanese, S., Pirunsan, U., Trott, P.: Adolescent standing postural response to backpack loads: a randomised controlled experimental study. BMC Muscoskel. Disord. **3**, 10 (2002)
19. Cottalorda, J., Rahmani, A., Diop, M., Gautheron, V., Ebermeyer, E., Belli, A.: Influence of school bag carrying on gait kinetics. J. Pediatr. Orthop. B **12**(6), 357–364 (2003)
20. Golriz, S., Walker, B.: Can load carriage system weight, design and placement affect pain and discomfort? A systematic review. J. Back Musculoskelet. Rehabil. **24**, 1–16 (2011)
21. AlShayhan, F.A., Saadeddin, M.: Prevalence of low back pain among health sciences students. Eur. J. Orthop. Surg. Traumatol. **28**(2), 165–170 (2018). https://doi.org/10.1007/s00590-017-2034-5. Epub 2017 Aug 30
22. Ben Ayed, H., Yaich, S., Trigui, M., Ben Hmida, M., Ben Jemaa, M., Ammar, A., Jedidi, J., Karray, R., Feki, H., Mejdoub, Y., Kassis, M., Damak, J.: Prevalence, risk factors and outcomes of neck, shoulders and low-back pain in secondary-school children. J. Res. Health Sci. **26**(19), e00440, 1 2019
23. Bettany-Saltikov, J., Cole, L.: The effect of frontpacks, shoulder bags and handheld bags on 3D back shape and posture in young university students: an ISIS2 study. Stud. Health Technol. Inf. **176**, 117–121 (2012)
24. Biagi, C., Nunzio, M.D., Bordoni, A., Gori, D., Lanari, M.: Effect of adherence to mediterranean diet during pregnancy on children's health: a systematic review. Nutrients **11**(5), 997 (2019). https://doi.org/10.3390/nu11050997
25. Birrell, S.A., Hooper, R.H.: Initial subjective load carriage injury data collected with interviews and questionnaires. Mil. Med. **172**, 306–311 (2007)
26. Calvo-Munoz, I., Kovacs, F.M., Roque, M.: Risk factors for low back pain in childhood and adolescence: a systematic review. Clin. J. Pain **34** (5), 468–484 (2018) https://doi.org/10.1097/AJP.0000000000000558
27. Chen, Y.L., Mu, Y.C.: Effects of backpack load and position on body strains in male schoolchildren while walking. PLoS ONE **13**(3), e0193648 (2018). https://doi.org/10.1371/journal.pone.0193648
28. Labbe, D., R Fries, L., Ferrage, A., Lenfant, F., Godinot, N., Martin, N.: Right sizing: sensory-based product design is a promising strategy to nudge consumers toward healthier portions. Nutrients **10** (10), E1544 (2018). https://doi.org/10.3390/nu10101544 pii."
29. Gong, W., Lee, S., Kim, B.: The comparison of pressure of the feet in stance and gait by the types of bags. J. Phys. Ther. Sci. **22**(3), 255–258 (2010). https://doi.org/10.1589/jpts.22.255
30. Kim, K., Kim, C.J., Oh, D.: Effect of backpack position on foot weight distribution of school-aged children. J. Phys. Ther. Sci. **27**(3), 747–749 (2015). https://doi.org/10.1589/jpts.27.747
31. Ramadan, M.Z., Al-Tayyar, S.N.: Development and experimental verification of an ergonomic backpack. BioMed Res. Int. **15** (2020) https://doi.org/10.1155/2020/1437126.
32. Sahli, S., Rebai, H., Ghroubi, S., Yahia, A., Guermazi, M., Elleuch, M.H.:. The effects of backpack load and carrying method on the balance of adolescent idiopathic scoliosis subjects. Spine J. **13** (12), 1835–1842 (2013). https://doi.org/10.1016/j.spinee.2013.06.023

33. Hadid, A., Gozes, G., Atoon, A., Gefen, A., Epstein, Y.: Effects of an improved biomechanical backpack strap design on load transfer to the shoulder soft tissues, J.C.N. Lee et al. Int. J. Ind. Ergonom. **84**(2021),103166 (2018) Jul 25 J. Biomech. **76**, 45–52 (2018). https://doi.org/10.1016/j.jbiomech.2018.05.016. Epub 2018 May 18.
34. Kim, M.H., Yi, C.H., Kwon, O.Y., Cho, S.H., Yoo, W.G.: Changes in neck muscle electromyography and forward head posture of children when carrying schoolbags. Ergonomics **51**(6), 890–901 (2008). https://doi.org/10.1080/00140130701852747
35. Endo, K., Yamazaki, N., & Tsuru, S.: Effect of shoulder strap width on trapezius muscle activity and subjective discomfort during load carriage. J. Phys. Ther. Sci. **28**(8), 2310–2313 (2016)
36. Wettenschwiler, P., Lorenzetti, S., Stüssi, E., Brühwiler, P. A.: The influence of shoulder strap width on subjective perception of load and trunk muscle activity during backpack carriage. Appl. Ergon. **43**(2), 343–347 (2012). https://doi.org/10.1016/j.apergo.2011.06.004
37. Ramadan, M.Z., Al-Tayyar, S.N.: Comparison between the physiological responses and subjective ratings of a group of male students to three backpack designs. Int. J. Environ. Res. Publ. Health **16**, 4104 (2019). https://doi.org/10.3390/ijerph16214104

59. Hadid, A., Gozes, G., Atoun, A., Gefen, A., Epstein, Y.: Effects of an improved biomechanical backpack strap design on load transfer to the shoulder soft tissues. J. Biomech. 76, 45–52 (2018). https://doi.org/10.1016/j.jbiomech.2018.05.016. Epub 2018 May 18
60. [illegible]

Part III
Smart Manufacturing Systems and Industry 4.0

Chapter 12
Systematic Approach of Lean Implementation Using Multi-level Value Stream Mapping: A Case Study

S. Gowtham, A. Prabukarthi, and V. D. Gogul

Abstract Value stream mapping (VSM) is a lean manufacturing (LM) technique used to examine the flow of materials and information for a particular product family. This paper aims to discuss the significance of VSM in an LM context in the gear manufacturing industry. The cycloidal speed reducer is one of the products that was chosen. There are 18 sub-components required for assembling the cycloidal speed reducer. The purpose of this research suggests a new approach of preferred multi-level VSM mapping based on manufacturing setup time reduction found in the gap analysis of the present state map. Using this idea, the study was conducted to determine the areas in which work in process, lead time, setup time, and cycle time need to be improved as part of developing a future state map and implementing Kaizen on the current state map. The results obtained from the study are as follows: The production lead time is reduced from 1131 to 1028 min, VA time was reduced from 323 to 308 min, and the NVA time was reduced from 671 to 582 min.

Keywords Lean manufacturing · Value stream mapping · Multi-level VSM · Gear manufacturing industry

S. Gowtham (✉) · A. Prabukarthi · V. D. Gogul
Department of Mechanical Engineering, PSG College of Technology, Coimbatore, Tamil Nadu, India
e-mail: gowthamcloud@gmail.com

A. Prabukarthi
e-mail: apk.mech@psgtech.ac.in

V. D. Gogul
e-mail: 22mf01@psgtech.ac.in

J. Maiti et al. (eds.), *Recent Advances in Industrial and Systems Engineering*, Lecture Notes on Multidisciplinary Industrial Engineering,
https://doi.org/10.1007/978-981-96-8323-9_12

12.1 Introduction

The gear manufacturing industry is one of the most active, employing continuous improvement tactics to reduce costs and enhance product quality to stay competitive in their business. Companies are also using a variety of strategic tools for lean manufacturing to maintain a balance between product quality and cost. As a result of increased competition, manufacturing businesses have been forced to modify their production systems. In today's world, the industrial scene has been confronted with greater problems as a result of increased international competitiveness. Lean manufacturing (LM) philosophy has played an important role in improving company performance, not only at the operational level but also at the business level. This truth was primarily emphasized in the research as an agreement that the acceptable application of LM would leverage businesses. LM is a well-known management concept in many industrial firms. LM has been implemented in both the private and public sectors. There are numerous success stories of lean initiatives that result in shorter production lead times, higher quality, and lower prices. Several organizations have imposed or attempted to adopt the principles and objectives of LM throughout the previous decade. Some firms have implemented a few LM strategies, while others have implemented a whole spectrum of LM elements. LM is a collection of approaches for achieving leanness in a manufacturing company. Black† [1] LM was designed by the Toyota Production System, which strives to decrease inventory, remove waste, increase throughput, and motivate employees to point out issues and recommend fixes [2, 3]. Value stream mapping (VSM) is a method for locating waste in a production system. VSM is used in the LM unit to help in visualizing the entire production process by showing each piece of information and material flow. The purpose is to identify all types of waste within the VSM and to impose the required measures to remove them. Many researchers have used the VSM tool to reduce waste in production systems. VSM is a way of connecting quality improvement projects and lean thinking in order to deliver the best total profit to a firm through waste removal and value-saving activities.

During these activities, a process can be broken down into various work components for simpler measurement. These tasks can be divided into value-added (VA) activities that increase the value of the product, such as machining the sub-components and assembling the main components, and non-value-added (NVA) activities that do not affect the product's value, such as loading and unloading, machine cleaning, material handling, dimension checking, unused tool movement, and operator movement. Using the time study and work study approaches outlined above, the measurement of these activities may be determined. In this study, we created a framework for the approach that is suggested for applying value stream mapping to increase the efficacy and efficiency of the case study under consideration. The gear manufacturing facility that is involved in the manufacture and assembly of the primary components is the basis for the case study presented in this article. A flowchart illustrates the specific steps in the procedure. The research study begins with executing the VSM analysis and a performance review of the present production process design in order to meet the goal, i.e., cycle time reduction. It covers the actual field visit scanning and planning of process objects, as well as takt and cycle time calculations. The current value stream map may be created using various steps. Manupati et al. [4] Value stream design begins with a process cycle time reduction when true optimization is feasible. Following the cycle time decrease, the future value stream map is created using the newly computed data. Thus, by presenting a production process design with a markedly reduced cycle time and improved transparency in production control, the study provides validation of the methods adopted.

The paper is structured as follows: Sect. 12.2 contains an extensive overview of the literature. In Sect. 12.3, there is a discussion of the methodological framework. Section 12.4 is presented the case study on the use of VSM in a manufacturing company that produces gears and creates a map of the case firm's current and future states. The innovative concepts used to increase the effectiveness of various production techniques are discussed in Sect. 12.5, and future value stream maps are used to illustrate the improvements. In Sect. 12.6, the changes that have been made are evaluated and discussed. The appropriate calculations from the findings of the present and future value stream maps are made in Sect. 12.7.

12.2 Literature Review

This VSM literature review's goal is to provide information on the investigated areas of lean manufacturing where VSM has been used. Singh et al. [5] investigated that VSM is a useful tool for making recommendations about how to reduce manufacturing waste, lead times, and rework. The timeline comparing VA and NVA timings made it evident that there were significant amounts of waste, leading to the long lead times, and further current and future state maps illustrated the production process. According to the study's final findings, processing time was decreased overall by 16.22%, lead time by 37.56%, and inventory by 30%. The VA ratio increased by 33.65%. Sharma and Virmani [6] implemented the lean production system through the VSM technique, as well as cellular manufacturing, Kanban, 5S, TPM, ALS, supermarkets, online inspection, and strain index, which were all merged into the future state map to get rid of various types of waste. Manupati et al. [4] used logical methods to investigate how small efforts in the direction of putting lean principles into practice will help you reach your waste reduction goals, which will increase component manufacturing efficiency by reducing cycle times. Work elements were divided into activities with and without value addition as a result of work measurement. The analysis of the production line's wastes and non-value-added activities was followed by the application of various alternative approaches to address the wastes discovered. Vinodh et al. [7] have discussed an example where VSM and other lean principles were applied at the Indian Camshaft Organization in the process sector. Gurumurthy and Kodali [8] Researchers have proposed integrating simulation and VSM to present a case study of a company that employs a job shop production system to manufacture doors and windows in order to illustrate the usage of virtual simulation modeling (VSM) in lean manufacturing system (LMS) design. For the case organization, simulation models were created using the Queuing Event Simulation Tool for various situations, such as "before LM" and "after LM," to show how the case organization will evolve after incorporating certain LM aspects. Vamsi Krishna Jasti and Sharma [9] examined the VSM to highlight the beneficial effects on process ratio, takt time, process inventory level, line speed, total lead, and process time, and decreased manpower. In terms of quality, price, and delivery, it assists the business in ensuring customer satisfaction. Mishra et al. [10] investigated the value stream mapping, modeling, and sustainability software were all integrated into the research using a lean and green strategy. Several lean methods, including Ishikawa, FMEA, and Poka-Yoke, were also employed. A break-even analysis was also performed to show the plans' financial advantages. Vinodh et al. [11] defined lean manufacturing techniques as a type of VSM focused on the micro-analysis of manufacturing processes for waste quantification, increase in labor productivity, quality improvement, and throughput time reduction in the context of a global market. Abideen and Mohamad [12] have implemented value stream mapping (VSM) and discrete event simulation (DES) combined in a Malaysian pharmaceutical manufacturing warehouse. By first creating a supply chain value stream map (both present state and future state) and integrating its data into a DES modeling and simulation software to dynamically visualize the changes in future state value stream map, the primary focus is shifted toward reducing the warehouse supply chain lead time. Ebrahimi vet al. [13] Researchers have used 3SM in a real-world setting, and the results demonstrate its applicability as well as its faster setup times and improved sustainability standards. It also broadens the scope of SMED to include sustainability improvement by demonstrating what/how setup activities/tasks at setup workstations and factory-wide levels affect sustainability requirements. Vinodh et al. [14] describe it as the methodical elimination of waste, and two types of validation are being done utilizing a questionnaire-based methodology. The executives' practical receptivity to implementing the recommendations was gathered in the first step. Statistical validation has been done in the second step. The study's findings included quantification of waste, an increase in labor productivity, improved quality, and a decrease in throughput time.

12.3 Methodology

First, examine the literature on lean manufacturing and VSM. VSM was chosen as the tool to help the case organization solve its problem, then followed by data collection and further calculations based on the information gathered. The shop floor's micro-level examination revealed a variety of activities. Activity categorization was done based on the collected data to precisely determine the valuable activity and waste. Then, for the specified product line, a current state map was created. Then, in consultation with the organization's management, a brainstorming session was held to examine the current state map and come up with some improvement recommendations. These suggestions for improvement included a new approach to multi-level VSM mapping based on manufacturing setup time reduction in the gaps found in the analysis of the present state map. Using this idea, a future state map was used to conclude. Finally, the case organization executed the prepared future state map. The findings are based on a pilot implementation of the projected future state map. The detailed flow of the current research work is presented in Fig. 12.1.

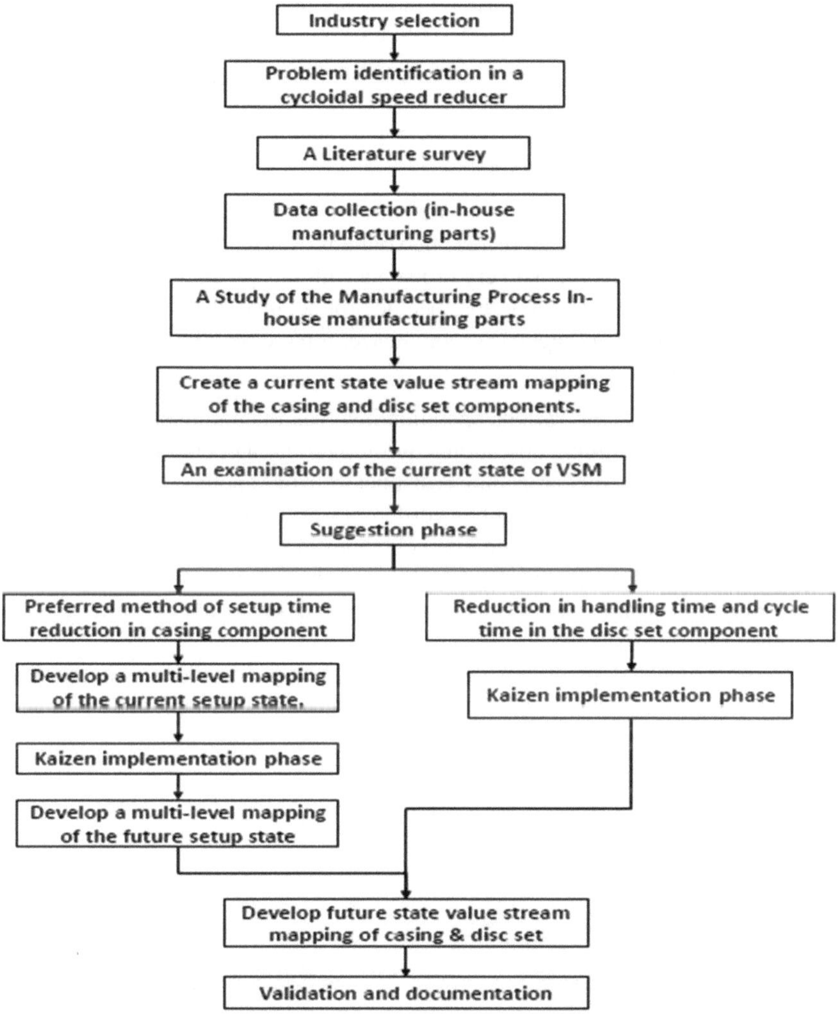

Fig. 12.1 Flowchart of field study

12.3.1 Product Family Selection

The gearbox manufacturing industry studied produces a number of different types of gearboxes of various sizes from its machine shop. They are widely used in robots, machine tools, and equipment in the process industry. Before starting the VSM, it is essential to select a product family. The cycloidal speed reducer was one of the products selected for this study. It consists of a total of 18 components, namely casing, input flange, planet pin, disk set, output flange, taper roller, spacer, and HS shaft which are in-house manufacturing components, and outer race, oil seal 1, oil seal 2, cage, needle roller 1, needle roller 2, shaft plug, and nut are vendor manufacturing components. The exploded view of the casing component of the cycloidal speed reducer is illustrated in Fig. 12.2.

In this study, the cycloidal speed reducer, the casing, and disk set are chosen for VSM based on the following factors:

1. It is the biggest product family in terms of turnover for the organization studied.
2. The customer had indicated increasing volumes over the following years.
3. The company was facing delivery problems with this product.
4. The manufacturing process of this specific product involved the use of all major equipment in the sheet metal division; hence, it increased the scope and impact of the VSM improvement.

Figure 12.3. depicts the various operation involved in components of the casing and disk set. Each machining operation requires a different operator. The timeframe of each operator's particular procedure is distinct from one another. Depending on the production schedule, each operator has a different workstation assigned to them. The goal of this research is to determine the cycle times of all currently in-use processes using time study and work study approaches and then to optimize those processes wherever possible using lean tools to shorten the process's total cycle time and plan the production appropriately.

After selecting and justifying the product's family considered for the VSM improvement project. The main objectives to be achieved through the VSM study are as follows:

1. Reducing production lead time.

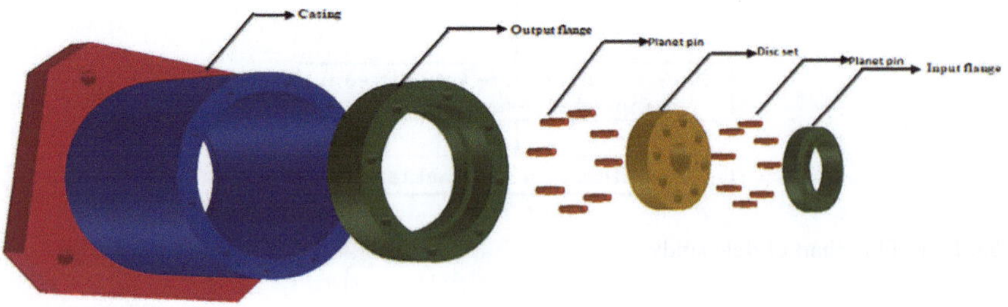

Fig. 12.2 Subcomponents of cycloidal speed reducer

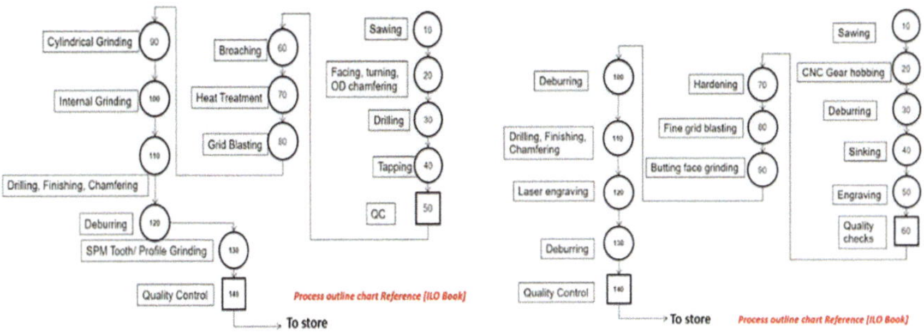

Fig. 12.3 Flow diagram of casing and disk set components

2. Reducing in-process and finished goods inventory.
3. Reducing process bottlenecks with respect to customer demand.

12.4 Current State Value Stream Mapping

The major two components of cycloidal speed reducer considered for this study are the casing and the disk set. This section has a wide range of casing and disk set production options. The length and shape of the components vary depending on the needs of the customer. All types of casing and disk set manufacturing procedures are the same. The characteristics of each manufacturer were acquired by manual time observation. C/T, C/O, VA time, NVA time, number of operators, and WIP were among the measurements taken. Except for the number of operators and WIP, recorded timings on VSM are based on average time. For each procedure in the U-bolt portion, the time available for work and uptime were calculated. This information was also gathered from shop floor employees and foremen. The overall production time is 21 h. There is a 30-min lunch break and two 15-min tea breaks.

- Break and lunch time/shift = 1 h.
- First shift timing = 8.00 am to 4:30 pm = 8:30 h.
- Second shift timing = 4:30 pm to 1: 00 am = 8:30 h.
- Third shift timing = 1:00 am to 8:00 am = 7 h.
- Total available time = 21 h = 1260 min.
- Customer demand per month = 30.
- Takt time $= \frac{\text{Available time}}{\text{Customer demand}} = \frac{1260}{30} = 42\ \text{min}$.

The available time for production is 1260 min. This production time is consistent throughout the manufacturing process.

Uptime = actual operation time/available time of work.

Uptime of cutting operation = (actual operation time—changeover time)/ available time off work.

$$= (1260 - 80)/1260$$

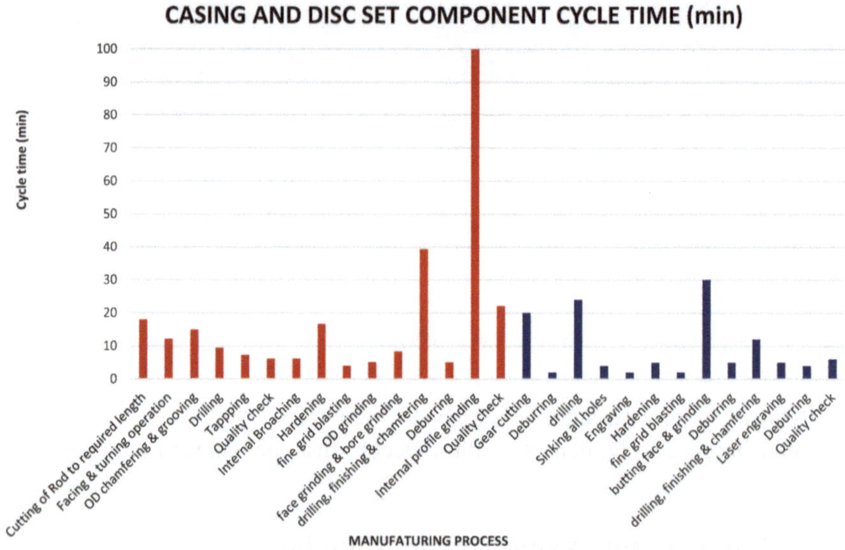

Fig. 4 Cycle time of casing and disk set component

$$\text{Uptime of cutting operation} = 93.6\%.$$

The demand arrives at the end of the month via its supplier to the marketing department, which then releases the order to PPC and sends the requirement to the purchasing department. Then, the purchase department issues a work order (WO). The PPC department personally distributes their daily planning document on the machine floor. They create a batch card for material according to their planning sheet. Following that, RM (raw material) travels from the RM yard to the cutting operation. From cutting operations to a variety of processes such as facing, turning, grinding, and drilling, as well as other details such as C/T, WIP, C/O, and production lead time, the VSM of the current state (Fig. 12.4.) is shown. After studying the VSM, the actual C/T of the casing and disk set is 877 s. The casing and disk set are the two components machining the operation in parallel. So, the total lead time for the product is 864 min.

12.4.1 Analysis of Current State Value Stream Mapping

Figure 12.4 depicts the current condition of the VSM casing and disk set section, which has sufficient information on each of the casing and disk set production processes as shown in (Tables 12.1 and 12.2) this allows us to identify the NVA process, bottleneck operations, and their underlying causes. VSM's concept depicts current status issues and acts as a reference for filling in the gaps. As a result, it helps in understanding how things will work once certain modifications have been implemented, as well as in the preparation of future state VSM.

From the present state map VSM of casing and disk set, the following gap locations were determined.

- C/T variation is higher, resulting in production inconsistency and WIP on the shop floor. The grinding operation had the highest C/T, as illustrated in Fig. 12.4. The main cause of line imbalance is this. As a result, there is a lot of WIP inventory on the remaining processes.
- WIP inventory varies greatly among manufacturing processes—WIP inventory is extremely high on turning machines, grinding machines, and heating activities as compared to other operations.
- The setup time of the XYZ machine was high due to three different machining operations of turning, drilling, and chamfering performed one after the other.

Table 12.1 Summary of casing component

Operations	Cycle time (min)	Setup time (min)	Value-added time (min)	Available time (min)	Uptime (%)	No of operators
Sawing	17.95	80	17.95	1260	93.6	1
Facing and turning operation	12.16	45	12.16	1260	96.4	1
OD chamfering and grooving	14.96	45	14.96	1260	96.4	1
Drilling	9.46	30	9.46	1260	97.6	1
Tapping	7.27	10	7.27	1260	99.9	1
Quality check	6.17		6.17	1260		1
Internal broaching	6.2	30	6.2	1260	97.6	1
Hardening	-		-	1260		1
Fine grid blasting	4.08		4.08	1260	99.3	1
OD grinding	5.16	15	5.16	1260	98.8	1
Face grinding and bore grinding	8.37	20	8.37	1260	98.4	1
Drilling, finishing, and chamfering	39.42	150	39.42	1260	88	1
Deburring	5.07	5	5.07	1260	99.9	1
Internal profile grinding	100.26	90	100.26	1260	92.8	1
Quality check	22.15		22.15	1260		1

Table 12.2 Summary of disk set component

Operations	Cycle time (min)	Setup time (min)	Value-added time (min)	Available time (min)	Uptime (%)	No of operators
CNC gear hobbing	20	60	20	1260	95	1
Deburring	2	8	2	1260	99.3	1
Drilling	24	45	24	1260	96.4	1
The sinking of all holes	4	15	4	1260	98.8	1
Engraving	2	8	2	1260	99.3	1
Quality check	5		5	1260		1
Hardening				1260		1
Fine grid blasting	2	8	2	1260	99.3	1
Butting face grinding	30	20	30	1260	98.4	1
Deburring	5	8	5	1260	99.3	1
Finishing and chamfering	12	45	12	1260	96.4	1
Laser engraving	5	60	5	1260	95	1
Deburring sharp corners	4	8	4	1260	99.3	1
Final quality checks	20		20	1260		1

- Low value-added ratio—the current state of the VSM VA ratio is quite low.
- Uptime—Because of the high C/O time and the high setting time of the machining process, the uptime at the drilling, finishing, and chamfering operation in the casing component is quite low.
- Differences between VA and NVA activities—Some NVA activities do not require a product but do require a process because the production cost of the product is relatively high, so this needs to be improved.

12.4.2 Suggestions Phase

From the existing state VSM, it was discovered that the time spent was excessive and that there was scope for further improvement by eliminating NVA processes and reducing waste. The issue was addressed at the ground level through brainstorming with the casing and disk set section of operators and cell leaders. This brainstorming process also helps in the identification of improvements and suggestions. Table 12.3 illustrates the improvement method and problem-solving recommendations.

Table 12.3 Suggestions for improvements

Operations	Problem identification	Suggestions	Improvements
Casing			
Drilling, finishing, and chamfering	Setup time is higher compared to other operations, so the uptime is low	Need to simplify the design of setup time reduction	Setup time is reduced
Disk set			
Gear hobbing	Handling time is higher	Simply the process of handling by introducing the Kaizen proposal	Handling time is reduced
Cylindrical grinding	Cycle time is high	Simply the process of grinding operation by introducing Kaizen proposal	Cycle time is reduced

12.4.3 *Preferred Method of Setup Time Reduction in Casing Component*

This research suggests a new approach to multi-level VSM mapping based on manufacturing setup time reduction in light of the gaps found in the analysis of the present state map. Applying VSM has the dual goals of highlighting waste causes and finding ways to make them go away. Traditionally, VSM has been used to map the manufacturing flow from raw materials into the hands of the consumer as well as the design flow from concept to launch. It also seeks to improve the efficiency of labor, machinery, and materials. The strategy suggested in this article assists in obtaining a comprehensive understanding of setup activities and analyses the setup's effects on the whole manufacturing process.

The traditional approach to reducing setup time concentrates on setup tasks and makes an effort to shorten setup time. As a result, the preferred multi-level VSM maps all setup operations at the activity and task levels on the timeline. Three goals are being pursued by this technique:

1. To clearly view and assess the internal and external activities of the setup process.
2. Transferring a few internal operations to external ones.
3. Making both more efficient.

12.4.4 *Step-by-Step Procedure*

This section elaborates on the preferred new methodology used in this article. Figure 12.1 includes a summary of the procedure.

Step 1: Develop the multi-level mapping of the current setup state of the casing component.

The process mapping is completed on two levels, as shown in the figure.

Level 1: Mapping of mainstream manufacturing processes: All relevant manufacturing processes of casing and disk set are mapped in the current state.

Level 2: As illustrated in Fig. 12.6, a focused setup stream map of the XYZ machine identifies all actions of the under-study setup.

Mapping the manufacturing processes at level 1 demonstrates the linkages between the setup workstation and the other manufacturing operations. To support the process map at level 1, key data like cycle time, setup time, operational and labor capacity, inventory level, and material requirement planning (MRP) data are required. At level 1, a comprehensive picture of the production processes aids in understanding the impacts of cycle time and setup time. At level 2, mapping the setup activities and their specific tasks aids comprehension of the internal and external setup activities and their related tasks. The duration of the setup operations and tasks is mostly quantified in terms of time. In terms of distinguishing internal and external activities/tasks, moving the internal activities/tasks to external ones, simplifying them, and evaluating their effects, this is very important for any improvement activity.

12.4.4.1 Step Stream Map of XYZ Machine

The setup procedure for the XYZ machine consists of 58 activities, and each activity can perform three different machining operations of turning, drilling, and chamfering one after the other. Setup activities and three different operations are briefly outlined in the appendix.

First, a casing setup stream map is created to show the value-added time, non-value-added time, and overall manufacturing lead time of the XYZ machine. The XYZ machine's current setup state map consists of 58 activities, each of which can perform three distinct machining operations—turning, drilling, and chamfering—one after another. The setup of the XYZ machine is the main topic of this paper, as can be seen in Fig. 12.5. After that, all of the XYZ machine setup's activities are plotted on a timeline Fig. 12.6, where IA, EA, and operator allocations are shown. This paper illustrates the external setup activity data box differentiated with a dashed line to visually distinguish between IA and EA.

12.4.4.2 Kaizen Implementation

LM success will be determined not just by the use of LM tools and procedures, but also by top management involvement and leadership, as well as employee attitudes toward the organization. Following discussions with management and relevant persons in the casing division, the following implementations in casing have been carried out on the shop floor section.

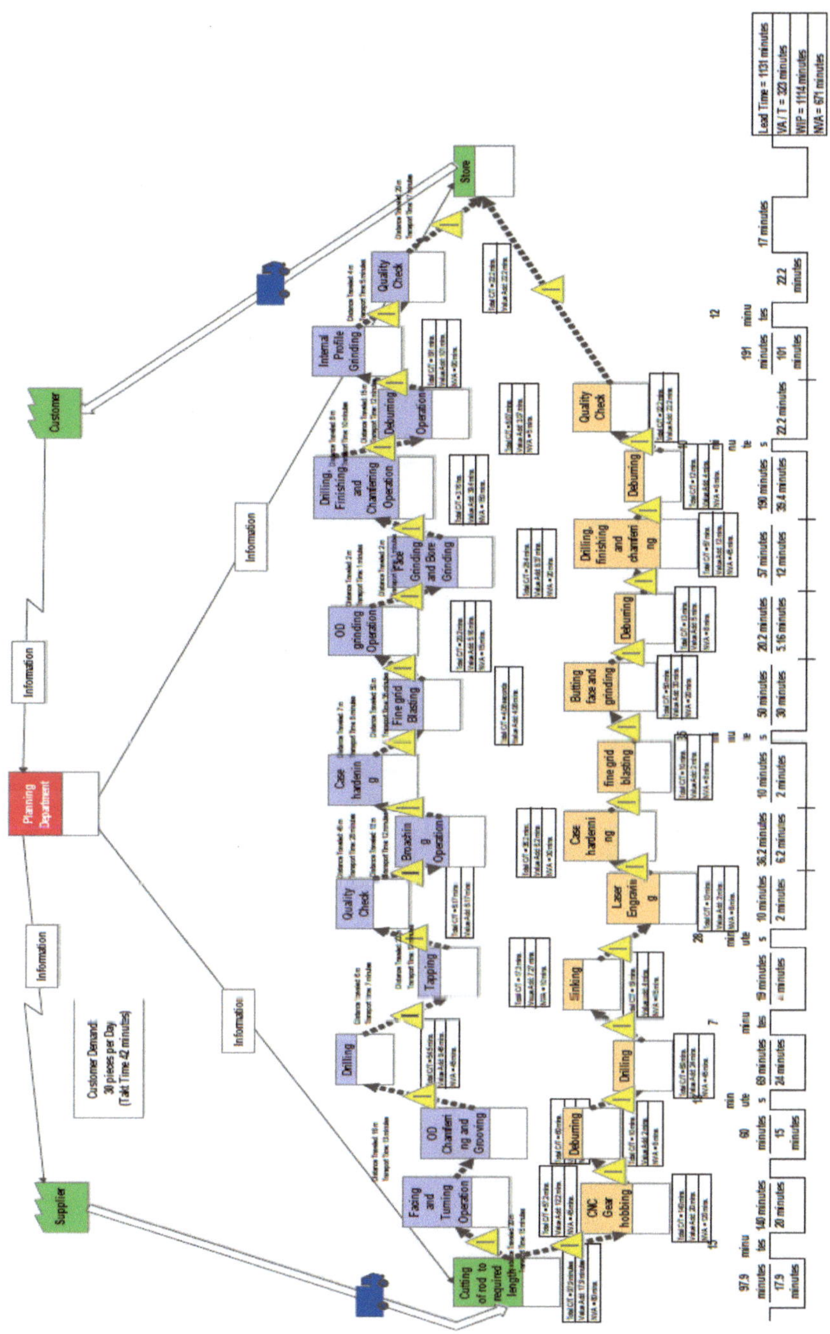

Fig. 12.5 Current state value stream map

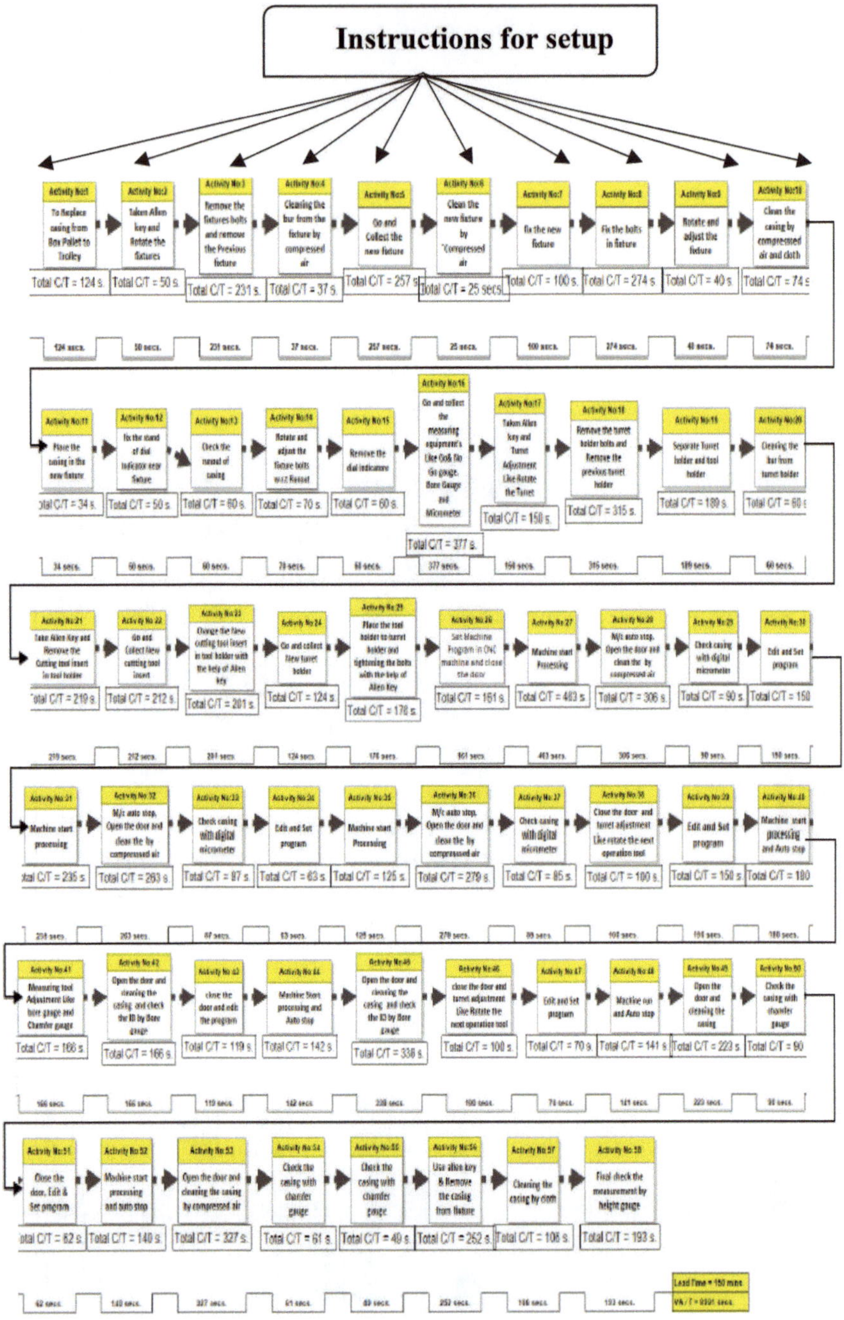

Fig. 12.6 Multi-level current state value stream map

Firstly, the entire setup was videotaped in order to document the current state of a typical research setup. A setup study sheet was produced, which included all of the tasks that had to be completed in order to finish the setup process. According to the current setup process, it will take 8182 s to get from the previous good piece of the previous run to the first good piece of the next run. After analyzing the process setup state map, the three ideas were identified to facilitate the improvement in setup time reduction such as the concurrent execution approach, normalization approach, and water spider approach.

For applying the concurrent execution approach setup process, preplanning is more important and a process that is carried out to increase the number of trained operators and to provide the capacity to complete the setup chores concurrently with the designated activities (activities 1, 6, and 42). Some of the aforementioned activities are parallelized in this regard. Amazingly, using a parallel technique results in a massive reduction in setup time, as seen in Table 12.4.

For applying the normalization approach, the previous job man and machine utilization report a batch of 30 components (not including setup time for the previous job). For every new job, machining starts till the first job OK starts for production, and for the remaining 29 components, the man idleness is high (70–80%). So, now is the time to figure out the next component action. Based on the next component operation, the tool holder and the new cutting tool insert are fixed and ready to changeover the next operation. This period allowed time to figure out the next component activity, such as the preparation work for activities 21 and 23, saving 420 s as seen in Table 12.4.

For implementing the water spider approach, for every new component, the operator goes and collects a new fixture and fixes it in the machine, then collects the tool holder from the shelf and cleaned it with compressed air, then collects the turret holder from another shelf and cleaned it compressed air, and then goes to the store and new cutting tool insert with that the machine to be idle, and this activity will take 799 s. By introducing the water spider concept whose main job is to make sure that materials are supplied to where they are needed, mainly the operator is spared from spending his time on transportation tasks; thus, water spider minimizes transportation waste as seen in Table 12.4.

12.4.4.3 Develop the Multi-Level Mapping of the Future Setup State

Based upon the improvements made in the setup process of the XYZ machine, perform the three distinct machining operations—turning, drilling, and chamfering—one after another. The result illustrates the setup time reduction from 8182 to 6155 s, representing a 32% improvement, and the machine idle time was reduced by 1219 s.

Table 12.4 Reduction in setup time

Activity no	Activity	Time (s)	Before SMED	Justification	Benefits
1	To replace the casing from the box pallet to the trolley	124	It was an internal activity	Converted into external activity	Parallel operation
5	Go and collect the new fixture	257	It was an internal activity	Converted into external activity (Kaizen proposal)	Introducing the water spider concept, the machine idle time is reduced by 799 s
16	Go and collect the measuring equipment: Like-Go and No-Go gauges, bore gauges, and micrometers	377			
22	Go and collect the new cutting tool insert	212			
24	Go and collect the new turret holder	124			
6	Clean the new fixture with compressed air	5	It was an internal activity	Converted into external activity	Parallel operation
21	Take the Allen key and remove the cutting tool insert from the tool holder	219	It was an internal activity	Converted into external activity (Kaizen proposal)	The machine idle time is reduced by 420 s
23	Change the new cutting tool insert in the tool holder with the help of the Allen key	201			
42	Measuring tool adjustments like bore gauge and chamfer gauge	166	It was an internal activity	Converted into external activity	Easy, fast, and parallel operation

12.5 Handling and Cycle Time Reduction in Disk Set Component

Disk set is one of the critical components in the manufacturing industry and performs various machining operations such as CNC gear hobbing, deburring, drilling, sinking, laser engraving, hardening, fine grid blasting, face grinding, finishing, and chamfering. With that, the CNC gear hobbing operations take more handling time compared to other operations. So that to simplify the handling time, the Kaizen proposal was introduced Likewise, the face grinding operations can also be simplified by introducing the new Kaizen proposal.

KAIZEN PROPOSAL SHEET

Location	Internal	Objective	To reduce the cycle time in CNC gear hobbing machine	Date	20/05/2021	Shift	1
BEFORE				AFTER			
				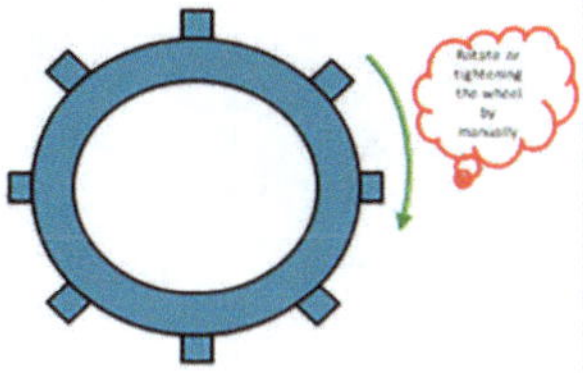			
PROBLEM				EFFECTS			
For Every new component, the operator mounts the workpiece in the fixture by loosening the wheel by using the C-clamp. It takes around 4 min.				**The proposal is to modify the wheel for tightening or loosening the fixture within a minute without using another clamp.**			

Fig. 12.7 Kaizen proposal sheet

CNC Gear Hobbing in Disk Set Manufacturing

During gear hobbing for every new component, the operator mounts the disk set in the fixture by loosening the wheel by using a C-clamp and it takes an average of 4 min. The handling time reduction of CNC gear hobbing in disk set using this proposal is illustrated in Fig. 12.7.

For easy mounting and dismantling of the disk set in the fixture, this proposal may suitable for tightening or loosening the fixture within a minute without using the C-clamp instead of manual tightening or loosening may occur.

Face Grinding Operation in Disk Set Component

In the cylindrical grinding machine, the operator may perform the grinding operation one by one of the disk sets. The tapered mandrel is used for mounting the disk set for grinding operation; changeover time takes an average of 5 min, and the cycle time is 25 min.

The cycle time reduction of face grinding operation in disk set using this proposal is illustrated in Fig. 12.8. In the proposed method, at a time two-disk sets are mounted in the stepped mandrel and the operator performs the grinding operation at the edge of the disk set. The changeover time is reduced from 5 to 2 min.

12.6 Future State Value Stream Mapping

Lean thinking offers principles for developing a future state of the map and an approach to drawing the map. Among these rules are the following: Production rate should be determined based on TAKT time; production schedule should be based on bottleneck operation; try to maintain continuous flow and improve overall process

KAIZEN PROPOSAL SHEET

Location	Internal	Objective	To reduce the cycle time in cylindrical grinding operation	Date	20/05/2021	Shift	1
BEFORE				AFTER			
				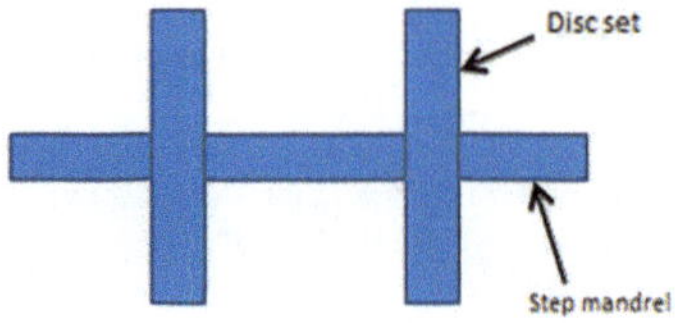			
PROBLEM				EFFECTS			
The conventional cylindrical machine is used for grinding operation at the edge of the disc set. The tapered mandrel is used for mounting the workpiece. The change over time takes 3 min for every new component.				**The proposal by, at a time two-disc sets are mounted in the stepped mandrel, and grinding operation is done at the edge of the disc set. The change over time is reduced by half.**			

Fig. 12.8 Kaizen proposal sheet

improvement through cycle time improvement or changeover time reduction, etc., as shown in (Fig. 12.9).

The future state map has been established by generating numerous enhancement recommendations following the rigorous research of the casing and disk set product. Figure 12.10. shows the suggested future state map. The goal of future state VSM is to meet customer demand while lowering C/T, C/O, manufacturing lead time, and WIP.

The main proposals developed for "casing and disk set" at cycloidal speed reducer are listed below.

1. In the context of the casing component, the three Kaizen ideas like concurrent executive approach, normalization approach, and water spider approach have to be implemented in the XYZ machine to effectively reduce the setup time.
2. In the context of disk set component, for quick mounting and dismantling of the disk set in fixtures at CNC gear hobbing machine, the Kaizen proposal was introduced for handling time reduction. Likewise, an improvement was observed in face grinding operations.

12.7 Result and Discussion

Following the implementation of the lean activities as outlined in future state map, the potential improvements are recognized. The lead time, value-added time, WIP time, and non-value-added time improvements can be compared between the current and future value stream maps, which can also be used to note the changes in these parameters of the production system. To make these changes, a variety of tools are used, including setup time reduction, Kaizen implementation, work study, and time

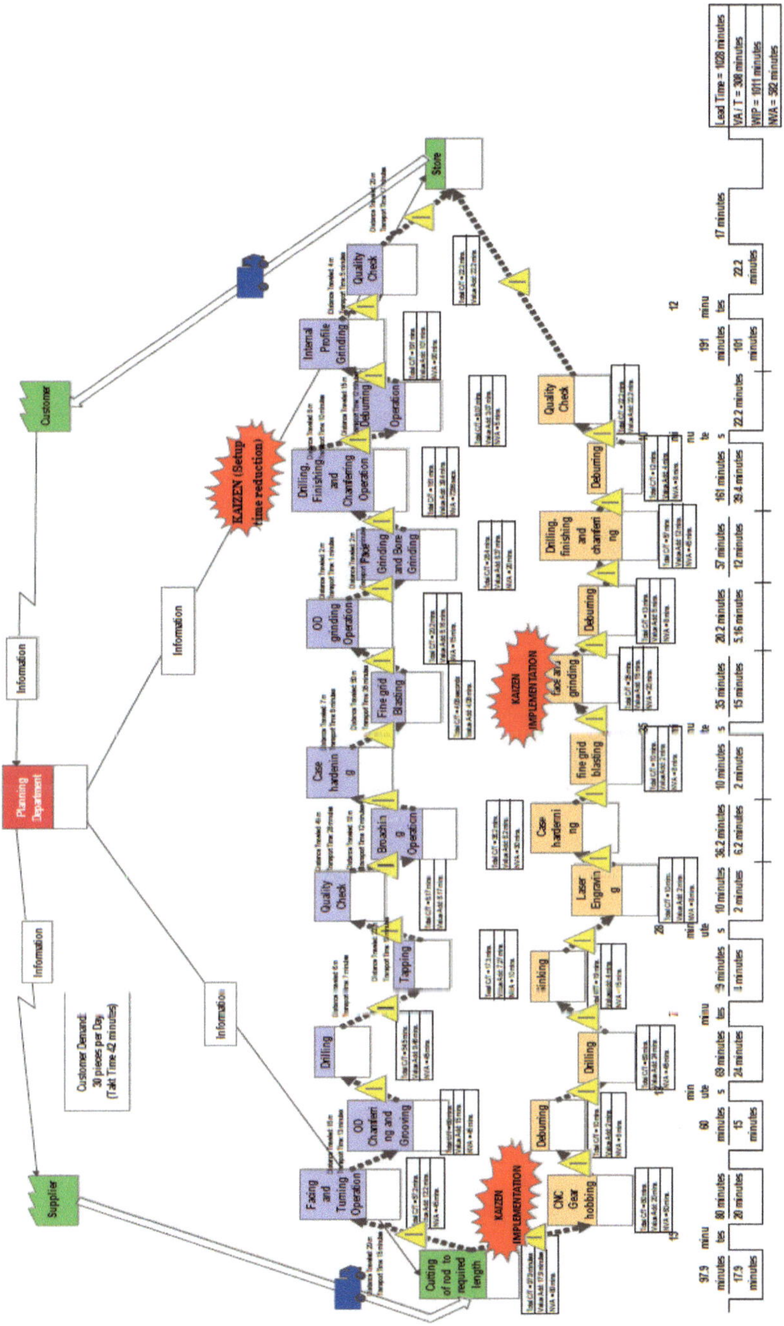

Fig. 12.9 Future state value stream map

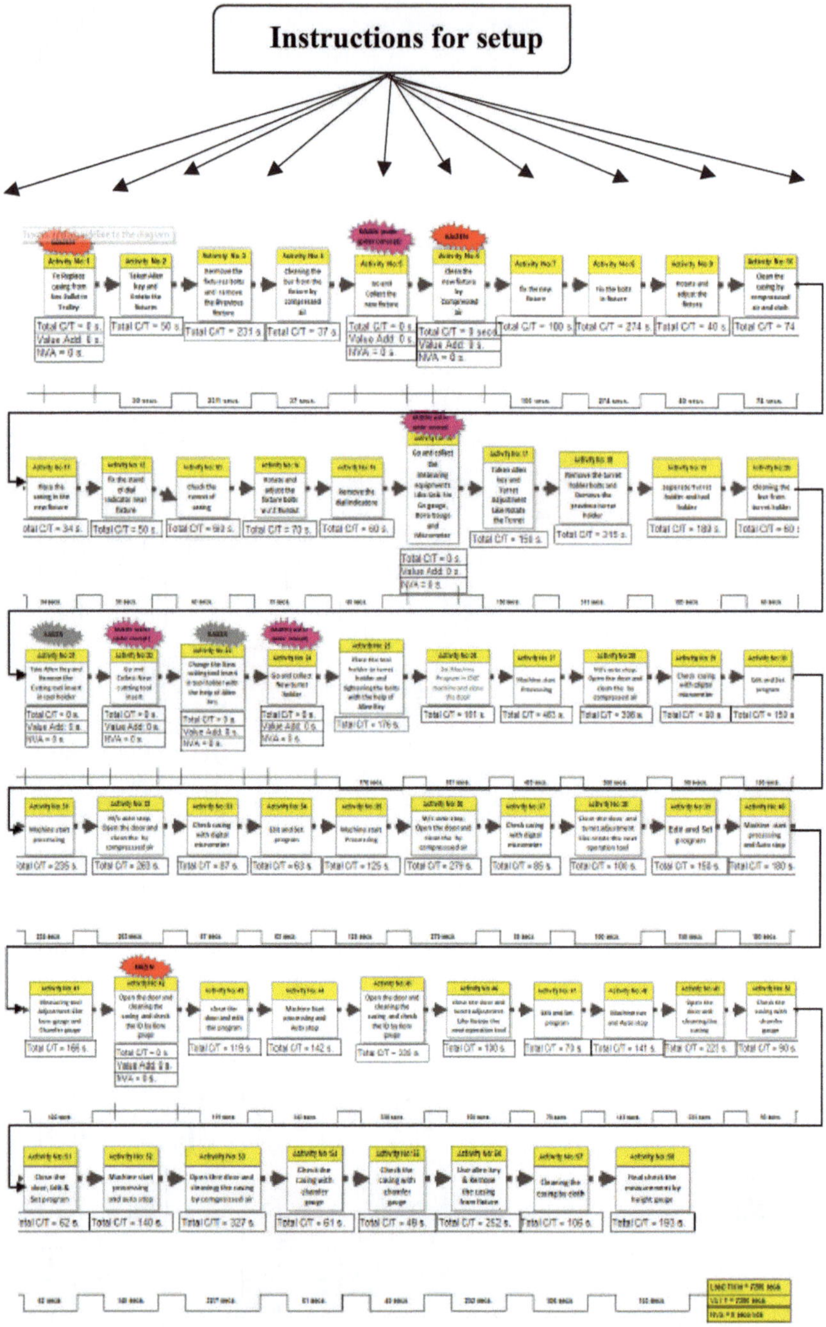

Fig. 12.10 Multi-level future state value stream map

Table 12.5 Evaluating performance by comparing the current and future state map

Description	Current state map	Future state map	Improvements (%)
Production lead time (min)	1131	1028	9.1
VA time (min)	323	308	4.6
WIP time (min)	1114	1011	9.2
NVA time (min)	671	582	13.2

study strategies for locating waste. Thus, these tools contributed to the elimination of non-value-added operations and the reduction of waste, which helped to improve production efficiency (Table 12.5).

The following details are included in future state map:

1. Lead time is reduced from 1131 to 1028 min.
2. Value-added time is reduced from 323 to 308 min.
3. Work in progress time is reduced from 1114 to 1011 min.
4. Non-value-added time is reduced from 671 to 582 min.

12.8 Conclusion and Theoretical Implications

Modern industrial companies face an onslaught of new competition. As a result, modern businesses have been pushed to change their manufacturing patterns from mass production to lean production. Lean manufacturing focuses on reducing waste and reconfiguring manufacturing processes to achieve shorter lead times and lower costs. VSM, which focuses on the micro-analysis of manufacturing processes, is one of the most essential lean manufacturing strategies. The results are illustrated as the overall production lead time is 9.1%, improvements are made, and the non-value-added time is 13.2%. This case study was conducted in this context for the development of VSM for the gear manufacturing business. At every stage of the procedure, the improvement is measured and documented. Due to its ability to adjust to variations in demand, the suggested approach has shortened lead times and streamlined the manufacturing process. As the suggested strategy incorporates lean principles and sustainability criteria, it offers more adaptability and flexibility, paving the way for the organization's continued success. Supervisors must ensure that the senior leadership group is fully supportive of and aligned with the VSM project during its implementation. Above all, decision-makers, from operators to upper management, coordinate. This study shows how the LM tool VSM can be used practically in a manufacturing unit. For brand-new researchers in this area, this will be helpful.

12.8.1 Limitations and Future Research Direction

This study clearly demonstrates that LM is one of the long-term manufacturing philosophies that can require time and money to yield the desired benefits for any organization. Furthermore, one of the key limitations of VSM as a static tool is that it frequently concentrates on a single product line in manufacturing. Additionally, the study's findings indicate that while VSM can identify various lean wastes, it is extremely complex and challenging to capture every event that occurs during the manufacturing of a cycloidal speed reducer. This study specifically focuses on and analyzes one of the bottleneck components in cycloidal speed reducer manufacturing. One of the evaluations that can be conducted in future is the level of maturity assessment for lean manufacturing. Such evaluations can help find the current level of process in SWOT (strengths, weaknesses, opportunities, and threats). In addition to one of the future studies, researchers can do multi-level value stream mapping for different products in various industrial manufacturing organizations.

References

1. Black†, J.: Design rules for implementing the Toyota production system. Int. J. Prod. Res. **45**(16), 3639–3664 (2007)
2. Hines, P., Rich, N., Bicheno, J., Brunt, D.: Value stream management. Int. J. Logist. Manag. **9**(1), 25–42 (1998)
3. Hines, P., Rich, N.: The seven value stream mapping tools. Int. J. Oper. Prod. Manag. **17**(1), 46–64 (1997)
4. Manupati, V.K., Varela, M., Gangal, C., Mobin, M.: Application of value stream mapping for cycle time reduction in production of link and roller assembly. Int. J. Serv. Operat. Manage. **33**(2), 135–158 (2019)
5. Singh, J., Gandhi, S.K., Singh, H.: Assessing lean thinking using value stream mapping in a manufacturing unit—a case study. Int. J. Bus. Excell. **17**(3), 306–335 (2019)
6. Sharma, V., Virmani, N.: Development of lean production system using value stream mapping approach: a case study. Int. J. Prod. Qual. Manag. **30**(2), 168–185 (2020)
7. Vinodh, S., Arvind, K., Somanaathan, M.: Application of value stream mapping in an Indian camshaft manufacturing organisation. J. Manuf. Technol. Manag. **21**(7), 888–900 (2010)
8. Gurumurthy, A., Kodali, R.: Design of lean manufacturing systems using value stream mapping with simulation. J. Manuf. Technol. Manag. **22**(4), 444–473 (2011)
9. Vamsi Krishna Jasti, N., Sharma, A.: Lean manufacturing implementation using value stream mapping as a tool. Int. J. Lean Six Sigma, **5**(1), 89–116 (2014)
10. Mishra AK, Sharma A, Sachdeo M, K.J.: Development of sustainable value stream mapping (SVSM) for unit part manufacturing. Int. J. Lean Six Sigma, **11**(3), 493–514 (2019)
11. Vinodh, S., Selvaraj, T., Chintha, S.K., KEK, V.: Development of value stream map for an Indian automotive components manufacturing organization. J. Eng. Des. Technol. **13**(3), 380–399 (2015)
12. Abideen, A., Mohamad, F.B.: Improving the performance of a Malaysian pharmaceutical warehouse supply chain by integrating value stream mapping and discrete event simulation. J. Model. Manag. **16**(1), 70–102 (2020)

13. Ebrahimi, A., Khakpour, R., Saghiri, S.: Sustainable setup stream mapping (3SM): a systematic approach to lean sustainable manufacturing. Prod. Plann. Cont. 1–19 (2021)
14. Vinodh, S., Somanaathan, M., Arvind, K.: Development of value stream map for achieving leanness in a manufacturing organization. J. Eng. Des. Technol. **11**(2), 129–141 (2013)

Chapter 13
Convergent Technologies Empowering Human–Computer Interaction in Programmable Logic Controllers for an Affordable Industry 4.0 Integration

Saptarshi Das, S. Roychowdhury, and S. Ghosh

Abstract In this article, we introduce an innovative solution for improving human–computer interaction (HCI) in programmable logic controllers (PLCs). By integrating Python, web APIs, sockets, and the pymodbus library, we've developed a web-based Human–Machine Interface (HMI) tailored for industrial controllers. The system employs web sockets for real-time connectivity between the HMI and PLC, enabling swift data exchange. Additionally, it incorporates REST web services supporting JSON messages, enhancing communication efficiency. The approach is compatible with various SCADA network protocols, providing a versatile solution for robust controller-HMI connections. The primary goal is to offer a cost-effective and customizable technology, promoting the adoption of Industry 4.0 principles among Small and Medium-sized Enterprises (SMEs). Illustrated through a case study in a medium-sized beverage bottling plant, the solution automates Cleaning in Place (CIP) processes. It provides enhanced reliability and security, addressing critical aspects of industrial operations. Unlike desktop-based SCADA software, our web-based solution transforms a PC into a server. It allows platform-independent access to any number of users via browsers on devices like laptops, Android mobiles, and tablets. The technology's distinctive feature lies in its ability to facilitate creative and multi-layered monitoring systems, leveraging the Modbus communication protocol in PLCs. In summary, this article presents a comprehensive HCI approach in PLCs, integrating Python, web APIs, sockets, and pymodbus. The solution not only tackles real-time data exchange challenges but also emerges as a cost-effective, versatile tool applicable to Industry 4.0 adoption, cybersecurity testing, high-speed simulations,

S. Das (✉) · S. Roychowdhury
Department of Industrial and Systems Engineering, Indian Institute of Technology, Kharagpur, Kharagpur, India
e-mail: saptarshi.das@iitkgp.ac.in

S. Ghosh
Department of Instrumentation and Control, Calcutta University, Calcutta, India

J. Maiti et al. (eds.), *Recent Advances in Industrial and Systems Engineering*, Lecture Notes on Multidisciplinary Industrial Engineering,
https://doi.org/10.1007/978-981-96-8323-9_13

and animated visualizations of mechatronic systems. Its potential to reshape digitization services by enhancing efficiency, flexibility, and accessibility for SMEs is evident.

Keywords HMI · PLC · Pymodbus · Digitization · REST-API · SME · Industry 4.0

13.1 Introduction

The realm of HCI within the domain of PLC has garnered significant attention in recent times. The increasing popularity of web-based HMI can be attributed to their adaptable nature and user-friendly design. This conference article introduces an innovative approach to HCI in PLCs by harnessing Python, web API, socket, and pymodbus. The primary goal of this proposed system is to provide a web-based HMI tailored for industrial controllers, particularly within the context of single-page applications. Proprietary SCADA software is primarily desktop-based and entails substantial annual license fees. This solution, however, is not confined to a single system and offers a more flexible user access. Traditional systems lack the flexibility to implement variable user role-based control for read and write access. Furthermore, the integration of an SMS/WhatsApp business gateway for sending critical alerts is often restricted in such systems. While some newer versions incorporate browser-based access, their potential for additional feature enhancements, such as periodic data cloning into a cloud or a redundant server, remains limited. A web-based IoT framework for SCADA preparation provides a cost-effective solution for SMEs to embrace digitization. It additionally enables the utilization of various external APIs and triggers to enhance the scope of monitoring and control. This, in turn, is used to implement workflow automation by integrating email services, chatbots, statistical tools, project management, customer relationship management, payment gateways, and a lot more. In this paper, we have detailed the system overview in Sect. 13.2. Section 13.3 details its implementation, Sect. 13.4 presents the practical results and the conclusion is presented in Sect. 13.5.

13.1.1 Literature Review

In this section, we have discussed relevant literature that explores similar concepts in various systems. The article focusing on smart building applications details the utilization of Modbus protocol and HTTP REST interface in building management systems (BMS) [1]. The design of an open communication protocol for the industrial internet of things (IIoT) also showcases the application of message queuing telemetry transport (MQTT), REST, and MTconnect [2]. The research on a water distribution testbed clearly outlines the use of Modbus serial communication to retrieve measurement data from analyzers [3]. This application employs the pymodbus library through a Python script and RESTful web services to transmit the extracted data to a visualization platform. In the article on SCADA cybersecurity awareness, the authors demonstrate the use of pymodbus to transmit a malicious Modbus TCP/IP frame, manipulating the register of a PLC with a false value [4]. Furthermore, in the article on IoT machine-to-machine architecture, pymodbus is employed to connect IoT sensors and actuators to a BMS [5]. In their work on the implementation of a low-cost SCADA system, the authors have used an Arduino UNO board to accumulate sensor data [6]. Modbus protocol is programmed in the board, which seamlessly exchanges the data with the SCADA software. The article on low-cost SCADA mentions that the real-time monitoring of a PLC-based machine is confined within the factory premises [7]. It has further presented a case study of an IIoT-based web framework to connect a PLC-supported machine to a cloud-based network. Another work on open-source SCADA for industrial process monitoring has used an OPC-UA server to host the application. Different Python libraries like pymodbus, pycan, and pycomm have been used to allow incoming communication from RS-485, CAN, and RS-232, respectively.

13.2 System Description

Industrial communication encompasses a mix of fieldbus systems, Ethernet-based approaches, and certain wireless solutions. Currently, Real-Time Ethernet (RTE) has gained widespread adoption as a standard in the field of industrial automation. However, it's worth noting that there is an absence of a single standard, but rather various mutually incompatible implementations. Key requirements for effective industrial automation include reduced latencies, determinism, independence from the physical layer, fault tolerance without the need for additional hardware, enhanced security, and interoperability among solutions from different manufacturers. The challenge of incompatibility in industrial networking has been extensively discussed in the existing literature, presenting a range of potential solutions [8, 9]. Among these solutions, OPC Unified Architecture (OPC-UA) has emerged as a notable technology for facilitating machine-to-machine communication in industrial automation. OPC specifies the communication of real-time plant information across control devices from different manufacturers. In this study, we have used a FreeOpcUA project to devise a versatile solution for ensuring communication compatibility within the SCADA system. It is equipped with several features, including user management through accounts, real-time write/read capabilities for multiple variables with a maximum latency of 10 ms, scalability to support various communication protocols (such as CAN, Modbus, RS-232, RS-485, and RTE), and database connectivity through MySQL. To enhance flexibility, additional communication protocols can be seamlessly integrated into the OPC server by incorporating available Python libraries. The proposed SCADA system serves as an OPC client, allowing access to the data

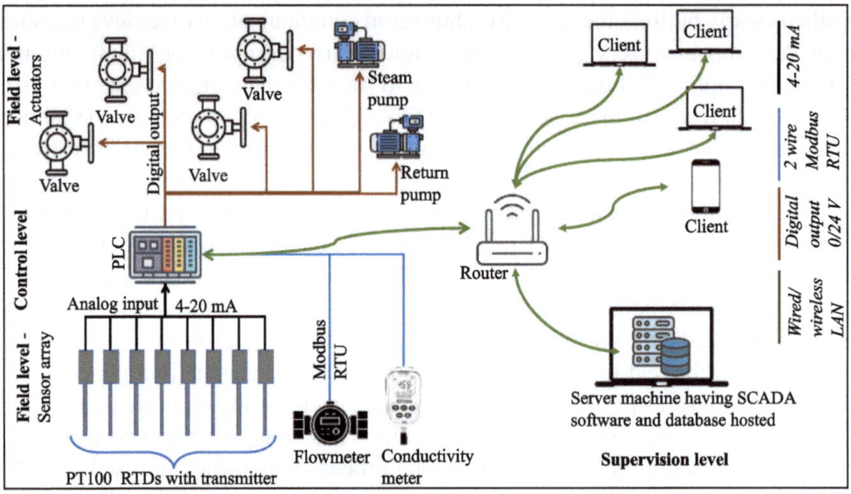

Fig. 13.1 System architecture

stored in the OPC-UA server. This communication capability effectively addresses the interaction between layers two and three of the CIM reference model. We have presented an overview of the system in Fig. 13.1, emphasizing its reliance on open-source platforms. The sensor array of the CIP system, where we have implemented the solution, contains PT-100 type RTDs, electromagnetic flowmeters, a conductivity meter, and a level sensor. The actuation includes multiple butterfly valves with a pneumatic actuator. These are operated by pneumatic solenoid valves, which are triggered by the digital output of the PLC. We have used pymodbus for the data exchange between the PLC and the server PC. The web framework has a react-based front end along with a PHP-based backend. There are two tables inside the MySQL database one of which is dedicated to user credentials and read–write permissions while the other is for logged data. Other system parameters, like logging interval and duration, are chosen from user inputs and stored in a JSON file.

13.3 System Implementation

The computer-integrated manufacturing (CIM) reference model categorizes a process into five key levels:

1. Field level
2. Control level
3. Supervisory level
4. Manufacturing execution system (MES)
5. Enterprise resource planning (ERP)

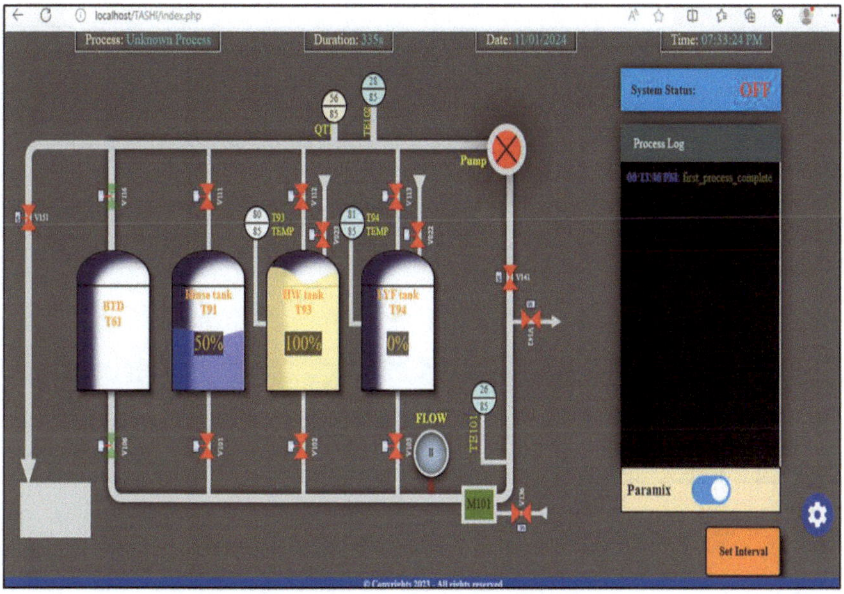

Fig. 13.2 Visual representation of the web-based SCADA software

In this paper, we have presented an architectural framework designed to facilitate interaction among the initial three levels. The field-level instruments, comprising the sensor array and actuators, establish connections through analog input (4–20 mA), Modbus communication input, or digital output of the PLC. The PLC functions as the control layer, housing the automation program in sequential ladder logic. The supervision layer involves the server machine and client devices, responsible for system visualization. The PLC is linked to the server via Modbus TCP/IP over the local area network (LAN) using a router. Client device visibility is expanded through HTTP request and response. Figure 13.2 provides a visual representation of the SCADA software, while Fig. 13.3 displays the logged data, which can be exported to Excel. Figure 13.4 includes a code snippet demonstrating the use of the pymodbus library to read data from the PLC. Additionally, Fig. 13.5 showcases the control page for executing programmed CIP operations or manually actuating individual valves and pumps (or motors). The code snippet in Fig. 13.6 illustrates how control page button presses are translated into write requests to the PLC using Modbus.

13.4 System Implementation

We have presented the evaluation of the system components as well as the different functionalities of the web-based SCADA/HMI.

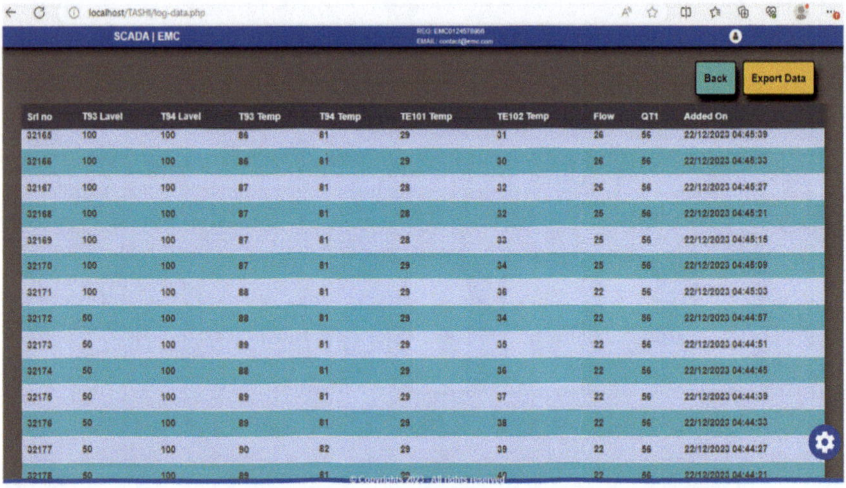

Fig. 13.3 Logged data with a time stamp and ready to be exported in a spreadsheet

Fig. 13.4 Code snippet representing the import of data from the PLC using its IP

```
import os
from pymodbus.client import ModbusTcpClient
import requests
import random
import time

# Constants for MODBUS connection
PLC_IP = '192.168.1.50'  # PLC's IP address
PLC_PORT = 502

# Function to read a holding register value
def read_holding_register(client, address):
    try:
        # Read a single holding register value
        response    =    client.read_holding_regis-
ters(address, 1, unit=1)
....
```

13.4.1 System Operation

Figure 13.1 provides a detailed overview of the system framework. The entire structure, ranging from the field layer to the supervision layer, is highly modular. We have prioritized ease of troubleshooting at every stage, enabling quick communication re-establishment in case of errors. Interoperability has been a key focus in developing this solution. The Python program's API communicates with the SCADA software through a GET request, receiving a JSON-structured response. This JSON response is adaptable for integration into any third-party system if data importation is needed for other applications. Data from the PLC is imported at a rapid interval of 500 ms, allowing users the flexibility to log values at any interval above 0.5 s, considering the criticality of the operation. The CIP automation involves multiple steps, and it's crucial for the SCADA system to identify the process name. We achieve this

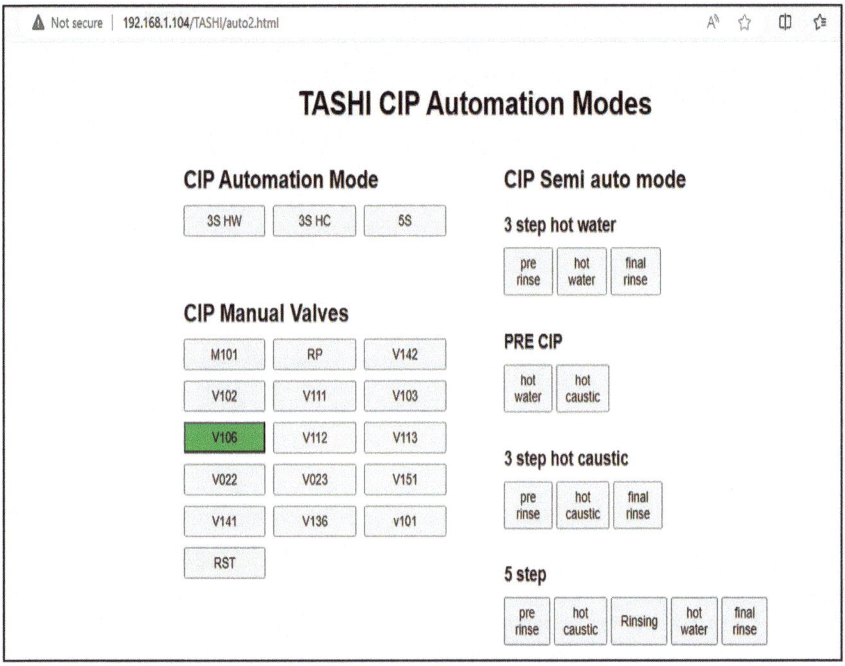

Fig. 13.5 Overview of the SCADA control page

by mapping different read coil memories of the PLC to the process name variable. As the Python program reads PLC memories, it discerns which associated process memory is HIGH (or 1), pointing to the corresponding process name in its dictionary. The concept is elaborated in detail in the code snippet in Fig. 13.7

13.4.2 Remote Access

The system not only provides access to devices within the same LAN network for monitoring and control but also extends its reach for remote operations through the integration of the ZeroTier application. This innovative approach eliminates the necessity to host the SCADA application on an external server. In addition, it mitigates concerns related to potential disruptions in internet connectivity, safeguarding the continuity of SCADA data logging. By leveraging the ZeroTier application, a virtual LAN layer is established across the WAN. This means that a client device, regardless of its location worldwide, can seamlessly access the software through the same ZeroTier account, essentially becoming part of the same network. This strategic use of ZeroTier enhances the system's flexibility and accessibility for remote monitoring and control.

Fig. 13.6 Code snippet representing how control page button presses are translated to write requests to the PLC

```
...

app = Flask(__name__)
CORS(app)

# Modbus TCP connection configuration
modbus_host = '192.168.1.50'  # Modbus TCP host -- PLC
modbus_port = 502  # Modbus TCP port

client = ModbusClient(modbus_host, port=modbus_port)

# Initialize a dictionary to store the state of each
button
button_states = {}

@app.route('/button_press', methods=['POST'])
def handle_button_press():
    data = request.get_json()
    button_number = data.get('value')
    button_state = data.get('state')
    print(f"Received button number: {button_number},
state: {button_state}")
...
....
......

else:
# Determine the coil value based on the button state
                coil_value = True if button_state ==
'ON' else False

# Write the coil value to the specified coil address
(button number)
                client.write_coil(button_number,
coil_value)

.........

            return jsonify({"message": f"Button press
received.   Setting   coil   {button_number}   to
{coil_value}."})
```

The SCADA software's performance is evaluated based on the features outlined in Table 13.1.

```
# Define the process titles and associated coil addresses
    process_title_coil_addresses = {
            "HW Pre Rinse": 2173,
            "Hot Water": 2174,
            "Hot Water Final Rinse": 2175,
            "Pre CIP Hot Water": 2176,
            "Pre CIP Hot Caustic": 2177,
            "Hot Caustic Pre Rinse": 2178,
            "Hot Caustic": 2179,
            ...
             ...

        }
....
# Determine the process title based on the coil values
        process_title = "Unknown Process"
       for title, address in process_title_coil_addresses.items():
            if read_coil(client, address):
                process_title = title
                break
```

Fig. 13.7 Code snippet depicting memory mapping for process name identification

Table 13.1 Overview of the system's features

Feature	Result
Visualization of process variable	Satisfied
Data storage and logging at preset interval & duration	Satisfied
Alarm notification	Not required
Actuation control	Satisfied
Export of logged data in simple spreadsheet	Satisfied
Globalization of visibility and access control	Satisfied
Platform-independent access from device with any operating system	Satisfied

13.5 Conclusion

High-level automation protocols or middleware, such as OPC-UA, prove essential for addressing automation needs in terms of functionality and data models. These protocols, grounded in widely adopted Internet technologies for communication services, offer an efficient and cost-effective solution. This cost-effectiveness is particularly advantageous for Small and Medium Enterprises (SMEs), where budget constraints often play a crucial role in technology adoption. Furthermore, the integration of internet technologies like web services or modern IoT-specific protocols with the supervision layer, in conjunction with OPC-UA communication linking the control and field levels, contributes to increased system performance. The versatility of the system's OPC-UA server in managing various communication protocols enhances cost-effectiveness, making it an ideal choice for integrating existing processes. This adaptability is crucial for SMEs looking to optimize their operations without incurring excessive expenses. Additionally, the incorporation of different APIs facilitates seamless data exchange within the system. The ability to integrate with diverse APIs ensures smooth interoperability with various software and devices, enhancing overall operational efficiency. This feature is particularly beneficial for SMEs with existing systems, as it streamlines the integration process and minimizes potential disruptions during the implementation of the CIM reference model architecture. Moreover, the system's capability to provide insights and critical alerts is invaluable. MODBUS-based access has some security concerns in terms of cyberattacks [4]. To mitigate these risks, the implementation of an intermediate microcontroller between the server and the PLC has been proposed. This microcontroller is designed to receive commands from the server through MQTT and subsequently relay these commands to the PLC via MODBUS.

This way, the direct MODBUS connection between the server and the PLC can be avoided, thereby reducing the exposure to vulnerability. Real-time insights enable proactive decision-making, while critical alerts ensure timely responses to potential issues, minimizing downtime and optimizing resource utilization. This aspect is crucial for SMEs seeking to maintain operational continuity and competitiveness in their respective industries. In conclusion, the combined features of cost-effectiveness, seamless integration with different APIs, and the provision of insights and critical alerts make the described system a robust and beneficial solution, particularly tailored for the needs of SMEs aiming to enhance their automation processes within a CIM reference model framework.

References

1. Jayasankar, M., Giridharan, A.: Specification of REST API services for modbus protocol using formal technique. EAI Endorsed Trans. Cloud Syst. **3**(10) (2017)
2. Hsiao, C.-H., Lee, W.-P.: OPIIoT: design and implementation of an open communication protocol platform for industrial internet of things. Internet Things **16**, 100441 (2021)
3. Adepu, S., Mishra, G., Mathur, A.: Access control in water distribution networks: a case study. In: 2017 IEEE International Conference on Software Quality, Reliability and Security (QRS), pp. 184–191 (2017)
4. Puys, M., Thevenon, P.-H., Mocanu, S., Gallissot, M., Sivelle, C.: SCADA cybersecurity awareness and teaching with hardware-in-the-loop platforms. J. Wirel. Mob. Netw. Ubiquit. Comput. Depend. Appl. **13**(1), 4–32 (2022)
5. Babayigit, B., Sattuf, H.: An IIoT and web-based low-cost SCADA system for industrial automation. In: 2019 11th International Conference on Electrical and Electronics Engineering (ELECO), pp. 890–894 (2019)
6. Allafi, I., Iqbal, T.: Low-cost SCADA system using arduino and reliance SCADA for a standalone photovoltaic system. J. Sol. Energy **2018**, 1–8 (2018)
7. Merchan, D.F., Peralta, J.A., Vazquez-Rodas, A., Minchala, L.I., Astudillo-Salinas, D.: Open source SCADA system for advanced monitoring of industrial processes. In: 2017 International Conference on Information Systems and Computer Science (INCISCOS), pp. 160–165 (2017)

8. Abbas, H.A., Mohamed, A.M.: Review on the design of web based SCADA systems based on OPC da protocol (2015). arXiv preprint arXiv:1506.05069
9. Guo, Y., Cheng, C., Xin, X., Xie, Y.: OPC communication protection method based on access control and anomaly traffic detection. In: 2016 IEEE International Conference on Internet of Things (iThings) and IEEE Green Computing and Communications (GreenCom) and IEEE Cyber, Physical and Social Computing (CPSCom) and IEEE Smart Data (SmartData), pp. 732–737 (2016)

Chapter 14
A Machine Tool Digital Twin to Implement Smartness in Discrete Production Environment

Soumen Paul, Nayan Manna, Supriyo Roy, and Tarun Kanti Jana

Abstract Rapid digitalization of manufacturing activities envisioned in Industry 4.0 is realized through Smart Manufacturing (SM) and Cyber Physical Production Systems (CPPS) to cope with the ever-increasing business challenges. The integration of physical systems and processes with the virtual world is accomplished by Digital Twin (DT), which acts as the key enabler to realize data-driven manufacturing. A DT is an abstraction of physical world into cyber space for shop-floor data collection, analysis and improved decision making such as simulation, optimization, prognostic health monitoring (PHM) and prediction for exercising better control of processes to enhance the overall performance of plant and machinery. In this paper, basic building blocks and functionalities of a DT are presented. Finally, a Machine Tool Digital Twin (MTDT) is proposed together with a CPPS, which is capable of improved decision making through optimization, simulation, and prediction of various Performance Indicators of metal cutting processes and machine tools.

Keywords Cyber physical production systems · Industry 4.0 · Machine tool digital twin · Machine tool 4.0 · Smart manufacturing

14.1 Introduction

Since the beginning of civilization to the current age, the industries have undergone various technological developments and reforms starting from the water and steam power generations, exploration of natural resources and development of machine

S. Paul
Department of Information Technology, Haldia Institute of Technology, Haldia, India

N. Manna
Department of Electrical Engineering, Haldia Institute of Technology, Haldia, India

S. Roy · T. K. Jana (✉)
Department of Mechanical Engineering, Haldia Institute of Technology, Haldia, India
e-mail: tarun.jana2000@gmail.com

J. Maiti et al. (eds.), *Recent Advances in Industrial and Systems Engineering*, Lecture Notes on Multidisciplinary Industrial Engineering,
https://doi.org/10.1007/978-981-96-8323-9_14

tools to the present day's smart products and services to satisfy the rising industrial demands and to improve the quality of life, as captured through the different industrial revolutions [16, 33, 46]. Accordingly, the manufacturing industries have also evolved through different phases, each one having its unique focus, characteristics, related themes, and thrust areas [49]. The entire transition of manufacturing paradigm is enabled by three distinct technologies: (i) electro-mechanical and mechanical hardware based automated systems (from 1900–1970) compatible to mass production, (ii) intelligent and mechatronic systems based on programmable automation (from 1970–2010) cater to batch manufacturing, and finally, (iii) data-intensive ICT and web-based autonomous systems (2010 onwards) to address the present day's growing requirements of mass customization and mass personalization.

On the heels of ever-increasing business challenges across the globe in the 21st Century, the industry is now experiencing the wave of fourth industrial revolution, Industry 4.0, to bring about radical changes in controlling and improving the activities and operations of manufacturing industries to address the critical challenges of mass customization [28, 40, 51]. Industry 4.0 underpins the use of advanced automation such as sensor technology, RFID, embedded systems, IoT devices in physical equipment coupled with ubiquitous computing in cyber space to realize massive digitalization in manufacturing to facilitate improved decision making and better control the process in an unprecedented manner. The state-of-the-art facilities and synergistic integration between physical shop-floor and its virtual counterpart through dynamic data and information sharing lead to the development of smart manufacturing (SM) [53, 55] and cyber physical production systems (CPPS) [18, 26, 41], which are iconic representations of Industry 4.0.

Smart manufacturing is a data-driven manufacturing paradigm which relies on real-time data capturing, modelling, and processing for continuous control of the various activities of manufacturing ecosystems through smart products and activities [23, 34]. Although a lot of commonalities exists between earlier intelligent manufacturing and present-day's smart manufacturing, the major distinguishing feature between these two is that the former predominantly utilizes mechatronics systems based on intelligent programmable controllers, whereas the later focuses more on real-time data capturing, processing, and dynamic decision making, and finally, acts on it.

A CPPS emphasizes on integration aspects between physical world and the cyber world. It is characterized by smartness (autonomous), connectedness, and responsiveness and manifests "Self-X" capabilities, e.g., self-awareness, self-prediction, self-optimize, self-organize, etc. [24, 54]. Both SM and CPPS are intertwined with the vision of Industry 4.0, and quite often, these are used interchangeably. The integration of physical world and the virtual world is accomplished through communication, computing, and control (3C) [21] that helps realize and fulfill the underlying philosophies of SM and CPPS.

A digital twin is a mirror image or digital abstraction of physical shop-floor and products, which relies on real-time data collection to carry out improved decision making such as simulation, prediction, optimization, etc. using varieties of software tools in cyber space such that meaningful information resulted from various analysis are used to better regulate the process in the physical world to yield better output and exhibit improved performances [4]. A DT is the embodiment of digital manufacturing and acts as main driver of SM and CPPS. The data is collected by various sensors and IoT devices from product and equipment in the physical shop-floor, which is then transferred to digital world through communication network for analysis. Useful information generated by processing raw data using varieties of techniques such as Artificial Intelligence and Machine Learning (AI & ML), big data analytics (BDA), Cloud Computing (CC), edge computing, etc. for the purpose of modelling, simulation, optimization, prediction in the digital space leads to actionable conclusions such as design validation, improved process control, health monitoring of machine critical components (PHM), etc. in the physical world. The twinning is, therefore, a cyclic process connecting the physical world with the cyber world through data transfer and cyber to physical world through action-oriented information propagation in such a way, so that virtual copy becomes the mirror image of the physical world and vice versa. Both physical equipment and its virtual part co-evolve dynamically throughout their entire lifecycle in synchronization with each other.

Over the last couple of years, with the advent of emergent problem solving techniques spearheaded by digitalization, the DT has gained added momentum and attracts the attention of both academia and industries. However, widespread implementation in actual factory is still an open and challenging issue considering varieties of constraints: (i) The difficulties associated with the seamless data and information transfer considering wide disparate sources, devices, and platforms such as sensors, CNC machine tool controller, Gateways like MTConnect, etc. are yet to be resolved to realize DT in its true perspective and in cost-effective manner. (ii) The most important aspect of DT is perhaps computing and decision making, which has hitherto not been captured adequately by the published articles. There is a lack of elaboration regarding the use of exact technique(s), enabling technologies, and appropriate software tools for analysis, simulation, prediction, etc. in different problem contexts, considering relevant operating variables/parameters vis-à-vis responses/objectives.

The present work, therefore proposes a generalized and systematic approach to implement DT functionalities and required technologies for improved decision making in machine tool systems in smart manufacturing environment. The work also develops a CPPS comprising of machine tool systems and a Machine Tool Digital Twin (MTDT) for execution of higher level decision making in metal cutting environment.

The rest of the paper is arranged as follows. Section 14.2 discusses the DT origin, taxonomy, characteristics, and related work. The DT components, functional requirements, and associated technologies and platforms are presented in Sect. 14.3. The proposed CPPS and MTDT for improved decision making in metal cutting are covered in Sect. 14.4. Finally, a short discussion and concluding remarks are provided in Sect. 14.5 to complete the article.

14.2 Origin of DT, Taxonomy, Characteristics, and Related Work

This section presents the origin of DT and review of the subsequent DT definitions by a number of researchers, along with the taxonomy, major characteristics, and hierarchy of DT. Such an endeavour is extremely important to understand the DT framework and it facilitates identifying the entities present in a DT along with their role. Further, it provides the roadmap to model a DT to fulfil its intended objectives and functionalities.

14.2.1 DT Origin and Definitions

DT is the technology to implement digitalization in manufacturing through autonomous behavior of the system entities [35]. A DT is the virtual entity, which is instrumental in digital manufacturing and acts as a key enabler of SM and CPPS [14, 43]. The early concept of DT dates back to 2003 and credited to Michael Grieves [10], who conceived the idea of virtual representation of a physical product in a very humble and modest way by collecting data manually using paper-based system in the context of Product Life-Cycle Management (PLM). According to Grieves, three distinct activities (use-case) namely conceptualization, comparison, and collaboration are the foundation of the problem solving and innovation in future manufacturing. The concept remained in its infancy over long period due to lack of technological developments, which could be befitting for implementation in its true spirit. The first major breakthrough in DT application was reported by National Aeronautics and Space Administration (NASA) and the US Air Force in 2012 for health maintenance and predicting the residual life of aerospace vehicles [9]. With the evolution of information and communication technology (ICT) in the recent past, the concept of DT has matured enough and its scope has flourished by leaps and bounds so as to embrace it as a potential platform in realizing SM and CPPS in the context of Industry 4.0 [7, 22]. The number of research publications in DT has increased exponentially in the last five to six years [6, 14, 27]. The scope of DT being highly attractive and appealing, leading industries have already adopted it [52], and this increasing trend will boost its further growth in the years to come.

A DT is multi-dimensional in nature and in the absence of any general consensus, a wide number of definitions are put forward by various researchers [8, 17, 25]. Further, Kritzinger et al. [17] and Fuller et al. [8] also distinguished DT from other similar, but its simpler versions such as Digital Model (DM) and Digital Shadow (DS) based on limited provisions in automatic data and information exchange and integration. Semeraro et al. [38] also studied a wide number of DT definitions by various researchers chronologically between 2002 and 2019 and categorized them into five different clusters. Numerous definitions augment the scope and flexibility of DT applications.

14.2.2 DT Taxonomy

DT taxonomy is vital to identify the components and their contribution to realizing DT activities. A DT primarily has three components: Physical object (entity), Virtual object, and Connection between them for to and from data and information transfer, as identified by Grieves [10]. Tao and Zhang [42] identified four components of a DT namely Physical Shop-floor (PS), Virtual Shop-floor (VS), Shop-floor Service System (SSS), and Shop-floor Digital Twin Data (SDTD) to offer the potential services a DT can provide by utilizing classified dataset as well as fused data. Tao et al. [44] further included 'Connection' as a vital component so as to yield five components of a DT. Jones et al. [14] described the key terminologies associated with DT, while conducting the state of the survey with focus on characterizing the DT, identification of gaps in domain knowledge, and future research directions. The authors eventually mapped their findings to the 'CIRP Encyclopaedia 8-dimension model for Digital Twin'.

14.2.3 DT Characteristics

Once the DT components are established, it is pertinent to identify the DT properties, which enable accomplishment of the twinning and thereby, doing justice to its name and functions.

As the data transfer from physical world to virtual world takes place dynamically, the status of various measured attributes in both physical and virtual world simultaneously at any instance is defined as state [12, 31]. The values of state variables are utilized to carry out processing and dynamic decision making and these values keep on changing with time similar to a thermodynamic process. The rate at which data synchronization between the physical world and the cyber world takes place is defined as twinning rate. The success of twinning is determined by number of attributes and the accuracy of the corresponding data level, which are fed to the virtual world. The number of parameters to fully describe the process/object along with the closeness and accuracy of the associated data set are instrumental in realizing true mirroring or twinning and referred to as fidelity [37, 38]. High fidelity data set are essential to represent the physical world in the virtual environment to ensure reliable twinning. Schleich et al. [37] and Semeraro et al. [38] have also studied other salient properties such as scalability, dynamicity, interoperability, and modularity, which are also crucial for DT modelling.

14.2.4 DT Hierarchy

DT hierarchy implies the classification of DT based on the level of applications and therefore, it bears significant relevance and understanding for selecting the most appropriate one for DT modelling considering the scope and domain of work precisely.

Based on the complexities and purview of work, it can be implemented in three-tier configuration: (i) unit level, (ii) system level, and (iii) system of systems (SoS) level [32]. The most fundamental and basic atomic level of application is known as unit level such as a Machine Tool or a Coordinate Measuring Machine (CMM). For example, a unit-level DT can be created for the machine tool process and components. As the process continues, the real-time data can be transferred to digital space for analysis and decision making to better control the process in physical world. Both physical and virtual entity keep on interacting and synchronizing with time. When a number of unit-level DTs cooperate and collaborate, integration of data and services across all these DTs are required so as to move up to the next level of hierarchy, known as system-level DT. For example, logical grouping of CNC machine tools, robots, AGVs, etc. leads to the realization of a factory, which can be regarded as a system. Multi-Agent approach can be of immensely beneficial to accrue the potential benefits in system level. Finally, conglomeration of multiple system-levels DTs can augment the scope of network-based activities leading to the development of system of systems (SoS) levels DTs, which is envisaged as the highest level of digitalization. For example: different plants, design, purchase, etc. can be combined to realize an enterprise. Combination of different phases of product life-cycle is also an example of SoS.

14.2.5 Related Work

This sub-section reviews some of the prominent work related to DT in a nutshell, which helps to explore new opportunities to strengthen the DT capabilities.

Alam and Saddik [2] proposed a DT architecture for cloud-based CPS (C2PS) to describe analytically the three major activities identified by 3C. The work utilized Bayesian Belief Network to implement smart interaction controller to consider modified situations dynamically. Finally, a telematics-based prototype driving assistance application for the vehicular domain of C2PS, VCPS proved the performance of the architecture reference model. Cimino et al. [5] carried out an extensive literature survey on implementation of DT in varieties of applications, and identified the major gaps in the integration of the DTs with the control systems in regard to Manufacturing Execution Systems (MES), where the production relies on the Automation Pyramid and the various services it can offer. Finally, the authors implemented a DT in a MES-equipped assembly line in a lab scale with a focus on data acquisition and historical records for product tracking. Adamenko et al. [1] proposed data-based digital twin, system-based digital twin, and their combinations as possible approaches to design of DT with relative advantages and disadvantages and their suitability for a specific application. Liu et al. [20] conducted extensive survey on DT literatures and categorized the articles into three distinct groups: (i) DT definition, (ii) technology and software tools used, and (iii) industrial applications in the areas such as design, manufacturing, service and retire phase. The work is concluded with the summary of literatures and recommendations for enrichment. Jiang et al. [13] applied discrete event system (DES) modelling theory based on Grieves three-dimensional DT model using seven basic elements: controller, executor, processor, buffer, flowing entity, virtual service node, and logistics path in a DMS. The physical manufacturing cells are mapped into virtual environment through logical, geometric and data models. The proposed modelling techniques is applied to real shop-floor producing aerospace structural items to simulate different shop-floor activities such as production plan, process monitoring and scheduling. VanDerHorn and Mahadevan [47] proposed a comprehensive and generalized definition along with a detailed characteristics of DT to avoid ambiguity arises from other digital technologies. Furthermore, implementation of DT is presented through a case study, where DT is utilized to support the asset integrity management of a naval vessel. The work also raises the future challenges and opportunities of DT. Sharma et al. [39] studied the research gap between the actual DT concept and implementation, the role of machine learning and big data in DT, the barriers to the DT realization, area-specific implementation, DT components (essential and imperative) and properties, and compared three case studies in the light of stated properties and components.

DT being a rapidly evolving and promising fertile field, research on DT; particularly on definition, characteristics, key components, architectures, enabling technologies, design and functional issues, potential applications, and gaps is an ongoing continuous process. Few prominent articles covering the aforementioned topics are further credited to Hribernik et al. [11], Qi et al. [30] and Zhuang et al. [56].

14.3 Digital Twin Functionalities and Enabling Technologies

The five-dimensional (5D) model proposed by Tao et al. [44] is widely accepted by a number of researchers. However, in the present work, a six-dimensional (6D) model of DT is proposed, where Software Tools are considered additionally, and it is represented by Eq. 14.1.

$$\mathrm{DT} = [PE, VE, S, D, C_n, ST] \tag{14.1}$$

where, PE represents physical entity, VE represents virtual entity, S are the services offered by DT, D represents data and information, C_n are the connections, and ST refers to various software tools used for modelling and analytical purposes.

The various services (S), which any DT is expected to offer primarily depend on scope and significance, such as optimization, prediction, etc. Such higher level decision making activities (services) require judicious blending of knowledge and intellectual skills of the user, refer to as capability. Nevertheless, the work can't be executed without requisite STs, since analysis are essentially carried out using varieties of software tools and therefore, STs are considered as a distinct and indispensable entity of a DT to yield a 6D model. The proposed DT is shown in Fig. 14.1.

14.3.1 Digital Twin

A DT is a digital representation of physical objects converted into a digital version of its environment. It is a digital replica of a physical real-world object for decision making based on real-time data. The useful information resulted from analysis is used to regulate the process in the physical world.

14.3.2 Physical Object of DT

The physical entity can have three levels such as unit level, system level, and system of systems level (SoS). A basic stand-alone physical entity is considered as unit-level DT. In the system level, a group of interrelated objects of unit level interacts to form a unified whole following a set of rules. A system of systems (SoS) is the collection and integration of multiple, independent systems in regard to a larger and more complex system.

14.3.3 Data Acquisition and Communication

A DT comprises various physical elements of real world and virtual elements of cyber world. It provides communication link of flow of collected data from physical world to the virtual world and the information or result from the virtual to the physical world. The behavior of physical object is translated into a virtual object that is interconnected with a high level of integration and communication between them. This data communication link enables data transmission and exchange from physical object to virtual object in CPPS through required networks. It allows to integrate and simulate the behavior of physical product.

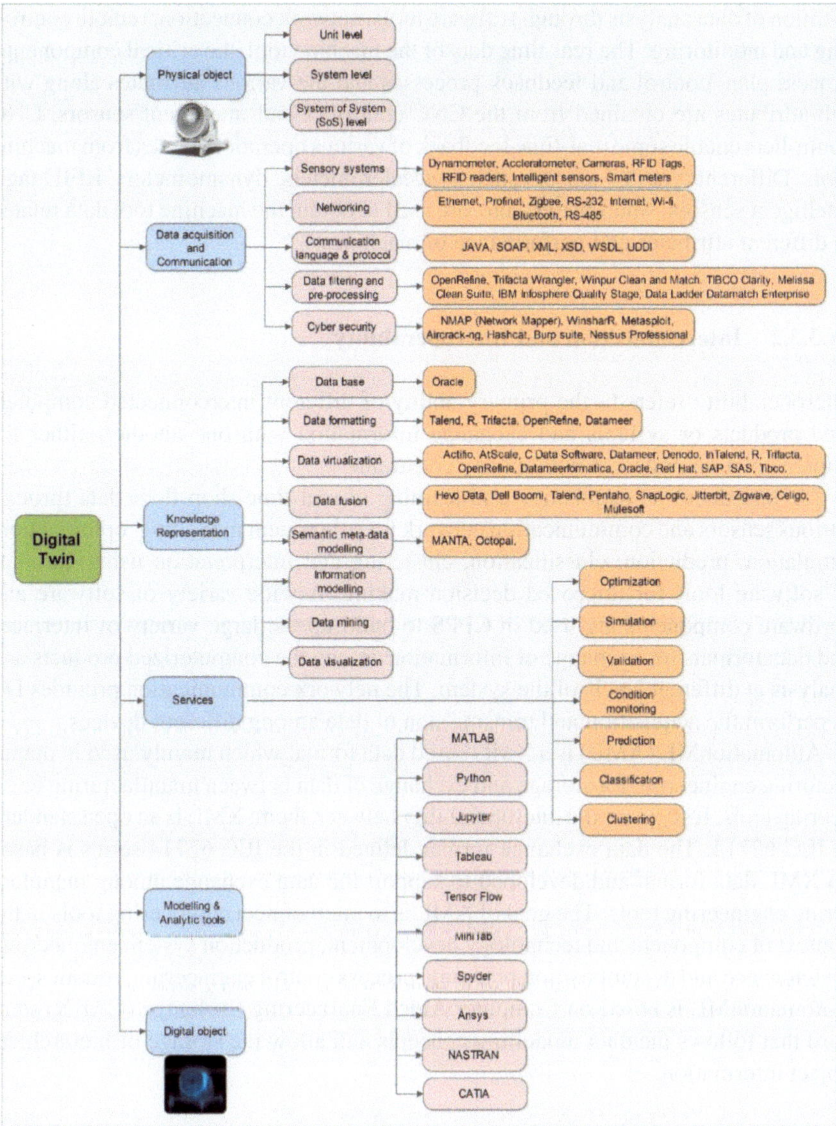

Fig. 14.1 6-D model of digital twin, functionalities, and enabling technologies

14.3.3.1 Data Acquisition

Data acquisition (DAQ) is defined as the process of digitizing the real world physical data from multiple resources through various applications that is stored, analysed, and displayed in a human resource interface. Data acquisition system in DT includes the addition of data analysis through software tools, network connection, remote controlling and monitoring. The real-time data of the machine tool, the critical components, process plan, control and feedback processes and the various attributes along with sub-attributes are obtained from the CNC controller and intelligent sensors. CNC controllers enable some real-time feedback of various operation modes from machine tools. Different types of sensors such as accelerometers, dynamometers, RFID tags, intelligent sensors, smart meters, etc. are used to obtain the machine tool data related to different attributes and performance of machines.

14.3.3.2 Interconnection and Interoperability

Interoperability refers to the primary ability of different interconnected computerized products or systems and exchange information with one another, either for implementation or access without any constraint.

DT introduces an acquisition and learning of real-time shop-floor data through various sensors and communication network for subsequent modelling, optimization, simulation, prediction, classification, clustering and interpretation using varieties of software tools for improved decision making. A wide variety of software and hardware components are used in CPPS to build up the large variety of interfaces and data formats for exchange of information among the computerized products and analysis at different levels of the system. The network communication provides DT to perform the acquisition and transmission of data among different devices.

AutomationML (AML) is a XML based data format which mainly used in manufacturing engineering for storage and exchange of data between manufacturing engineering tools. It supports the interoperability between them. XML is an open standard in IEC 62714. The data exchange format defined in the IEC 62714 series is based on XML data format and developed to support the data exchange among manufacturing engineering tools. The goal of AML is to interconnect engineering tools in the context of component and technology development, production system engineering, maintenance and decomposition planning, process control engineering, robotics, etc. AutomationML is based on Computer Aided Engineering Exchange (CAEX) standard that follows the data modelling concepts and allow the storage of hierarchical object information.

14.3.3.3 Communication Between Physical and Virtual World

The machine tools are usually supported by LinuxCNC software on Linux operating system. The PN532 NFC with Arduino Uno microcontroller board can read the RFID tags and sends the machine tool data to the MTConnect through proper network connection. MTConnect is a gateway that connects physical world to the virtual world through one-way data transfer. PN532 NFC is a highly integrated electronic devices, i.e., transceiver, which provides wireless communication at 13.56 MHz. A transceiver is transmitter and receiver in radio communication. It can both send and receive messages through wired or wireless connection. The signals acquired by the accelerometer, dynamometer and various other intelligent sensors are transferred to a PXI 1031 data acquisition platform. The collected signals are pre-processed and can be filtered by LabView software. The pre-processed sensor data is then possible to transfer by LabView to the MTConnect through TCP/IP protocol.

TCP/IP protocol uses networks layer that furnishes the machine tool with advanced connectivity such as 5G that offers more speed, lower latency, and ability to connect large numbers of smart devices and sensors within a network. It connects the physical machine tool with MTConnect through transmission of the real-time data from the CNC controllers and data acquisition devices. Based on the requirement of communication; various types of networks such as Ethernet, Wi-Fi, Internet, Ethernet, Bluetooth, I2C (Inter-Integrated Circuit), Zigbee, RS 232, etc. are implemented to realize the desired and reliable data communication between physical world and digital world. MTConnect receives all the real-time data set from the networks layers and assigns the unique ID to each data item. It timestamps each data and transforms all these data items into Simple Hierarchical Data Representation (SHDR) format. MTConnect eliminates the duplication of data. Information model is used over the filtered data in MTConnect.

The information model is the technology used for knowledge representation in digital twin which represents digital version of the physical and functional characteristics of any built object. *In CPPS,* Information models are a type of application data model that is used in historical database for reporting, slicing, and dicing the pre-processed data set. The information model is an XML file that represents the logical structure and relations of all machine tool data. After conversion to SHDR data; MTConnect establishes the relation of each data item with its predefined allied component of the information model using unique data IDs. Thus, the data from heterogeneous sources are clubbed together leading to data fusion. Data fusion is the mechanism that integrates the multiple data sources to yield more consistent, scalable, accurate, and useful information for comprehensive representation the status of certain components. Due to the heterogeneity sources of machine tools and systems, the collected data have the heterogeneous characteristics and formats. The unique data format and interface must to be rightly defined, so that collected data can be recorded, processed and maintained. Every machine or device within network connection can interact with each other through the different communication protocols such as SPI, I2C, CAN, HART, TCP/IP, etc.

MTConnect may work as a web server and formats all the real-time data into XML format defined by MT Connect standard and sends these XML files for processing either to local server or cloud server. This server contains various software tools such as Jupyter, Spyder, MATLAB, Tableau, and Minitab, etc. for improved decision making. These XML files are sent following HTTP requests through TCP/IP protocol. These historical data of machine tool in XML format can then be accessed by other applications to provide value-added services and self-configuration. In cloud-based platform the cloud layer contains various cloud-based applications that access the MTConnect or separate web server through the internet and provide cloud-based services.

However, remote process monitoring on digital and smart devices such as smart phones, tablets, smart watch, drones, etc. can be accomplished following the request of XML files directly from the MTConnect device for analysis and visualization of data. In MT Connect various big data analytics tools can be used to access and process the historical manufacturing data for improved decision making in CPPS. Thus, the cloud-based service has led to sustainable product developments, production planning, and enterprise management using big data analytics, artificial intelligence, machine learning and modern communication technologies.

14.3.4 Data Filtering and Pre-Processing

Data filtering is the process of sorting large set of data to obtain specific smaller set of data based on defined problem and solution criteria. Data filtering process enables to focus on specific sample set of data and discard the remaining portion of original data set that are not relevant with the problem definition.

Data pre-processing is the first and crucial process of preparing filtered data such that it is suitable for machine learning or deep learning model. In CPS, the machine tool data may not be in an understandable format. Often raw data can be inconsistent, incomplete, and may contain many errors such as noises, missing data values, and may be in unusable format that cannot be used directly used in machine learning models. Thus, to improve the quality of the data set and to simplify it by a method is called data pre-processing. Only then this pre-processed data are further progressed through different data analytics tools. Data pre-processing methods are divided into the four categories: Data cleaning, Data integration, Data transformation, Data reduction.

Data cleaning is the method of removing noise and correcting inconsistencies of data. Data integration involves grouping of data from multiple sources. Data transformation, also known as normalization, is the process of organizing the database to remove redundancy. Data reduction is the process of reducing data size by eliminating redundant features.

14.3.5 Cyber Security

Cyber physical systems (CPS) comprise various interconnected systems based on digital twin. It can manipulate and monitor all the physical and virtual objects and their processes. All the systems are firmly related to Internet of Things (IoT). The interconnection of cyber and physical worlds raises the issue of massive security challenges. The vulnerabilities of CPS security such as threats and attacks are reviewed and ensure the prevention of malicious attack or cyber-attack on network communication between different devices and the unauthorized access.

14.3.6 Knowledge Representation

Knowledge representation is one of the most important activities of CPS for analysis and decision making through a series of functionalities. Knowledge representation is the process that is used to organize and formalize the knowledge in the knowledge base following IF–THEN-ELSE rules. The knowledge representation includes information about the way of knowledge represented in DT. Knowledge graphs, and the various query languages are used for extraction of knowledge discovery from historical data of machine tools. Various data representation tools, knowledge graph along with relational and non-relational databases are processed with structured query languages (SQL).

14.3.6.1 Knowledge Graph

A knowledge graph is a collection of interlinked details of entities, objects, relationships, situation, events and concepts that constitutes the relationship between them. By storing all these features and context as contextual information, it is visualized as a structure of graph that directly reflects the result, analysis, and interpretation of task. In CPS knowledge graph provides a framework for sharing the information among various devices in physical and cyber world. It also furnishes the data integration, analysis and unification of data in DT.

14.3.6.2 Ontology

Ontology represents the formal semantics in context of knowledge graph. It is a semantic data model that defines the types and properties of things that exist in a particular problem. Ontology is a generalized data model that ensures the modelling over the entities sharing certain properties and doesn't include the specific individual of the entity. In smart manufacturing where CPS is deployed, ontology supports the sharing of information for interaction among new product modules quoted by different stakeholders.

14.3.6.3 Database and Data Formatting

The role of database and database management systems ensures the centralized control and automation in CPS scenario. Database management system provides the real-time control issue into a computation of unique data modelling and becoming the repositories in the form of logical structure of data of physical objects. It also stores the historical data of machine tools and the decision making information resulted by data analysis. The large variety of data formats and types are necessary for build up the uniform data format and exchange of data in context of process information in the shop floor.

14.3.6.4 Data Virtualization and Visualization

Data virtualization and data fusion technology gives users and programmers to do the fast access, combine, transform, and analysing the datasets effectively through databases, big data sources, cloud internet and IoT systems.

Data visualization is the practice of translating information into a visual context, such as knowledge graph, map to manipulate, interpret, and identify the patterns, trends and outliers in large data set in a most efficient manner.

14.3.6.5 Semantic Data Modelling

Semantic data modelling is highly essential in CPPS as the semantic model is a meta-data model that contains physical database objects in an abstract form and modified into logical dimensions. The semantic model is designed for the analysis of formatted data set according to the structure of the physical objects.

14.3.6.6 Information Model

Information model represents the way of interaction of physical components interact and simulates the behavior of different components in a collaborative manner. Decision making algorithm such as machine learning and deep learning algorithm ensures the information model being capable of evaluation and validation.

14.3.6.7 Data Mining

Data mining refers to large number of techniques for detecting the anomalies, pattern, and correlations in formatted data of CPS so that the data to be acknowledged easily. The mining extracts the knowledge about the historical data instead of extracting data itself. The data mining is often termed as Knowledge discovery in database (KDD). Seven steps are normally used for KDD process. These are 1. Data cleaning, 2. Data integration, 3. Data selection, 4. Data transformation, 5. Data analysis, 6. Pattern evaluation and extraction, and 7. Knowledge representation.

14.3.6.8 Formal Concept Analysis

Formal Concept Analysis (FCA) refers to a mathematical concept of hierarchies mainly used for the analysis of data, i.e., for deriving logical relationship in the dataset between objects and attributes in the context of knowledge representation. This technique takes the input of binary data matrix with set of objects as rows and set of attributes as column. FCA finds out the inner concept of binary data matrix presented in data items. It is mainly used to cluster the data set through organizing the semi ordered matrix data set. A triplet (O, A, R) is represented as formal context, where O and A be the non-empty sets. O is the set of objects and A is the set of attributes, where R is the binary relation between O and A such that $R \subseteq O\ X\ A$. The formal context (O, A, R) is an input matrix dataset of m rows and n columns. It consists of a set of objects defined as $O = \{Ob_1, Ob_2, Ob_3, \ldots, Ob_m\}$, and set of attributes defined as $A = \{Atr_1, Atr_2, Atr_3, \ldots, Atr_n\}$. Relation $\mathbf{R}$ is defined as $\{Ob_i, Atr_j\} \subseteq R$, where the ith row and the jth column of formal context is not null.

14.4 Proposed CPPS and MTDT

This section presents the proposed CPPS, the associated MTDT, and varieties of decision making functions for improved machine tool activities.

14.4.1 Proposed Cyber Physical Production System

In the present work, a CPPS comprising a group of CNC machine tools (Turning centres and Machining centres) is proposed, which are engaged in discrete part production as shown in Fig. 14.2. Each machine tool is equipped with varieties of sensors, connected in a network, and having decision making abilities to exhibit adequate smartness. These machine tools are fourth generation machine tools, popularly known as Machine Tool 4.0 (MT 4.0), which are smart, connected over a network, adaptive, and manifests autonomous behavior to demonstrate Self-X, and therefore, regarded as a vital entity of a CPPS [19, 50].

The group of machine tools are envisaged to work as a Multi-Agent Systems (MAS), which rely on negotiation and cooperation based work to offer higher flexibility, better adaptability, and quick responsiveness [48]. The agent based approach is a potential platform for efficient Manufacturing Execution System (MES) control such as distributed scheduling. Although MAS approach is practiced over long time, it is very much compatible to CPPS and therefore, has got added relevance and momentum in smart manufacturing [29].

Varieties of sensors such as dynamometers, accelerometers, cameras, RFID tags, smart meters, etc. are deployed to collect the cutting zone parameters, e.g., cutting force, temperature, vibration, etc., which are sent to the cyber space through IOT devices for analysis maintaining communication protocol such as MTConnect or OPC-UA or AutomationML. The collected data is utilized for subsequent analysis using various application protocols and application tools accessed through cloud services.

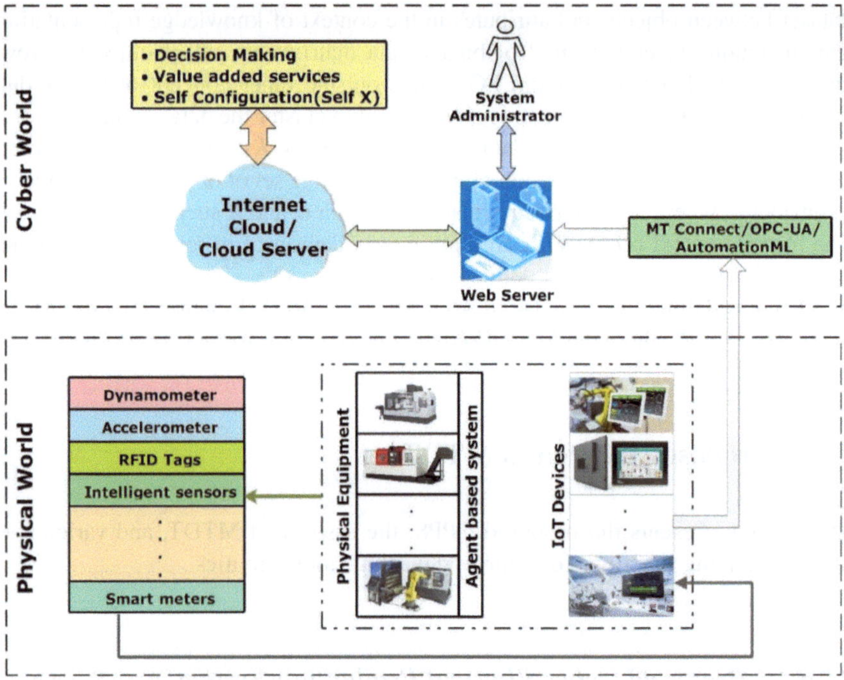

Fig. 14.2 Proposed cyber physical production system

Web server receives the machine tool data in digital format, preserves and maintains it in the data base, and utilize the same for analysis. It also considers information security (IS) using software tools, technologies, and processes. The network connection is protected using firewall and the different hardware and software components to prevent unauthorized access to the network as well as from other network. The system administrator acts as a human element in the loop and analyse the data through internet cloud using various software tools for improved decision making.

The results of analysis enable higher level decision making for better control the process and to accomplish other value-added functions such as simulation, prediction, health prognosis. Such enhanced activities eventually lead to self-configuration, e.g., self-aware, self-optimize, self-predict, etc., together called Self-X.

14.4.2 Machine Tool Digital Twin

A MTDT represents machine tool and machining processes in virtual environment for varieties of improved decision making activities [3]. To this effect, a MTDT is developed as unit-level DT considering a single machine tool and presented in

Fig. 14.3. The MTDT is at the core of the CPPS and instrumental in realizing smart manufacturing activities.

The major components of the MTDT are shop-floor data, information model, database, and various algorithms and analytic tools for smart decision making as shown in Fig. 14.3. Wide varieties of data such as process or operating variables, cutting zone parameters of machine tools, operation time, production rate, uptime, downtime, etc. are collected for analysis.

Some prominent functions which are carried out includes process parameter optimization for maximizing productivity, minimizing surface roughness, minimizing cost of production. Additional enhanced functions such as condition monitoring, prediction, simulation, etc. are also undertaken for further improvement in performance. The requisite data analytic tools are accessed from the cloud for the purpose of analysis and decision making.

The Human Machine Interfaces (HMIs) are used to regulate the processes and equipment depending on the outcome of decision making. Few popular HMIs are smart phone, Tablets, PCs, etc. and used for wide access to the data and functions supported by MTDT.

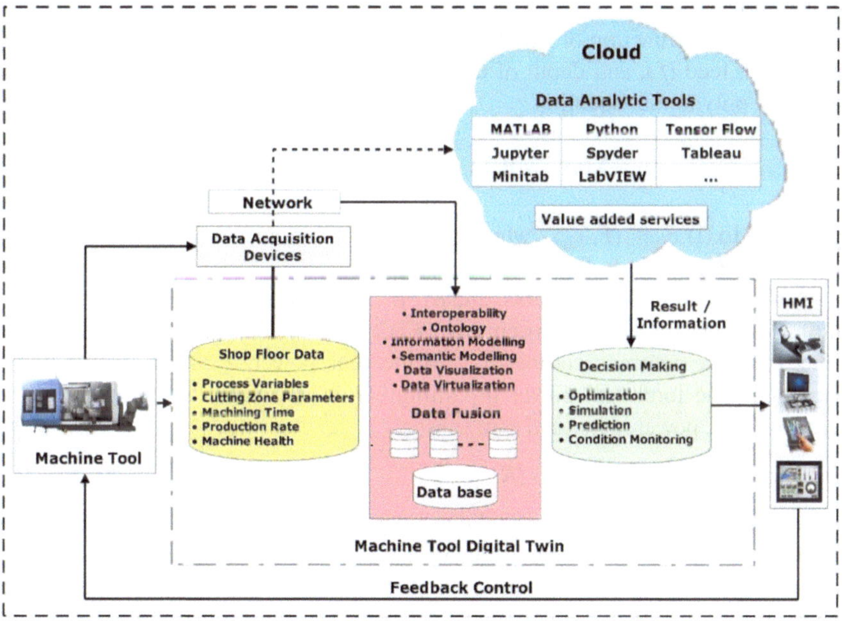

Fig. 14.3 Proposed machine tool digital twin (MTDT)

14.4.3 Decision Making by DT

This sub-section presents wide varieties of improved decision making by MTDT such as optimization, condition monitoring, prediction, simulation, etc. for overall performance improvement of the metal cutting process.

14.4.3.1 Optimization of Various Responses

One of the major functions of the proposed MTDT is to optimize the salient performance indicators through adjustments of the process parameters. Some prominent optimizations such as maximizing productivity, minimizing surface roughness, minimizing cost of production are considered under the scope of present work.

Maximizing Metal Removal Rate (MRR)

Reduction of cycle time is one of the most fundamental objectives of any metal cutting process; since reduction in cycle time increases the productivity. This is accomplished by improving the MRR. High MRR is usually desirable during large stock removal.

The objective function of MRR in terms of the process variables namely cutting speed (V_c), feed (f), and depth of cut (t) in turning operation is expressed by the following polynomial equation.

$$\text{Max} \quad \text{MRR} = K_1 \cdot (V_c)^{\alpha_1} (f)^{\beta_1} (t)^{\gamma_1} \tag{14.2}$$

Subject to (i) $F_c \le (F_c)_{\max}$, where, $F_c = K'(f)^{\delta}(t)^{\eta}$ and
(ii) $P_c \le P_{\max}$, where $P_c = u_c \times V_c \times f \times t$.
where, K_1 and K' are constants, and α_1, β_1, γ_1, δ η are exponents. F_c symbolizes cutting force encountered during cutting, $(F_c)_{\max}$ is the maximum permissible cutting force, P_c is the power in cutting, $P_{\max}$ is the peak power, u_c is the specific power in cutting. The force and power in cutting can be measured by combinations of dynamometer, power meter and rpm sensor.

Minimizing Surface Roughness (R_a)

One of the major attributes in metal cutting is improving the surface quality of the finished component by reducing the surface asperities. The objective function of minimizing surface roughness in terms of the process variables can be expressed generically by the following equation.

$$\text{Min} \quad R_a = K_2 \cdot (V_c)^{\alpha_2} (f)^{\beta_2} (t)^{\gamma_2} \tag{14.3}$$

where, K_2 is a constant, and α_2, β_2, γ_2 are exponents.

Good surface finish is considered usually during final cut, when high MRR is not recommended. Since, feed is the most significant parameter that affects surface roughness; it can also be modelled alternatively as shown in the Eq. (14.4).

$$\text{Min} \quad R_a = K_3 \cdot f + K_4 \cdot f^2 \tag{14.4}$$

where, K_3 and K_4 are constants

Multi-Objective Optimization

Unlike special purpose machine tools, CNC machines are characterized primarily by flexibility of operations. High stock removal employing repetitive tool path therefore, is not a judicious proposition. This is why a single cut is often preferred to reduce operation time as well as to achieve good surface finish simultaneously. A single function Ψ is proposed by Eq. (14.5) for multi-objective optimization to take care of *MRR* and R_a concurrently.

$$\text{Max} \quad \Psi = w_1 \cdot MRR - w_2 \cdot R_a \tag{14.5}$$

where, w_1 and w_2 are the weights assigned to *MRR* and R_a respectively

Minimizing Machining Cost

Reducing cost of machining is another crucial criterion which needs to be addressed. Considering the highest possible values of feed and depth of cut without violating the force constraint, the total cost of machining (C_{total}) can be modelled by Eq. (14.6).

$$\text{Min} \quad C_{\text{total}} = C_{m/c} \cdot t_{m/c} + C_{\text{tool}} \cdot \frac{t_{m/c}}{T} + C_{m/c} \cdot \left[\frac{t_{m/c}}{T} \cdot t_s\right] + C_{m/c} \cdot t_h \tag{14.6}$$

where, $C_{m/c}$ is the direct operating cost of the machine tool/unit time including cost of labour, C_{tool} is the cutting tool cost, $t_{m/c}$ is the total time during which the tool is engaged in cutting and time required in indexing, t_s is the time required for tool changing and setting, t_h is the loading and unloading time, and T is the tool life.

Assuming the limiting values of feed, and depth of cut as restricted by the force constraint, Eq. (14.6) can be re-written as shown in Eq. (14.7).

$$\text{Min} \quad C_{\text{total}} = C_{m/c} \cdot K_5 V_c^{-1} + \left[C_{m/c} \cdot t_s + C_{\text{tool}}\right] \cdot K_6 V_c^{\frac{1}{n}-1} + C_{m/c} \cdot t_h \tag{14.7}$$

where, K_5 and K_6 are constants

Approaches such as statistical, machine learning algorithms can be applied using appropriate software tools for optimization of the above responses.

14.4.3.2 Condition Monitoring and Prediction

Prediction is one of the powerful activities of DT for determining the future course of actions. In the present work, two such situations are considered related to metal cutting.

Prediction of temperature rise

During metal cutting, the strain energy in deforming the material gets converted to heat and consequently temperature rise takes place. High temperature is detrimental for tool-work systems; since it accelerates diffusion process that causes premature tool failure and adversely affects the finished component. Prediction of chip-tool interface temperature (θ_i) during continuous cutting such as turning can be expressed by Eq. 14.8.

$$\theta_i = K_7.(V_c)^{\alpha_3}(f)^{\beta_3}(t)^{\gamma_3} \tag{14.8}$$

where, K_7 is a constant, and α_3, β_3, γ_3 are exponents.

Infra-red camera can be used to measure the temperature. Machine learning and statistical methods are very efficient approaches for prediction of temperature rise.

Prediction of Tool Wear

As the cutting progresses, the tool which is exposed to harsh condition, gradually wears out and loses its cutting ability. The life of a tool (T) is correlated to the process variables, which can be expressed by Eq. 14.9.

$$T = \frac{K_8}{(V_c)^{\alpha}(f)^{\beta}(t)^{\gamma}} \tag{14.9}$$

where, K_8 is a constant, and α, β, γ are exponents.

The constant and the exponents can be obtained statistically from experimental data. The tool life can then be predicted for any values of the process variables.

The residual useful life (RUL) of a cutting tool can also be predicted based on force and vibration sensors signals without requiring any physical model. Various machine learning techniques such as Neural Network, Support Vector Machine (SVM), Deep Learning, etc. can be efficiently used for model training and testing, and finally for prediction.

The constants and exponents appeared in Eq. 14.2 through Eq. 14.9 are assumed to be generic considering tool-work combinations and cutting conditions.

14.4.3.3 Simulation

Finite element packages like ABAQUS, ANSYS, and COMSOL are powerful tools for simulation of different aspects of machining process. For example, [36] predicted the cutting force and deflection of end milling operation based on cutting force model proposed by Tlusty et al. [45] and deflection of end mill model developed by Kivanc et al. [15]. The expression for tangential force (F_t) and radial force (F_r) are given by Eqs. 14.10 and 14.11 respectively in differential form.

$$dF_t = 2(F_u/f_t)h(\theta)d\theta \tag{14.10}$$

$$dF_r = 2(F_u/f_t)P_f h(\theta)d\theta \tag{14.11}$$

where, F_u is unit force, f_t represents feed per tooth, h is chip thickness, θ is cutting angle, and P_f is the proportional factor.

The deflection of the cutter (f_x) normal to the direction of feed is expressed by Eq. 14.12.

$$f_x = C\frac{F_x}{E}\left[\frac{L1^3}{D1^4} + \frac{\left(L2^3 - L1^3\right)}{D2^4}\right]^N \tag{14.12}$$

where, $D1$ is the cutter diameter, $D2$ is the shank diameter, $L1$ is the flute length, $L2$ is the overall length, F_x is the applied load normal to the direction of feed, E is the modulus of elasticity of tool material. C and N are constants, which depend on number of flutes of the cutter. Johnson–Cook theory was used for defining the material property.

A comprehensive decision making activities related to metal cutting along with the various sensory systems, together with the varieties of statistical and machine learning techniques and relevant software tools are presented in Table 14.1.

14.5 Discussion and Conclusions

The present work discusses the key features of SM and CPPS in the context of Industry 4.0 and focusses on digital twin, which is the key entity to realize smart objects and activities. DT opens up a rather new and potential platform for improved decision making through varieties of activities for efficient operations and overall performance improvement in a CPPS. A detailed study of DT definitions and characteristic features are considered under the ambit of the present work. The work then proposes the detailed functional requirements of a six-dimensional DT together with the enabling technologies and tools for its implementation. A CPPS comprising of machine tool systems along with the MTDT is also proposed to execute wide spectrum of activities like optimization, prediction, condition monitoring, and simulation of machining processes through appropriate mathematical model, relevant techniques and algorithms, and the software tools to suit one's own requirements.

Two different simulation results reported in the literatures are found to be worth mentioning in this context. The work of Saffar et al. [36] shows that the cutting forces

Table 14.1 Various decision making activities using MTDT along with related sensors, techniques, and software tools

Purpose	Response	Sensor used	Algorithms	Software used
Optimization	Metal removal rate	Dynamometer, Power meter, rpm sensor	Support vector regression (SVR), Response surface methodology (RSM), Heuristic algorithm	MINITAB, Weka, MATLAB, Anaconda
	Surface roughness			
	Machining cost			
Prediction	Chip-tool interface temperature	IR-Camera	Random Forest (RF), Genetic algorithm (GA), Polynomial regression, Back propagation Neural Network (BPNN)	Weka, MATLAB, Anaconda
	Residual useful life of cutting tool	Dynamometer, Accelerometer	Neural network, Support vector machine (SVM), Deep learning	Jupyter, Spyder, Tensor Flow
Condition monitoring	Tool condition monitoring	Accelerometer	K-Star algorithm	Microsoft Excel, Python
Simulation	Cutting force and tool deflection in end milling	Dynamometer, Accelerometer	Numerical modelling using Lagrangian, Eulerian and Arbitrary Lagrangian Eulerian (ALE) approach	ANSYS, ABAQUS, COMSOL

in end milling obtained from simulation are in good compliance with those found through experiments while machining AISI 1045 steel by uncoated HSS cutter; although calculated cutting force normal to the direction of feed (obtained using Eqs. 14.10 and 14.11) are higher than the experimental values, which implies the accuracy of the simulation. The calculated values of tool deflections are however, found to be in good agreement with those obtained from simulations in three different cutting conditions. In another experiment, Armendia et al. [3] investigated the twin-control modelling features in industrial environments namely in automotive and aerospace industries. The authors considered complete machining and full multi-body simulation (MBS) in an integrated manner as well as pure machining separately and implemented in FEA solver SAMCEF Mecano in two different use-cases, which yield satisfactory results.

Despite the highly attractive features of DT, its implementation in true sense is a challenging task. The interoperability requirement of different data sources is one of the major hindrances to its success. Another critical challenge, particularly for MT 4.0, is that majority of the CNC controllers are proprietary and closed, which do not allow any external feedback control to exercise. The use of open controller as an alternative, can be of great help to get rid of such crisis; since control commands can be implemented through Application Programming Interfaces (APIs). Additionally, since cutting environment is dirty due to presence of high temperature, fumes, chips, coolant, etc., it is not conducive for installation of sophisticated sensors. Adequate care must be taken to protect them from cutting zone hazardous environment. The high network communication speed, network security, synchronization between physical world and the digital world through real-time data and information propagation, availability of low-cost gateways, etc. are also some of the open issues, which are critical to be addressed properly to accrue the true potential benefits of digitalization.

The present work however, can be regarded as strong theoretical foundation for implementation of DT capabilities. Future endeavour will attempt for its implementation and investigate the performance in realizing a quintessential Machine Tool 4.0.

References

1. Adamenko, D., Kunnen, S., Pluhnau, R., Loibl, A., Nagarajah, A.: Review and comparison of the methods of designing the digital twin. Proc. CIRP **91**, 27–32 (2020)
2. Alam, K.M., Saddik, A.E.: C2PS: A digital twin architecture reference model for the cloud-based cyber-physical systems. IEEE Access **5**, 2050–2062 (2017)
3. Armendia, M., Cugnon, F., Berglind, L., Ozturk, E., Gil, G., Selmi, J.: Evaluation of machine tool digital twin for machining operations in industrial environment. Proc. CIRP **82**, 231–236 (2019)
4. Attaran, M., Attaran, S., Celik, B.G.: The impact of digital twins on the evolution of intelligent manufacturing and industry 4.0. In: Advances in Computational Intelligence, vol. 3(11), pp. 1–15. (2023)
5. Cimino, C., Negri, E., Fumagalli, L.: Review of digital twin applications in manufacturing. Comput. Ind. **113**(103130), 1–15 (2019)
6. D'Amico, R.D., Erkoyuncu, J.A., Addepalli, S., Penver, S.: Cognitive digital twin: an approach to improve the maintenance management. CIRP J. Manuf. Sci. Technol. **38**, 613 630 (2022)
7. Ding, K., Chan, F.T.S., Zhang, X., Zhou, G., Zhang, F.: Defining a digital twin-based cyber-physical production system for autonomous manufacturing in smart shop floors. Int. J. Prod. Res. **57**(20), 6315–6334 (2019)
8. Fuller, A., Fan, Z., Day, C., Barlow, C.: Digital twin: enabling technologies, challenges and open research. IEEE Access **8**, 108952–108971 (2020)
9. Glaessgen, E., Stargel, D.: The digital twin paradigm for future NASA and U.S. Air Force vehicles. *In:* Proceedings 53rd AIAA/ASME/ASCE/AHS/ASC Structures, Structural Dynamics, and Materials Conference, pp. 1–14. (2012) [Online]
10. Grieves, M.: Digital twin: manufacturing excellence through virtual factory replication. White Paper, pp. 1–7. (2014)
11. Hribernik, K., Cabri, G., Mandreoli, F., Mentzas, G.: Autonomous, context-aware, adaptive digital twins—State of the art and roadmap. Comput. Ind. **133**(103508), 1–12 (2021)
12. Hu, W., Zhang, T., Deng, X., Liu, Z., Tan, J.: Digital twin: a state-of-the-art review of its enabling technologies, applications and challenges. J. Intell. Manufact. Spec. Equipm. **2**(1), 1–34 (2021)
13. Jiang, H., Qin, S., Fu, J., Zhang, J., Ding, G.: How to model and implement connections between physical and virtual models for digital twin application. J. Manuf. Syst. **58**, 36–51 (2021)
14. Jones, D., Snider, C., Nassehi, A., Yon, J., Hicks, B.: Characterizing the digital twin: a systematic literature review. CIRP J. Manufact. Sci. Technol. **29**(A), 36–52 (2020)
15. Kivanc, E.B., Budak, E.: Structural modeling of end mills for form error and stability analysis. Int. J. Mach. Tools Manuf **44**, 1151–1161 (2004)

16. Koc, T.C., Teker, S.: Industrial revolutions and its effects on quality of life. PressAcademia Proc. **9**, 304–311 (2019)
17. Kritzinger, W., Karner, M., Traar, G., Henjes, J., Sihn, W.: Digital twin in manufacturing: a categorical literature review and classification. IFAC PapersOnLine **51–11**, 1016–1022 (2018)
18. Liu, C., Jiang, P.: Cyber-physical system architecture in shop floor for intelligent manufacturing. Proc. CIRP **56**, 372–377 (2016)
19. Liu, C., Xu, X., Peng, Q., Zhou, Z.: MT connect-based cyber-physical machine tool: a case study. Proc. CIRP **72**, 492–497 (2018)
20. Liu, M., Fang, S., Dong, H., Xu, C.: Review of digital twin about concepts, technologies, and industrial applications. J. Manuf. Syst. **58**, 346–361 (2021)
21. Liu, Y., Peng, Y., Wang, B., Yao, S., Liu, Z.: Review on cyber-physical systems. IEEE/CAA J. Automatica Sinica **4**(1), 27–40 (2017)
22. Lu, Y., Liu, C., Wang, K.I.-K., Huang, H., Xu, X.: Digital twin-driven smart manufacturing: connotation, reference model, applications and research issues. Robot. Comput. Integrat. Manufact. **61**(101837), 1–14 (2020)
23. Lu, Y., Xu, X., Wang, L.: Smart manufacturing process and system automation—A critical review of the standards and envisioned scenarios. J. Manuf. Syst. **56**, 312–325 (2020)
24. Monostori, L., Kádár, B., Bauernhansl, T., Kondoh, S., Kumara, S., Reinhart, G., Sauer, O., Schuh, G., Sihn, W., Ueda, K.: Cyber-physical systems in manufacturing. CIRP Ann. Manuf. Technol. **65**(2), 621–641 (2016)
25. Negri, E., Fumagalli, L., Macchi, M.: A review of the roles of digital twin in CPS-based production systems. Proc. Manufact. **11**, 939–948 (2017)
26. Oks, S.J., Jalowski, M., Lechner, M., Mirschberger, S., Merklein, M., Vogel-Heuser, B., Möslein, K.M.: Cyber-physical systems in the context of industry 4.0: a review, categorization and outlook. Inform. Syst. Front. 1–42 (2022)
27. Onaji, I., Tiwari, D., Soulatiantork, P., Song, B., Tiwari, A.: Digital twin in manufacturing: conceptual framework and case studies. Int. J. Comput. Integr. Manuf. **35**(8), 831–858 (2022)
28. Ortt, R., Stolwijk, C., Punter, M.: Implementing industry 4.0: assessing the current state. J. Manufact. Technol. Managem. **31**(5), 825–836 (2020)
29. Pulikottil, T., Estrada-Jimenez, L.A., Rehman, H.U., Mo, F., Nikghadam-Hojjati, S., Barata, J.: Agent-based manufacturing—Review and expert evaluation. Int. J. Adv. Manufact. Technol. **127**, 2151–2180 (2023)
30. Qi, Q., Tao, F., Hu, T., Anwer, N., Liu, A., Wei, Y., Wang, L., Nee, A.Y.C.: Enabling technologies and tools for digital twin. J. Manuf. Syst. **58**, 3–21 (2021)
31. Qi, Q., Tao, F., Zuo, Y., Zhao, D.: Digital twin service towards smart manufacturing. Proc. CIRP **72**, 237–242 (2018)
32. Qi, Q., Zhao, D., Liao, T.W., Tao, F.: Modelling of cyber–physical systems and digital twin based on edge computing, fog computing and cloud computing towards smart manufacturing. In: Proceedings of ASME 2018 13th International Manufacturing Science and Engineering Conference; MSEC2018, vol. 51357, pp. V001T05A018. ASME (2018b)
33. Qin, J., Liu, Y., Grosvenor, R.: A categorical framework of manufacturing for industry 4.0 and beyond. Changeable, agile, reconfigurable and virtual production. Proc. CIRP **52**, 173–178 (2016)
34. Radziwon, A., Bilberg, A., Bogers, M., Madsen, E.S.: The smart factory: exploring adaptive and flexible manufacturing solutions. Proc. Eng. **69**, 1184–1190 (2014)
35. Rosen, R., Wichert, G.V., Lo, G., Bettenhausen, K.D.: About the importance of autonomy and digital twins for the future of manufacturing. IFAC-PapersOnLine **48–3**, 567–572 (2015)
36. Saffar, R.J., Razfar, M.R., Zarei, O., Ghassemieh, E.: Simulation of three-dimension cutting force and tool deflection in the end milling operation based on finite element method. Simul. Model. Pract. Theory **16**(10), 1677–1688 (2008)
37. Schleich, B., Anwer, N., Mathieu, L., Wartzack, S.: Shaping the digital twin for design and production engineering. CIRP Ann. Manuf. Technol. **66**(1), 141–144 (2016)
38. Semeraro, C., Lezoche, M., Panetto, H., Dassisti, M.: Digital twin paradigm: a systematic literature review. Comput. Ind. **130**(103469), 1–23 (2021)

39. Sharma, A., Kosasih, E., Zhang, J., Brintrup, A., Calinescu, A.: Digital twins: state of the art theory and practice, challenges, and open research questions. J. Ind. Inf. Integr. **30**(100383), 1–15 (2022)
40. Suleiman, Z., Shaikholla, S., Dikhanbayeva, D., Shehab, E., Turkyilmaz, A.: Industry 4.0: clustering of concepts and characteristics. Cogent Eng. **9**(2034264), 1–26 (2022)
41. Suvarna, M., Yap, K.S., Yang, W., Li, J., Ng, Y.T., Wang, X.: Cyber-physical production systems for data-driven, decentralized, and secure manufacturing—A perspective. Engineering **7**(9), 1212–1223 (2021)
42. Tao, F., Zhang, M.: Digital twin shop-floor: a new shop-floor paradigm towards smart manufacturing. IEEE Access **5**, 20418–20427 (2017)
43. Tao, F., Qi, Q., Wang, L., Nee, A.Y.C.: Digital twins and cyber-physical systems toward smart manufacturing and industry 4.0: correlation and comparison. Engineering **5**, 653–661 (2019)
44. Tao, F., Zhang, H., Liu, A., Nee, A.Y.C.: Digital twin in industry: state-of-the-art. IEEE Trans. Industr. Inf. **15**(4), 2405–2414 (2019)
45. Tlusty, J., MacNeil, P.: Dynamics of cutting forces in end milling. CIRP Ann. **24**(1), 21–25 (1975)
46. Vaidya, S., Ambad, P., Bhosle, S.: Industry 4.0—a glimpse. Proc. Manufact **20**, 233–238 (2018)
47. VanDerHorn, E., Mahadevan, S.: Digital twin: generalization, characterization and implementation. Decis. Support. Syst. **145**(113524), 1–11 (2021)
48. Vogel-Heuser, B., Lee, J., Leitão, P.: Agents enabling cyber-physical production systems. Automatisierungstechnik **63**(10), 777–789 (2015)
49. Wu, D., Rosen, D.W., Wang, L., Schaefer, D.: Cloud-based manufacturing: old wine in new bottles? Proc. CIRP **17**, 94–99 (2014)
50. Xu, X.: Machine tool 4.0 for the new era of manufacturing. Int. J. Adv. Manuf. Technol. **92**, 1893–1900 (2017)
51. Yang, F., Gu, S.: Industry 4.0, a revolution that requires technology and national strategies. Complex Intell. Syst. **7**, 1311–1325 (2021)
52. Yao, J.-F., Yang, Y., Wang, X.-C., Zhang, X.-P.: Systematic review of digital twin technology and applications. Visual Comput. Indus. Biomed. Art **6**(10), 1–20 (2023)
53. Zheng, P., Wang, H., Sang, Z., Zhong, R.Y., Liu, Y., Liu, C., Mubarok, K., Yu, S., Xu, X.: Smart manufacturing systems for industry 4.0: conceptual framework, scenarios, and future perspectives. Front. Mech. Eng. **13**(2), 137–150 (2018)
54. Zheng, P., Xia, L., Li, C., Li, X., Liu, B.: Towards self-X cognitive manufacturing network: an industrial knowledge graph-based multi-agent reinforcement learning approach. J. Manuf. Syst. **61**, 16–26 (2021)
55. Zhong, R.A., Xu, X., Klotz, E., Newman, S.T.: Intelligent manufacturing in the context of Industry 4.0: a review. Engineering **3**, 616–630 (2017)
56. Zhuang, C., Miao, T., Liu, J., Xiong, H.: The connotation of digital twin, and the construction and application method of shop-floor digital twin. Robot. Comput. Integrat. Manufact. **68**(102075), 1–16 (2021)

Part IV
Quality Assurance, Reliability, and Safety Engineering

Chapter 15
Reliability-Centered Maintenance (RCM) Analysis of IOT-Based Weight Data Communication System Using Fuzzy Fault Tree Analysis (FFTA)

Saptarshi Das and Bijan Sarkar

Abstract In the era of Industry 4.0, the IOT (Internet of things) system is a prime constituent of various production systems for making the system sustainable as well as reliable. To overcome any malfunction in the IOT system working under an uncertain environment, appropriate failure data is required for fault diagnosis as well as to set up maintenance policy. Meanwhile, due to unavailability and insufficient failure data, this task becomes very laborious and erratic. This paper aimed to set up fuzzy fault tree analysis (FFTA) using experts' judgment for reliability enumeration. Based on the obtained results, reliability-centered maintenance (RCM) analysis has been applied for adaptation of proper maintenance policy. An IOT-based weight data communication system has been considered as a case study.

Keywords IOT · Industry 4.0 · Fuzzy fault tree analysis · Reliability-centered maintenance

15.1 Introduction

Modern industries are adopting the IOT (Internet of things) in order to survive in the competitive market under Industry 4.0. IOT is the integration of software, hardware, and network connectivity through which communications are set up between different components in the domain. Some web-enabled devices, i.e., processor, sensors, communication devices, etc. are integrated with a conventional system and its sub-systems from which essential data are collected from a sub-system and send to other sub-system in form of information so that the required action could be taken

S. Das · B. Sarkar (✉)
Production Engineering Department, Jadavpur University, Calcutta, India
e-mail: bijan.sarkar@jadavpuruniversity.in

S. Das
e-mail: saptarshidas1989@gmail.com

J. Maiti et al. (eds.), *Recent Advances in Industrial and Systems Engineering*, Lecture Notes on Multidisciplinary Industrial Engineering,
https://doi.org/10.1007/978-981-96-8323-9_15

according to that. The system and the sub-systems can do most of the work without human intervention, though people can interact with the devices, e.g., to set them up, give them instructions or access the data. It provides a real time look into how the system really works. It is an essential technology now a days which continues to advance as more businesses realize the potential of connected devices to keep them competitive. Like all other systems, it also requires maintenance for stability and readiness, for which RCM (Reliability-centered maintenance) is a good choice. It is generally a preventive maintenance approach that focuses on optimizing the reliability of a system or process. It is technically an authentic process which focuses on the root cause of the failure of the system and maintenance operation conducted by focusing on these failures along with mode of failures, criticality of the system, and potential consequences. The fundamental of RCM can be sub-divided into three phases namely (i) **Decision**, where the factors which influence a system has been identified, (ii) **Analysis**, where the root cause of the factors influence the systems are identified and appropriate maintenance are adopted accordingly and lastly (iii) **Outcomes and feedback**, where the result obtained from maintenance program are analyzed and based on analysis proper task are carried out for further improvement of the system. From the basics of RCM, it may conclude that sufficient failure data are required in order to adopt RCM strategy. However, due to the unavailability of sufficient failure data, RCM adaptation becomes unrealizable. FFTA (Fuzzy fault tree analysis) plays a revival role under these circumstances. Its emphasis on qualitative failure data from experts in the respective fields and converted them to numerical values for further evaluation. A detail discussion on FFTA and RCM along with detail literature reviews has discussed on the later part. But the basic question may arise that what is the significance of this study? At a glance, the prime moto of the FFTA is to identify and recognize the most vulnerable component of a system using FTA and convert the expert's elicitation regarding failure of the components and system into a suitable numeric value using the fuzzy logic. On the basis of the failure data, the reliability of the system has been calculated and said values are the prime sources of RCM technique for setting up a suitable maintenance strategy. The framework of the research can be understand using a Venn diagram as shown in Fig. 15.1. Several literature surveys have been conducted on both RCM and FFTA which has been discussed in the theory part. Through the literature survey, it has been observed as a noble and holistic approach which combine the both RCM and FFTA in a single framework and on the other hand it has also found that no such approach has been initiated on the IOT system in order to adopt the proper maintenance strategy under uncertain and unstructured environment. The practical implication of the study is to augment the system availability into a large extent in order to improve the growth and development of the respective organization as a whole. The basic idea of the system, considered as the case study, has been delved.

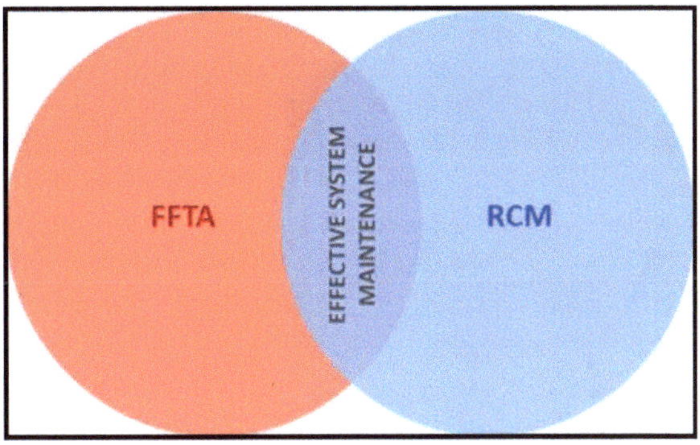

Fig. 15.1 Venn diagram of the proposed research work

15.1.1 System Overview

During data collection for research work, the concept of the said system has been adopted from a leading Indian manufacturing company of aluminum and its associated product. The data regarding weights carried by an ECL (Electrification Charpente Levage) crane, which is basically an overhead crane used in the aluminum smelting plant is available at a printer, set up inside the crane. There is no provision for digital data backup for future reference. Thus, during requirement of weight data for the requisite period is a difficult as well as laborious task and it leads to hampering the production system. Due to the massive height of the crane, the regular inspection as well as collection of data is a matter of safety concern. Considering the said factors an IOT-based weight data communication system has been adopted. This system systematically stores the data in the database system for any future reference and reflects the exact weight data information at the ground station. A typical block diagram of the system is pictured in Fig. 15.2. The weight information handled by the ECL crane is measured by the weight sensor and transmitted to the weight processor. This information is converted to a current signal having a range from 0 to 20 mAmp and received by a current loop converter. This device is attached either with a printer through a 25-pin serial connector for printing the weight value or with a display board for showing the weight. However, due to the availability of the weight data at the ground station as well as to store the digital data in the organizational database, the IOT system has been adopted. As Fig. 15.2 shows another current loop converter is connected in series with a weight processor. Likewise, the first one, it also received the current signal and transferred it into a pocket-sized computer which is the prime communication device of the system. The transmission of the signal is processed through a 25-pin serial to a 9-pin serial and then through a USB converter which is powered up by a 12 V external DC source. The data is transferred from the communication device to the ground station through Wi-Fi. A combination of programs written in Python script along with MySQL (My structured query language), a database management system set up the communication among the both cranes and the ground station. Both cranes are configured as Master–Master

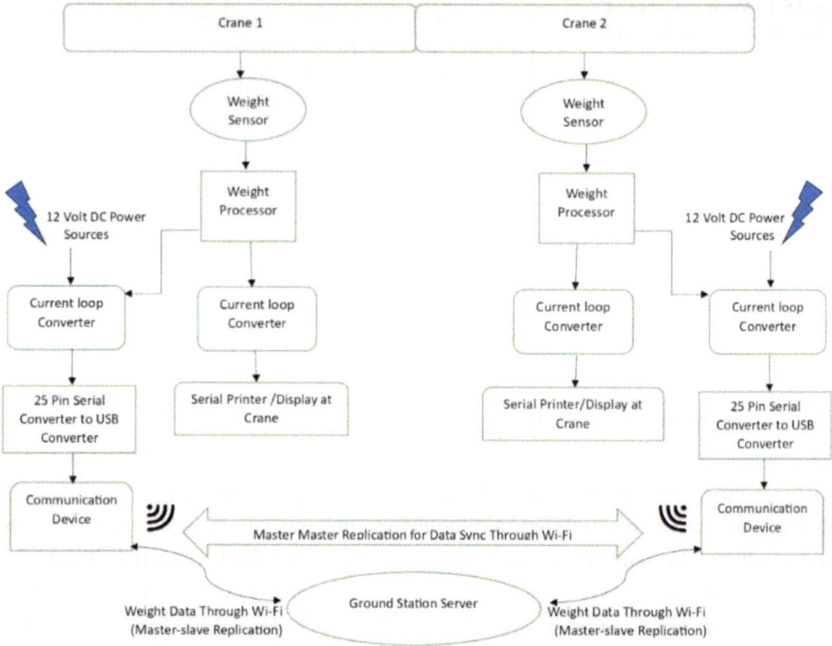

Fig. 15.2 Schematic diagram of IOT-based weight data communication system

replication in the MySQL database system. On the other side, the ground station is configured as Master–slave replication with either of the cranes in MySQL. In case of any failure in the wireless communication system, the ground station will get synchronized as the connection is ready. This leads to the elimination of data loss due to network failure. The ground station has an open-source PHP (Personal Home page currently known as Hypertext Pre-Processor) server which is programmed to fetch weight data from MySQL database. This system provides basic knowledge of an IOT system along with its predominance over conventional systems. But one thing has to be understood this is also an engineering system that comprises various types and modes of failures. While reviewing the system, the possible ground of failure has been identified and discussed in the next section.

15.1.2 FFTA

The concept of FFTA is based on two different topics. First weightage has been given to FTA which is a powerful tool to identify how the system may fail to function. It is a deductive or backward analysis as stated by Kabir [25]. The term "tree" implies the shape of the logic diagram which consists of nodes and branches [42]. This tool has extensive use to compute system reliability. It uses both qualitative and quantitative assessment to trace system failure. The causes of system failure are analyzed and gradually split into hierarchical levels until the effect of component failure can be identified [8]. Due to its versatility, it has been applied in different domains by researchers. Kang et al. [26] adopted FTA to find failure characteristics of semi-submersible floating offshore wind turbines. Both quantitative and qualitative assessments of the failure characteristics have been performed. Chi et al. [15] investigate the causes of fatal falls in the Taiwanese construction industry through FTA. They established a causal relationship among the events that cause accidents. Fazlollahtabar et al. [19] proposed an FTA-RBD-based integrated model for the reliability calculation of an automated guided vehicle (AGV) operating under a complex manufacturing system. This study has also evaluated system reliability of the manufacturing system using the minimal path and cut method (MPCM). Gachlou et al. [20] use FTA for risk assessment of a river basin to investigate undesirable events. In their work, the failure probability of the top events has been computed. Makajic-Nikolic et al. [33] apply the FTA method for risk assessment associated with infectious medical waste management. Chen et al. [14] proposed a method for fault diagnosis of an IOT system using FTA and fuzzy neural network (FNN). FNN is used to train the relationship mapping between fault symptoms and faults. The said research work majorly focused on quantitative data to enumerate failure probability. But in real-life situations, due to a shortage of useful failure data, the analysis of the effective life of a system becomes very sturdy. Under this uncertain environment, fuzzy set theory performs a key role by extracting calculable essence from qualitative data based on this idea. The idea of this theory was proposed by Prof. Zadeh [53]. This theory is extensively used in two situations, either for analysis of a complex system with an uncertain behavior or for a case where an inferential but rapid solution is required [36]. A fuzzy number or fuzzy probability is an impreciseness form of a real number. The weight of impreciseness varies from 0 to 1, which is known as the membership function. A fuzzy number represents itself in the probability space under a fuzzy set with a value ranging from 0 to 1. A triangular fuzzy number (TFN) and its membership function are illustrated in Fig. 15.3 along with its membership function [16]. $f(x)$ is the membership function of the triangular fuzzy no as drawn above and expressed as:

$$f(x) = \begin{cases} \frac{x-a}{b-a}, & a \leq x \leq b \\ \frac{c-x}{c-b}, & b \leq x \leq c \\ 0, & \text{otherwise} \end{cases} \tag{15.1}$$

The extraction of crisp values from qualitative parameters in the field of reliability analysis using fuzzy theory has been adopted by various researchers. Liu et al. [31] in their study, conducted an FTA of high-speed railway accidents. A fuzzy set theory has been applied to their research work to overcome uncertain characteristics of basic events in fault trees caused due to obscure information. Henceforth, this theory is considered a worthy choice for the computation of qualitative data resulting from FTA [11], and combining those studies, the concept of FFTA comes

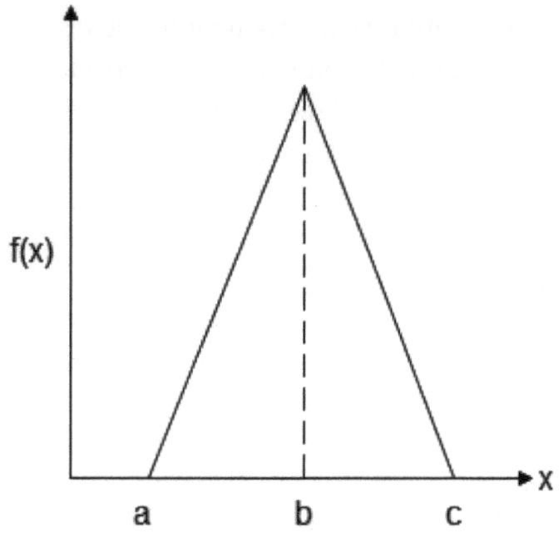

Fig. 15.3 TFN and its membership function

into the picture. Mahmood et al. [32] introduce a review of FFTA and reflect on its strengths, weaknesses, and application in their research work. Yazdi et al. [52] used FFTA to evaluate the failure probability (FP) of various basic events (BEs) in a chemical storage tank located at a chemical plant. Three experts from the respective field have been invited to provide qualitative information regarding the failure of basic events. Badida et al. [6] utilized FFTA to evaluate the FP of pipelines associated with oil and natural gas pipeline systems using expert opinion. Babaei et al. [5] performed FFTA of agricultural water conveyance and delivery systems to compute the FP of top events. Abdelgawad et al. [1] apply FFTA to the construction industry. A case study has been conducted on horizontal directional drilling in this regard. Kargar et al. [27] studied the risk assessment of Asymmetric Tandem Lifting operation on the basis of FFTA. A comprehensive approach has been taken by Tabesh et al. [44] for risk assessment of water treatment plant using FFTA. Gürgen et al. [21] have considered FFTA for identify the root causes of accidents resulting from the loss of a ship steering ability. Khajuria [28] has proposed FFTA on non-repairable system which contain uncertain information. Triangular intuitionistic fuzzy numbers (TIFNs) have been used for data uncertainty. Based on those literature reviews, it is found that FFTA comes out as an effective tool for FP calculation under uncertain environments. Based on the assumption, FFTA has been incorporated in this study for the calculation of FP of BEs of the said IOT system. A fault tree (FT) diagram of the IOT system has been illustrated in Fig. 15.4 to identify the elemental failure of the system. Table 15.1 discusses about particulars of different failures as drawn in FT.

15.1.3 Reliability-Centered Maintenance (RCM)

RCM is a mode of optimal maintenance [10] through which a system can carry out its designed function under a normal operating environment. Safety is the primary aim of

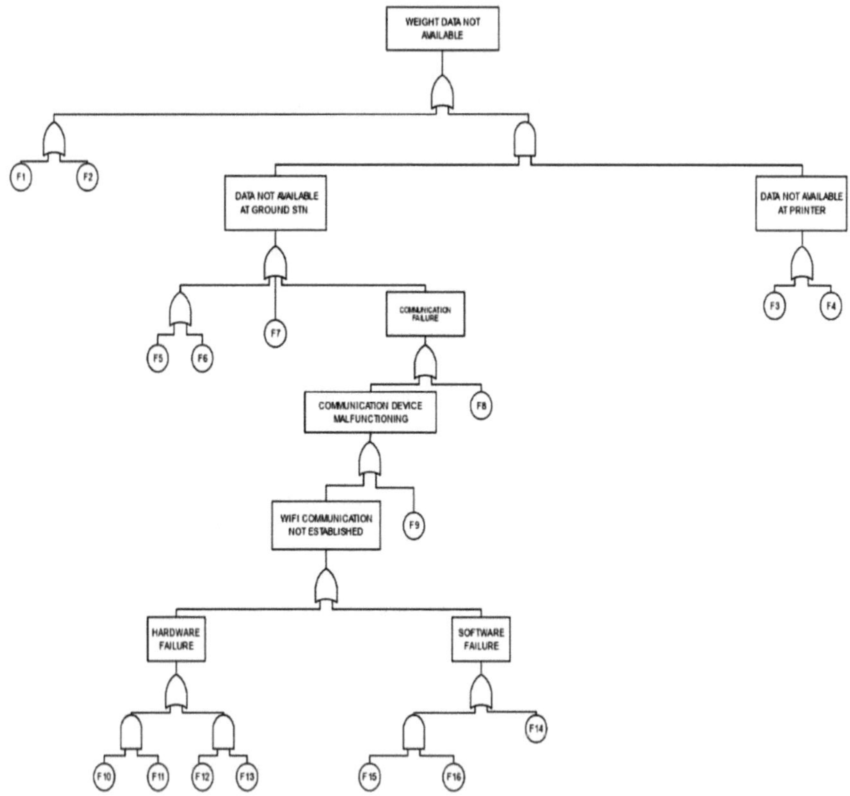

Fig. 15.4 Fault tree diagram of IOT-based weight data communication system

RCM followed by reliability, availability, and cost-effectiveness of the system [34]. Henceforth, this ensures the stabilization of the inherent reliability of the system. The selection of preventive maintenance strategies is based on the probability of failure modes and their consequences [37]. Preventive maintenance aims to secure the system before it fails [50]. It either follows a schedule or is determined by the condition of the system [2, 24]. RCM methodology put in the preventive maintenance action in a structured manner to extend the service life of the system [12, 30]. The concept along with a detailed methodological overview of RCM has been propounded by Tale [46]. The author has constructed a flow diagram of RCM methodology during this research work which was found very helpful during this study. This method found a unique paradigm over other maintenance strategies and was adopted in different engineering fields. Aji et al. [3] applied the RCM method on an onboard communications-based train control device and found this method useful. Fang et al. [18] exerted the RCM technique on the metro door system. Firstly, the sub-systems and their components have been identified. An RCM logic decision diagram has been set up and based on the diagram, the maintenance mode of each sub-system has

Table 15.1 Particulars of different failures in the FT

BEs	Particulars
F1	WT sensors failure
F2	WT processor failure
F3	Current loop converter failure
F4	Printer failure
F5	Current loop converter no. 2 failure
F6	12 Volt DC power source failure
F7	25 PIN serial to USB converter failure
F8	Power failure of communication device
F9	WiFi chip failure of communication device
F10	Ground station power failure
F11	Backup crane failure
F12	Ground station WiFi receiver failure
F13	Backup crane WiFi receiver failure
F14	Software failure due to malware attack
F15	Software failure: M-to-M synchronization
F16	Software failure: M-to-S synchronization

been chosen. Eriksen et al. [17] examined the applicability of RCM on Unmanned autonomous cargo ships. The limitation and necessary amendments of the RCM have been discussed in the article. Patil et al. [35] developed an optimized maintenance program using RCM. They choose steam boiler systems as their working domain. Implementation of RCM implies the reduction of maintenance costs by improving the availability of the system [23]. The RCM framework was also applied to distribution grids by Wang et al. [49] based on FTA. This study was found quite helpful because of the integration between RCM and FTA. Suryono et al. [43] considered the RCM technique to find the optimal inspection frequency PT. X used an automated machine. The average inspection time and average repair time based on a fixed reliability value have been computed based on the said method. Based on the RCM technique, Setiawan et al. [39] were able to increase fabric production by 13% by reducing downtime by up to 30% of an air jet loom (ALJ) machine. The above literature reviews signify the effectiveness of RCM in finding effective maintenance strategies based on the failure probability of the system as well as in minimizing downtime which signifies the improvement of system availability. Taha et al. [45] proposed a cyber physical system on electric busses to deals with the challenges of developing self-improvement and online risk control. The base of the online risk control approach used in this study is RCM which has been used for enhancement of system uptime. Shaheen and Németh [40] conducted a literature review focusing on Industry 4.0 and its effect on maintenance management systems. Tortorella et al. [48] have examined the impact of integrating Industry 4.0 on total productive maintenance. Based on the literature reviews of both FFTA and RCM, it is found that both FFTA and RCM are found useful in their respective domain and, considering their merits and demerits, a collaborative approach has been set up between them to select the correct maintenance strategy for the IOT-based data communication system during its operation.

15.2 Methodology

The basic methodology of this research work consists of two parts as stated before. A flowchart of the work has been constructed and portrayed in Fig. 15.5.

First, three steps of the block diagram were already covered by the introductory part. After constructing the FT diagram, the basic failure events have been identified as shown in Fig. 15.4. The FFTA has been conducted through the following stages as discussed.

15.2.1 Rating and Weightage Stage

A group of four experts from the relevant field having sound knowledge of the system were invited for their valuable opinions regarding the failure of respective IOT systems. The score rating criteria of the experts are represented in Table 15.2.

Based on the assigned score expressed in Table 15.2, the assigned score (x_i) of each expert has been computed and tabulated in Table 15.3. The sum of the assigned score $\left(\sum x_i\right)$ has also been evaluated to find the weightage associated with each expert.

Henceforth, the weightage, $W(E_u)$ of each expert has been presented in Table 15.4.

The linguistic rating of FP based on the expert's judgment has been collected at different periods for all BEs. In this study, the periods are considered as 2500 h, 5000 h, 7500 h, and 10,000 h, respectively. Experts' decisions corresponding to each period have been reflected in Tables 15.5, 15.6, 15.7 and 15.8.

15.2.2 Selection of Fuzzy Number and Membership Function

Numerous fuzzy numbers are used by different researchers working in different fields. But triangular fuzzy number (TFN) and trapezoidal fuzzy number (TZFN) are best suited for reliability analysis [32]. After studying different literature reviews, TFN finds a suitable choice for its simplicity as well as effectiveness [22]. A fuzzy set with seven linguistic representations has been considered in this study for precise calculation [13]. The TFN of each linguistic expression is tabulated in Table 15.9.

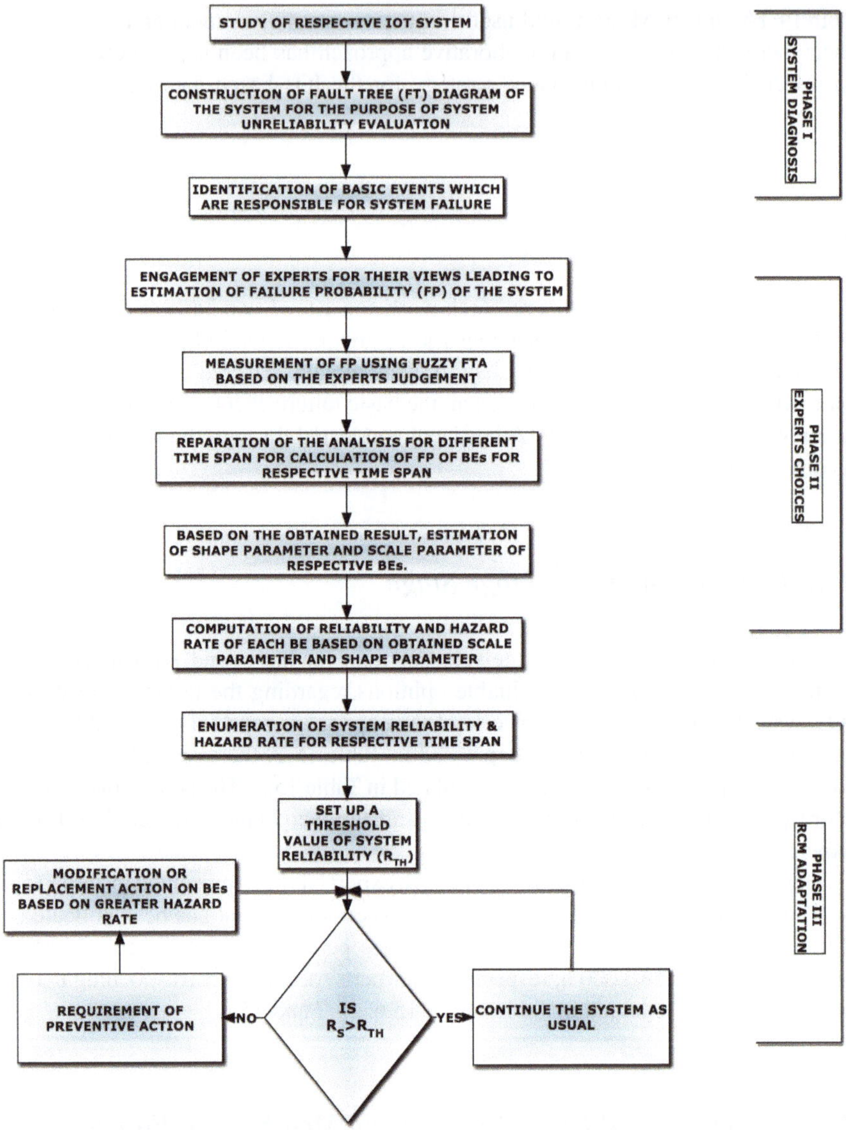

Fig. 15.5 Flow chart of the research work

The linguistic representations of fuzzy numbers have been reflected in Fig. 15.6. Based on the linguistic response proposed by the experts, aggregation of linguistic expressions has been conducted.

Table 15.2 Rating of experts based on criteria

Sl. no	Criterion	Category	Score
1	Designation	General Manager (GM)/Dy. GM/Professor	5
		AGM/Associate Professor	4
		Sr. Manager/Senior Researcher/Assistant Professor	3
		Assistant Manager/Foreman/Lecturer	2
		Junior Researcher/Supervisor	1
2	Job/research experience	Over 30 years	4
		21–30 years	3
		11–20 years	2
		10 and below 10 years	1
3	Qualification	PhD and above	4
		M.E/M. Tech/M.B.A	3
		B.E/B. Tech	2
		Diploma Engineering	1
4	Time of involvement with Job-shop activity	More than 50 h/week	4
		30 to 50 h/week	3
		15 to 29 h/week	2
		Less than 15 h/week	1

Table 15.3 Assign score of the experts

Expert (E)	Designation	Job/Research experience	Qualification	Time of involvement with job-shop activity	Assign score (x_i)
1	Dy. GM (R&D) (Score: 5)	25 years (Score: 3)	M.E (Score: 3)	18 h/week (Score: 2)	13
2	Professor (Industrial Engineering) (Score: 5)	32 years (Score: 4)	PhD (Score: 4)	6 h/week (Score: 1)	14
3	Senior Researcher (Score: 3)	5 years (Score: 1)	M. Tech (Score: 3)	36 h/week (Score: 3)	10
4	Sr. Manager (Score: 3)	16 years (Score: 2)	M. Tech (Score: 3)	55 h/week (Score: 4)	12
The sum of assign score ($\sum x_i$)					49

Table 15.4 Weightage of each expert

Expert (E_u)	1	2	3	4
Weightage $\left(\frac{x_i}{\sum x_i}\right)$	0.265	0.286	0.204	0.245

Table 15.5 Decision of exp. for 2500 h

BEs	E1	E2	E3	E4
F1	M	M	H	M
F2	M	M	M	H
F3	M	L	M	M
F4	VL	L	L	VL
F5	M	L	M	M
F6	M	M	M	M
F7	M	L	L	M
F8	VL	EL	EL	VL
F9	VL	EL	L	VL
F10	VL	VL	VL	VL
F11	L	L	VL	VL
F12	VL	VL	L	VL
F13	L	VL	L	VL
F14	L	M	L	VL
F15	EL	EL	EL	EL
F16	EL	EL	EL	EL

15.2.3 Aggregation of Basic Events

The steps associated with fuzzification and aggregation of linguistic expression in FFTA are discussed below.

Step 1: Enumerating the Degree of Similarity Between a Pair of Experts

It is defined as the closeness of judgment between a pair of experts out of j number of experts on the same event. Suppose an expert E_1 proposed a judgment $J_1 = (a_1, a_2, a_3)$ for an event and expert E_2 proposed a judgment $J_2 = (b_1, b_2, b_3)$ for the same event. The degree of similarity between the pair of experts, $S\ (J_1, J_2)$ is defined as:

$$S(J_1, J_2) = 1 - \frac{1}{J}\sum_{i=1}^{J}|a_i - b_i|, \text{ where } i = 1, 2, 3 \quad (15.2)$$

Table 15.6 Decision of exp. for 5000 h

BEs	E1	E2	E3	E4
F1	M	H	VH	H
F2	H	VH	H	H
F3	H	M	M	H
F4	L	L	M	L
F5	H	M	M	H
F6	H	H	H	M
F7	H	M	M	M
F8	VL	VL	VL	VL
F9	VL	VL	L	L
F10	L	VL	VL	L
F11	M	L	VL	L
F12	VL	L	L	VL
F13	L	L	L	L
F14	L	M	M	L
F15	EL	EL	EL	VL
F16	EL	EL	EL	VL

Table 15.7 Decision of exp. for 7500 h

BEs	E1	E2	E3	E4
F1	H	VH	VH	VH
F2	H	VH	VH	VH
F3	VH	H	H	H
F4	M	M	M	M
F5	VH	H	H	H
F6	VH	VH	VH	VH
F7	H	H	H	H
F8	L	L	VL	L
F9	L	L	M	L
F10	M	L	L	L
F11	M	L	L	M
F12	L	M	M	L
F13	M	L	L	M
F14	M	M	M	L
F15	VL	EL	EL	VL
F16	VL	EL	EL	VL

Table 15.8 Decision of exp. for 10,000 h

Bes	E1	E2	E3	E4
F1	VH	EH	EH	VH
F2	VH	EH	VH	EH
F3	VH	VH	VH	VH
F4	H	H	M	M
F5	VH	VH	VH	VH
F6	EH	VH	EH	EH
F7	VH	H	VH	H
F8	L	M	L	L
F9	M	M	M	M
F10	M	L	M	M
F11	M	M	L	H
F12	M	M	H	L
F13	H	M	M	M
F14	M	M	M	M
F15	VL	VL	VL	VL
F16	VL	VL	VL	VL

Table 15.9 Linguistic expression of TFN

Sl. no	Linguistic representation	TFN
1	Extremely low (EL)	(0, 0, 0.17)
2	Very low (VL)	(0, 0.17, 0.33)
3	Low (L)	(0.17, 0.33, 0.5)
4	Medium (M)	(0.33, 0.5, 0.67)
5	High (H)	(0.5, 0.67, 0.83)
6	Very high (VH)	(0.67, 0.83, 1)
7	Extremely high (EH)	(0.83, 1, 1)

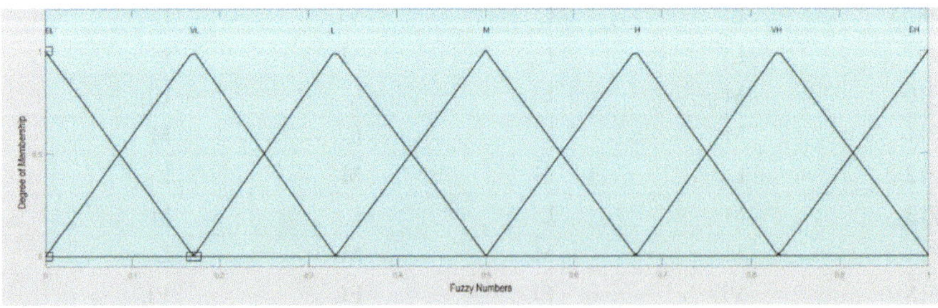

Fig. 15.6 Linguistic representation of failure probability

Step 2: Calculating the Average Agreement (AA) Degree of an Expert

This step computes the average value of the degree of similarity of an expert concerning the rest of the experts for the same event. Suppose there are four experts, namely E_1, E_2, E_3, and E_4 and their judgment on the same event is J_1 J_2 J_3, and J_4, respectively. The average agreement (AA) degree of E_1 can be written as:

$$\mathrm{AA}(E_1) = \frac{1}{j-1}[S(J_1, J_2) + S(J_1, J_3) + S(J_1, J_4)],$$
$$\text{where}\, j = \text{No. of experts}$$

The general expression of the average of agreement (AA) degree can be written as:

$$\mathrm{AA}(E_u) = \frac{1}{j-1} \sum_{\substack{u \neq v \\ v=1}}^{j} S(J_u, J_v) \tag{15.3}$$

Step 3: Computing Relative Agreement (RA) Degree of All Expert

The general expression of the relative agreement (RA) degree can be manifested as:

$$\mathrm{RA}(E_u) = \frac{\mathrm{AA}(E_u)}{\sum_{u=1}^{j} \mathrm{AA}(E_u)} \tag{15.4}$$

Step 4: Calculating the Consensus Coefficient (CC) of an Expert

While calculating the consensus coefficient (CC), a term called relaxation factor (γ) has to be considered. The range of (γ) varies from 0 to 1, i.e., $[0 \leq \gamma \leq 1]$. The consensus coefficient (CC) of an expert can be estimated as:

$$\mathrm{CC}(E_u) = \gamma \cdot W(E_u) + (1-\gamma) \cdot \mathrm{RA}(E_u) \tag{15.5}$$

It maintains a balance between weightage, $W(E_u)$, and relative agreement, $\mathrm{RA}(E_u)$ during the calculation of (CC). So, the precise value of γ should be considered during calculation. In the case of equal weightage, the value of γ considered as 0.5, and this consideration has also been adopted in this work.

Step 5: Computing Aggregated Result of Expert's Judgment, R_{AG}

The value of R_{AG} for each event can be computed as:

$$R_{\mathrm{AG}} = \sum_{u=1}^{j} \mathrm{CC}(E_u) \cdot J_u \tag{15.6}$$

The de-fuzzification of each BE would be started after enumerating the R_{AG} of each event.

15.2.4 Defuzzification and Computation of FP

Defuzzification implies the conversion of fuzzy numbers into crisp numbers. Several methods are available for the calculation of crisp value. In this work, the conversion of TFN to the corresponding crisp value is inspired by BAN et al. [7] as it was found more precise compared to others. Suppose the R_{AG} of an event is represented by (x, y, z) then the corresponding crisp value X' will be:

$$X' = \frac{(x + 2y + z)}{4} \tag{15.7}$$

The crisp value is vital for the computation of FP based on the following formulae used by Badida et al. [6].

$$k = \left[\left(\frac{1}{X'} - 1\right)^{\frac{1}{3}}\right] \times 2.301 \tag{15.8}$$

$$\text{FP} = \frac{1}{10^k} \tag{15.9}$$

Weibull distribution has been considered as a probability distribution function for all BEs, to compute reliability as well as hazard rate. Weibull distribution due to its indistinct characteristic shape as well as to fit experimental data that are not characterized by any distributions has a sound abidance [8]. FP of two-parameter Weibull distribution written as:

$$\text{FP} = 1 - \exp^{\left[-\left(\frac{t}{\alpha}\right)^{\beta}\right]} \tag{15.10}$$

where β, α are the shape parameter and scale parameters, respectively

t represents the time corresponding to FP.

To compute the values of the shape parameter and scale parameter, both FP and corresponding time are plotted in a Weibull probability graph paper as suggested by Kumar et al. [29]. The axis of the graph is:

$$x\,\text{axis} = \ln(t) \tag{15.11}$$

$$y\,\text{axis} = \ln\left[\ln\left(\frac{1}{1 - \text{FP}}\right)\right] \tag{15.12}$$

The values are put into an Excel sheet which constructs a best-fitted straight line as well as provides its equation through which both parameters have been reckoned.

15.2.5 Reliability (R) and Hazard Rate (λ) Calculation

Reliability (R) implies the probability of successful operation of a system for a specific period under a given operating condition. This indicates, $R(t) = (1 - \text{FP})$. Therefore, reliability in terms of formula can be written as:

$$R(t) = \exp^{\left[-\left(\frac{t}{\alpha}\right)^{\beta}\right]} \tag{15.13}$$

Hazard rate (λ) is defined as the instantaneous failure rate of the system per unit time. It can be written as:

$$\lambda = \frac{\beta t^{\beta-1}}{\alpha^{\beta}} \tag{15.14}$$

15.2.6 Adaptation of RCM Strategy

After enumerating the values of R and λ for all BEs for different periods, the system reliability (R_s) and system hazard rate (λ_s) have been computed and plotted against the said periods to observe the nature of the plotting and also to calculate intermediate values. The concept of Boolean algebra in terms of logic gates namely OR gate and AND gate is considered for the construction of FT. The equation of reliability and hazard rate for OR gate is given by Eqs. (15.15) and (15.16), respectively. Similarly, the equation of reliability and hazard rate for AND gate is presented in Eqs. (15.17) and (15.18), respectively [9].

$$R(t) = \prod_{i=1}^{m} R_i(t) \tag{15.15}$$

$$\lambda(t) = \sum_{i=1}^{m} \lambda_i(t) \tag{15.16}$$

$$R(t) = 1 - \prod_{i=1}^{m} [1 - R_i(t)] \tag{15.17}$$

$$\lambda(t) = \sum_{i=1}^{m} \lambda_i(t) \frac{R_i(t)}{1 - R_i(t)} \left(\frac{\prod_{j=1}^{m} \left(1 - R_j(t)\right)}{1 - \prod_{j=1}^{m} \left(1 - R_j(t)\right)} \right) \tag{15.18}$$

A threshold value R_{TH} of system reliability has been established by industrial experts, design engineers, maintenance engineers, and manufacturers' viewpoints. Preventive maintenance action would be immediately initiated once the threshold value found is greater than system reliability (R_S). The element responsible for BEs having the highest hazard rate value will be replaced with a new one for system sustainability. This results in the improvement of R_S and reduces λ_S.

15.3 Findings and Contributions

15.3.1 Result of FP Using FFTA

The result of FP has been obtained after a numerous calculation based on the equations stated above. For easy understanding of the reader, the result of each step for 2500 h is discussed below. Under this step, it can be observed from Table 15.5 that, both Expert E1and E2 has rated Basic Event F1 as "Medium" and the corresponding TFN of "Medium" has shown as (0.33,0.5,0.67) in Table 15.9. Following the Eq. 15.2, the degree of similarity of E1 and E2 for F1 has been calculated below and remaining results tabulated in Table 15.10.

$$S(\text{E1}, \text{E2}) = 1 - \frac{1}{3}[(0.33 - 0.33) + (0.5 - 0.5) + (0.67 - 0.67)] = 1$$

The average of the agreement (AA) degree of each expert furnished in Table 15.11 has been computed vide Eq. (15.3). Evaluation of AA (E1) for F1 has been shown as a reference.

$$\text{AA(E1)} = \left\{\frac{1}{(4-1)} * [1 + 0.833333 + 1]\right\} = 0.944444$$

After obtaining the AA of experts, the relative agreement (RA) degree of every expert is evaluated based on Eq. (15.4) and put into Table 15.12. A reference calculation of RA (E1) for F1 has been given below.

$$\text{RA(E1)} = \left\{\frac{0.944444}{(0.944444 + 0.944444 + 0.833333 + 0.944444)}\right\} = 0.257576$$

Based on the calculated value of RA from Table 15.12 and the weightage of the experts in Table 15.4, consensus coefficient (CC) values are evaluated by considering Eq. (15.5). The CC values of the experts are furnished in Table 15.13. CC of E1 for F1 has been describe below.

$$\text{CC(E1) for F1} = \{0.5 * 0.265 + (1 - 0.5) * 0.257576\} = 0.2613$$

Table 15.10 Degree of similarity between pair of experts

Bes	*S* (E1,E2)	*S* (E1,E3)	*S* (E1,E4)	*S* (E2,E3)	*S* (E2,E4)	*S* (E3,E4)
F1	1	0.833333	1	0.833333	1	0.833333
F2	1	1	0.833333	1	0.833333	0.833333
F3	0.833333	1	1	0.833333	0.833333	1
F4	0.833333	0.833333	1	1	0.833333	0.833333
F5	0.833333	1	1	0.833333	0.833333	1
F6	1	1	1	1	1	1
F7	0.833333	0.833333	1	1	0.833333	0.833333
F8	0.89	0.89	1	1	0.89	0.89
F9	0.89	0.833333	1	0.723333	0.89	0.833333
F10	1	1	1	1	1	1
F11	1	0.833333	0.833333	0.833333	0.833333	1
F12	1	0.833333	1	0.833333	1	0.833333
F13	0.833333	1	0.833333	0.833333	1	0.833333
F14	0.833333	1	0.833333	0.833333	0.666667	0.833333
F15	1	1	1	1	1	1
F16	1	1	1	1	1	1

Table 15.11 Average of agreement (AA) degree of each expert

BEs	AA (E1)	AA (E2)	AA (E3)	AA (E4)
F1	0.944444	0.944444	0.833333	0.944444
F2	0.944444	0.944444	0.944444	0.833333
F3	0.944444	0.833333	0.944444	0.944444
F4	0.888889	0.888889	0.888889	0.888889
F5	0.944444	0.833333	0.944444	0.944444
F6	1	1	1	1
F7	0.888889	0.888889	0.888889	0.888889
F8	0.926667	0.926667	0.926667	0.926667
F9	0.907778	0.834444	0.796667	0.907778
F10	1	1	1	1
F11	0.888889	0.888889	0.888889	0.888889
F12	0.944444	0.944444	0.833333	0.944444
F13	0.888889	0.888889	0.888889	0.888889
F14	0.888889	0.777778	0.888889	0.777778
F15	1	1	1	1
F16	1	1	1	1

Table 15.12 Relative agreement (RA) degree of each expert

BEs	RA (E1)	RA (E2)	RA (E3)	RA (E4)
F1	0.257576	0.257576	0.227273	0.257576
F2	0.257576	0.257576	0.257576	0.227273
F3	0.257576	0.227273	0.257576	0.257576
F4	0.25	0.25	0.25	0.25
F5	0.257576	0.227273	0.257576	0.257576
F6	0.25	0.25	0.25	0.25
F7	0.25	0.25	0.25	0.25
F8	0.25	0.25	0.25	0.25
F9	0.263378	0.242102	0.231141	0.263378
F10	0.25	0.25	0.25	0.25
F11	0.25	0.25	0.25	0.25
F12	0.257576	0.257576	0.227273	0.257576
F13	0.25	0.25	0.25	0.25
F14	0.266667	0.233333	0.266667	0.233333
F15	0.25	0.25	0.25	0.25
F16	0.25	0.25	0.25	0.25

Table 15.13 CC values of experts

BEs	CC (E1)	CC (E2)	CC (E3)	CC (E4)
F1	0.2613	0.2718	0.2156	0.2513
F2	0.2613	0.2718	0.2308	0.2361
F3	0.2613	0.2566	0.2308	0.2513
F4	0.2575	0.2680	0.2270	0.2475
F5	0.2613	0.2566	0.2308	0.2513
F6	0.2575	0.2680	0.2270	0.2475
F7	0.2575	0.2680	0.2270	0.2475
F8	0.2575	0.2680	0.2270	0.2475
F9	0.2642	0.2641	0.2176	0.2542
F10	0.2575	0.2680	0.2270	0.2475
F11	0.2575	0.2680	0.2270	0.2475
F12	0.2613	0.2718	0.2156	0.2513
F13	0.2575	0.2680	0.2270	0.2475
F14	0.2658	0.2597	0.2353	0.2392
F15	0.2575	0.2680	0.2270	0.2475
F16	0.2575	0.2680	0.2270	0.2475

The aggregated result of the expert's judgment (R_{AG}) is the value that is useful for the calculation of crisp value for all the BEs. The value of R_{AG} is obtained by following Eq. (15.6). From Table 15.5, it can be seen that E1, E2, E3, and E4 have opted for F1 as Medium, Medium, High, and Medium, respectively. Based on the respective TFN as shown in Table 15.9, the calculation of (R_{AG}) for F1 has been given below and the remaining values are tabulated in Table 15.14.

$$\begin{aligned}(R_{AG})\text{ of }F1 &= [(0.33 * 0.2613) + (0.33 * 0.2718) + (0.5 * 0.2156) \\ &\quad + (0.33 * \ 0.2513)] = 0.3667 \\ &= [(0.5 * 0.2613) + (0.5 * 0.2718) + (0.67 * 0.2156) \\ &\quad + (0.5 * 0.2513)] = 0.5367 \\ &= [(0.67 * 0.2613) + (0.67 * 0.2718) + (0.83 * 0.2156) \\ &\quad + (0.67 * 0.2513)] = 0.7045\end{aligned}$$

Finally, the FP values of BEs are derived considering Eqs. (15.7) to (15.9) and furnished in Table 15.15. The numerical framework of F1 has been given below.

$$X' = \left\{\frac{0.3667 + 2 * 0.5367 + 0.7045}{4}\right\} = 0.53612$$

Table 15.14 Aggregated result of expert's judgment (R_{AG})

BEs	Aggregation of triangular fuzzy number		
F1	0.3667	0.5367	0.7045
F2	0.3701	0.5401	0.7078
F3	0.2889	0.4564	0.6264
F4	0.0842	0.2492	0.4142
F5	0.2889	0.4564	0.6264
F6	0.3300	0.5000	0.6700
F7	0.2508	0.4159	0.5859
F8	0.0000	0.0859	0.2508
F9	0.0370	0.1599	0.3247
F10	0.0000	0.1700	0.3300
F11	0.0893	0.2541	0.4193
F12	0.0367	0.2045	0.3667
F13	0.0824	0.2475	0.4124
F14	0.1709	0.3359	0.5035
F15	0.0000	0.0000	0.1700
F16	0.0000	0.0000	0.1700

Table 15.15 FP values of BEs

BEs	X'	K	FP
F1	0.53612	2.19263	0.006417585
F2	0.53955	2.18256	0.006568177
F3	0.45701	2.43708	0.003655264
F4	0.24918	3.32349	0.000474796
F5	0.45701	2.43708	0.003655264
F6	0.50000	2.30100	0.005000345
F7	0.41709	2.57262	0.002675357
F8	0.10563	4.68994	2.042E−05
F9	0.17039	3.89993	0.000125912
F10	0.16750	3.92681	0.000118356
F11	0.25421	3.29402	0.000508134
F12	0.20308	3.62938	0.000234758
F13	0.24744	3.33379	0.00046367
F14	0.33653	2.88524	0.001302458
F15	0.04250	6.49880	3.17102E−07
F16	0.04250	6.49880	3.17102E−07

$$k = \left[\left(\frac{1}{0.53612} - 1\right)^{\frac{1}{3}}\right] \times 2.301 = 2.19263$$

$$\text{FP} = \frac{1}{10^{2.19263}} = 0.006417585$$

15.3.2 Estimation of Reliability (R) and Hazard Rate (λ)

The values of R and λ are obtained from Eqs. (15.13) and (15.14), respectively, which are insoluble due to obscure values of β and α. Under these circumstances, those values are tracked from the Weibull probability graph paper. Equations (15.11) and (15.12) indicate the x-axis and y-axis of the graph, respectively. To solve those equations, period, and adjacent FP of all BEs are required. Likewise, previous method, the FP of all BEs has been evaluated for 5000, 7500, and 10,000 h, respectively. For example, the FP of F1 for 5000 h has been evaluated as 0.014470 and similarly, for 7500 h and 10,000 h, the FP values have appeared as 0.0335832 and 0.0743145, respectively. An Excel sheet has been utilized to calculate and plot the graph, and it also indicates the equation of the trend line, and detailed calculation for F1 has been provided below.

$$x\,\text{axis} = \ln(2500) = 7.824046$$

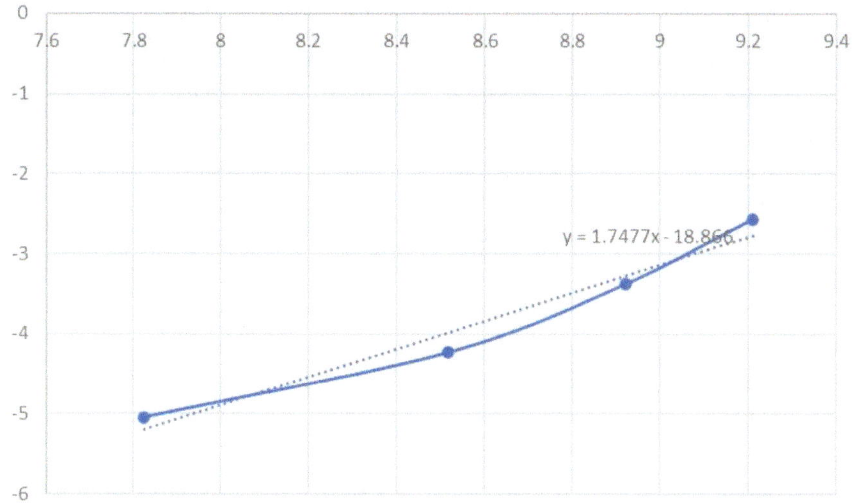

Fig. 15.7 Weibull distribution plotting for F1

$$y\,\text{axis} = \ln\left[\ln\left(\frac{1}{1 - 0.006417585}\right)\right] = -5.04549598$$

Continuing the same process for 5000, 7500, and 10,000 h as stated before, the Weibull distribution graph has been constructed by plotting the points. Considering the trend line, the equation has been generated from Microsoft Excel Software as stated above. Using the equation of the straight line, the values of β and α have been computed. Weibull distribution plotting for F1 has furnished in Fig. 15.7. β and α of other BEs are also calculated in the similar manner and tabulated in Table 15.16.

Using values of β and α as given in Table 15.16, the values of R and λ have been computed using Eqs. (15.13) and (15.14). Period-wise values of R and λ for all BEs are arranged in Tables 15.17 and 15.18, respectively. The reference calculation for evaluating R and λ under the time span of 2500 h has describe below for F1.

$$R(t) = \exp^{\left[-\left(\frac{2500}{48764.54671}\right)^{1.7477}\right]} = 0.994454088$$

$$\lambda = \frac{1.7477 * 2500^{(1.7477-1)}}{48764.54671^{1.7477}} = 6.5281 * 10^{-6}$$

Table 15.16 β,α values along with *R* square and adjusted *R* square values of BEs

BEs	Shape parameter (β)	Scale parameter (α)	*R* square (%)	*R* square (adjusted) (%)
F1	1.7477	48,764.54671	96.02	94.04
F2	1.7155	48,215.94525	98.02	97.03
F3	1.7672	64,400.94963	96.52	94.78
F4	2.149	90,131.72178	99.59	99.38
F5	1.7672	64,400.94963	96.52	94.78
F6	2.2084	30,447.64137	94.56	91.83
F7	1.6252	99,766.05886	98.96	98.45
F8	3.3525	65,684.44705	98.08	97.13
F9	2.632	81,450.50058	97.21	95.82
F10	2.5124	95,030.07831	98.94	98.41
F11	1.6392	266,380.0575	98.86	98.29
F12	2.2706	111,381.552	91.85	87.78
F13	1.8538	167,601.6193	96.96	95.44
F14	0.9662	2,393,985.949	99.90	99.85
F15	4.16	94,353.88178	98.92	98.37
F16	4.16	94,353.88178	98.92	98.37

Table 15.17 Period-wise values of R

BEs	Reliability (*R*)			
	2500 h	5000 h	7500 h	10,000 h
F1	0.994454088	0.981497049	0.96277512	0.939207194
F2	0.993779887	0.979717224	0.95975021	0.934918841
F3	0.996794704	0.989131377	0.97787493	0.963484852
F4	0.999549139	0.998001886	0.99523089	0.991168326
F5	0.996794704	0.989131377	0.97787493	0.963484852
F6	0.996003668	0.981663645	0.95570004	0.918025306
F7	0.997502863	0.992316742	0.98520314	0.976487612
F8	0.99998258	0.999822084	0.99930745	0.998184224
F9	0.999895796	0.999354232	0.99812379	0.996003697
F10	0.999892704	0.999387958	0.99830583	0.996512907
F11	0.999525411	0.99852243	0.99712992	0.995404658
F12	0.999819686	0.999130243	0.99781754	0.995810189
F13	0.999588618	0.998513857	0.99685126	0.994638747
F14	0.998683877	0.997430321	0.99620029	0.9949858
F15	0.999999724	0.999995072	0.99997338	0.999911899
F16	0.999999724	0.999995072	0.99997338	0.999911899

Table 15.18 Period-wise values of λ

BEs	Hazard rate (λ) in (failure/h)			
	2500 h	5000 h	7500 h	10,000 h
F1	3.88783E−06	6.5281E−06	8.83996E−06	1.09614E−05
F2	4.28157E−06	7.03056E−06	9.39687E−06	1.15446E−05
F3	2.2694E−06	3.86243E−06	5.27179E−06	6.57372E−06
F4	3.87648E−07	8.59648E−07	1.36978E−06	1.90636E−06
F5	2.2694E−06	3.86243E−06	5.27179E−06	6.57372E−06
F6	3.53727E−06	8.17397E−06	1.3342E−05	1.88885E−05
F7	1.62537E−06	2.50701E−06	3.23034E−06	3.86687E−06
F8	2.33604E−08	1.19303E−07	3.09676E−07	6.09292E−07
F9	1.09711E−07	3.40042E−07	6.59043E−07	1.05393E−06
F10	1.07834E−07	3.07633E−07	5.68006E−07	8.77628E−07
F11	3.11252E−07	4.84765E−07	6.28187E−07	7.55005E−07
F12	1.63783E−07	3.95146E−07	6.61453E−07	9.53337E−07
F13	3.0511E−07	5.51412E−07	7.79513E−07	9.96543E−07
F14	5.0899E−07	4.97204E−07	4.90436E−07	4.85691E−07
F15	4.58757E−10	4.10051E−09	1.47668E−08	3.66516E−08
F16	4.58757E−10	4.10051E−09	1.47668E−08	3.66516E−08

15.3.3 *RCM and Its Consequences*

System reliability (R_{sys}) and system hazard rate (λ_{sys}) have been evaluated from respective Eqs. (15.15) to (15.18) utilizing the values of Tables 15.17 and 15.18. Readers may go through the concept of reliability evaluation for series and parallel systems in this regard. Periodical values of R_{sys} and λ_{sys} are reflected in Table 15.19 and their periodical distribution curve as shown in Figs. 15.8 and 15.9, respectively.

A threshold value of system reliability, R_{th} which is the minimum operating criteria of the system has been chosen as 0.9. While tracing the R_{sys} vs time graph as shown in Fig. 15.8, it is found that R_{sys} is reaching R_{th} approximately at 8610 h. Among all the BEs, the hazard rate value of event F6 was found maximum with a value of 1.57636×10^{-05}/h. at that instant. After replacing the prime constituent responsible for F6, i.e., 12 V power source, the value of R_{sys} shifted to 0.902. This minute change

Table 15.19 Periodical values of R_{sys} and λ_{sys}

Time (h)	R_{sys}	λ_{sys} (per hour)
2500	0.988228409	2.00691E−05
5000	0.961098764	3.37697E−05
7500	0.921916181	4.63204E−05
10,000	0.872324394	5.93621E−05

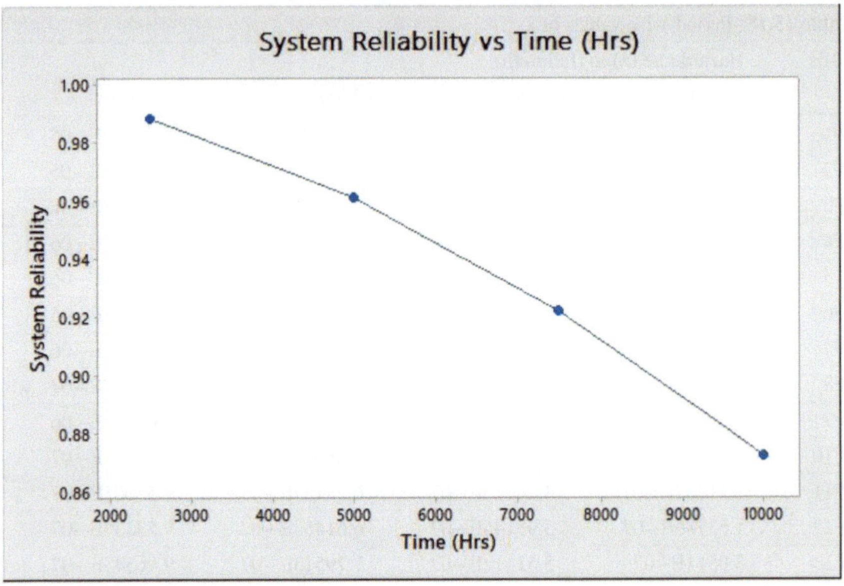

Fig. 15.8 Periodical distribution curve of R_{sys}

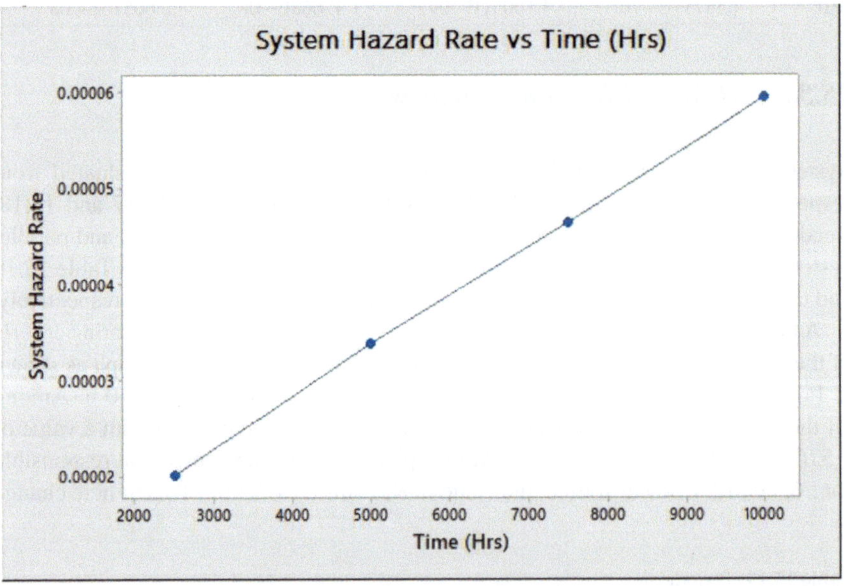

Fig. 15.9 Periodical distribution curve of λ_{sys}

is due to the parallel configuration of data transfer between the ground station and printer. The R_{sys} approaches R_{th} for the second time approximately at 8720 h. The hazard rate of event F2 has a maximum value of 1.04669×10^{-05}/h. on that period. Adopting the previous maintenance strategy for said BE, the R_{sys} improved from 0.9 to 0.9499. Table 15.20 displaying the effect of replacement on R_{sys}. Figure 15.10 emphasizes on improvement of R_{sys} due to the adaptation of the replacement strategy to make the system sustainable.

Table 15.20 Effect of replacement of components on R_{sys}

Time (h)	System reliability before replacement	System reliability after replacement action corresponding to F6	System reliability after replacement action corresponding to F2
2500	0.988228409	–	–
5000	0.961098764	–	–
7500	0.921916181	–	–
8610	0.901098936	0.902852737	–
8720	0.898928935	0.900766325	0.949988276
10,000	0.872324394	0.875300215	0.934380433

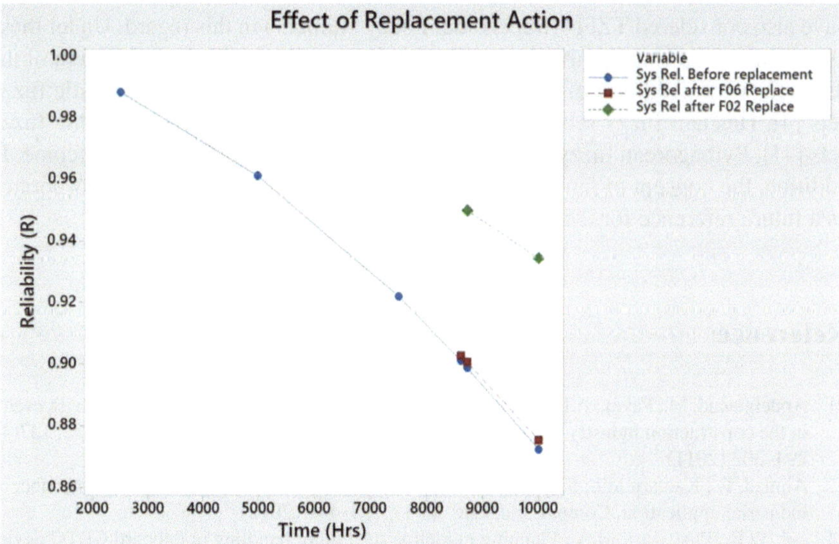

Fig. 15.10 Effect of R_{sys} due to replacement strategy

15.4 Conclusion

The ideology of an IOT-based data communication system has been adopted from the Industry 4.0 philosophy. This system has been adopted in an aluminum smelting plant as stated before, but can also be used in different manufacturing sectors depending on its necessity. Under these circumstances, this system needs to be reliable as well as available otherwise a huge loss of resources might be faced by the concerned firm. To identify the faults of the system under an uncertain environment, FFTA is used as a key tool and its significance has been reflected in this work. Despite of unavailable failure data of the IOT system, it has successfully disclosed the probability of component-level failures. Based on detected failures, RCM has been adopted for unwanted hamper in production work as well as loss of work-related data. The RCM methodology used in the research work emphasizes replacement strategies based on the hazard rate (λ_S) of the components. A significant improvement in system reliability (R_S) has been observed after the completion of maintenance work which implies betterment of system availability and sustainability. Based on the result obtained, it may be concluded to adopt this FFTA-based RCM methodology in a structured manner on different IOT-based systems for better stability and availability. The actual experimental result might slightly vary as fuzzy set theory portrays the results based on expert elicitation which has been found as the prime limitation of the study. On the other hand, only TFN has been considered for FFTA, though various studies have also considered TZFN (trapezoidal Fuzzy Number) in this regard. Under these circumstances, a comparative analysis has to be considered for the betterment of the research work. Addition of superior concept of fuzzy sets, i.e., Intuitionistic fuzzy sets [4], Hesitant fuzzy sets [47], Fermatean Fuzzy Sets [38], Neutrosophic fuzzy sets [41], Pythagorean fuzzy set [51], etc. might be effective for a better outcome. In addition, the concept of fault diagnosis and failure prognosis might be incorporated as a future reference for the health assessment of the system.

Reference

1. Abdelgawad, M., Fayek, A.R.: Fuzzy reliability analyzer: quantitative assessment of risk events in the construction industry using fuzzy fault-tree analysis. J. Construc. Eng. Manage. **137**(4), 294–302 (2011)
2. Ahmad, R., Kamaruddin, S.: An overview of time-based and condition-based maintenance in industrial application. Comput. Ind. Eng. **63**(1), 135–149 (2012)
3. Aji, W.B., Puspasari, M.A.: Planning guideline for failure handling in onboard CBTC device using RCM (reliability center maintenance) analysis
4. Atanassov, K.: Intuitionistic fuzzy sets. Fuzzy Sets Syst. **20**(1), 87–96 (1986). https://doi.org/10.1016/S0165-0114 [(86), 80034-3]
5. Babaei, M., Roozbahani, A., Shahdany, S.M.H.: Risk assessment of agricultural water conveyance and delivery systems by fuzzy fault tree analysis method. Water Resour. Manage **32**, 4079–4101 (2018)

6. Badida, P., Balasubramaniam, Y., Jayaprakash, J.: Risk evaluation of oil and natural gas pipelines due to natural hazards using fuzzy fault tree analysis. J. Natural Gas Sci. Eng. **66**, 284–292 (2019)
7. Ban, A.I., Ban, O.I., Bogdan, V., Popa, D.C.S., Tuse, D.: Performance evaluation model of Romanian manufacturing listed companies by fuzzy AHP and TOPSIS. Technol. Econ. Dev. Econ. **26**(4), 808–836 (2020)
8. Billinton, R., Allan, R.N.: Reliability Evaluation of Engineering Systems, vol. 792. Plenum Press, New York (1992)
9. Boland, P.J.: A reliability comparison of basic systems using hazard rate functions. Appl. Stochas. Models Data Anal. **13**(3–4), 377–384 (1997)
10. Bressi, S., Santos, J., Losa, M.: Optimization of maintenance strategies for railway track-bed considering probabilistic degradation models and different reliability levels. Reliab. Eng. Syst. Safety **207**, 107359 (2021)
11. Cheliyan, A.S., Bhattacharyya, S.K.: Fuzzy fault tree analysis of oil and gas leak-age in subsea production systems. J. Ocean Eng. Sci. **3**(1), 38–48 (2018)
12. Chen, M., Nakamura, S., Nakagawa, T.: Replacement and preventive maintenance models with random working times. IEICE Trans. Fundam. Electron. Commun. Comput. Sci. **93**(2), 500–507 (2010)
13. Chen, T.Y., Ku, T.C.: Importance-assessing method with fuzzy number-valued fuzzy measures and discussions on TFNs And TrFNs. Int. J. Fuzzy Syst. **10**(2) (2008)
14. Chen, Y., Zhen, Z., Yu, H., Xu, J.: Application of fault tree analysis and fuzzy neural networks to fault diagnosis in the internet of things (IoT) for aquaculture. Sensors **17**(1), 153 (2017)
15. Chi, C.F., Lin, S.Z., Dewi, R.S.: Graphical fault tree analysis for fatal falls in the construction industry. Accid. Anal. Prev. **72**, 359–369 (2014)
16. Dey, B., Bairagi, B., Sarkar, B., Sanyal, S.K.: Warehouse location selection by fuzzy multi-criteria decision making methodologies based on subjective and objective criteria. Int. J. Manage. Sci. Eng. Manag. **11**(4), 262–278 (2016)
17. Eriksen, S., Utne, I.B., Lützen, M.: An RCM approach for assessing reliability challenges and maintenance needs of unmanned cargo ships. Reliab. Eng. Syst. Saf. **210**, 107550 (2021)
18. Fang, F., Zhao, Z.J., Huang, C., Zhang, X.Y., Wang, H.T., Yang, Y.J.: Application of reliability-centered maintenance in metro door system. IEEE Access **7**, 186167–186174 (2019)
19. Fazlollahtabar, H., Niaki, S.T.A.: Fault tree analysis for reliability evaluation of an advanced complex manufacturing system. J. Adv. Manuf. Syst. **17**(01), 107–118 (2018)
20. Gachlou, M., Roozbahani, A., Banihabib, M.E.: Comprehensive risk assessment of river basins using fault tree analysis. J. Hydrol. **577**, 123974 (2019)
21. Gürgen, S., Yazır, D., Konur, O.: Fuzzy fault tree analysis for loss of ship steering ability. Ocean Eng. **279**, 114419 (2023)
22. Herrera, F., Martínez, L., Sánchez, P.J.: (2005). Managing non-homogeneous information in group decision making. Eur. J. Oper. Res. **166**(1), 115–132
23. Jena, M.C., Mishra, S.K., Moharana, H.S.: Integration of Industry 4.0 with reliability centered maintenance to enhance sustainable manufacturing. Environ. Prog. Sustain. Energy **43**(2), e14321 (2024)
24. Jimenez, V.J., Bouhmala, N., Gausdal, A.H.: Developing a predictive maintenance model for vessel machinery. J. Ocean Eng. Sci. **5**(4), 358–386 (2020)
25. Kabir, S.: An overview of fault tree analysis and its application in model based dependability analysis. Expert Syst. Appl. **77**, 114–135 (2017)
26. Kang, J., Sun, L., Soares, C.G.: Fault tree analysis of floating offshore wind tur-bines. Renew. Energy **133**, 1455–1467 (2019)
27. Kargar, V., Jahangiri, M., Alimohammadlu, M., Kamalinia, M., Mirazahossieninejad, M.: Risk assessment of mobile crane overturning in asymmetric tandem lifting (ATL) operation based on fuzzy fault tree analysis (FFTA). Results Eng. **16**, 100755 (2022)
28. Khajuria, R.: Intuitionistic fuzzy fault tree analysis of PCBA using novel arithmetic operations. J. Qual. Maint. Eng. **29**(4), 822–841 (2023)

29. Kumar, A.R., Krishnan, V.: A study on system reliability in Weibull distribution. Methods **5**(3), 23–28 (2017)
30. Liu, G., Chen, S., Jin, H., Liu, S.: Optimum opportunistic maintenance schedule incorporating delay time theory with imperfect maintenance. Reliab. Eng. Syst. Safety **213**, 107668 (2021)
31. Liu, P., Yang, L., Gao, Z., Li, S., Gao, Y.: Fault tree analysis combined with quantitative analysis for high-speed railway accidents. Saf. Sci. **79**, 344–357 (2015)
32. Mahmood, Y.A., Ahmadi, A., Verma, A.K., Srividya, A., Kumar, U.: Fuzzy fault tree analysis: a review of concept and application. Int. J. Syst. Assur. Eng. Manag. **4**, 19–32 (2013)
33. Makajic-Nikolic, D., Petrovic, N., Belic, A., Rokvic, M., Radakovic, J.A., Tubic, V.: The fault tree analysis of infectious medical waste management. J. Clean. Prod. **113**, 365–373 (2016)
34. Okwuobi, S., Ishola, F., Ajayi, O., Salawu, E., Aworinde, A., Olatunji, O., Akinlabi, S.A.: A reliability-centered maintenance study for an individual section-forming ma-chine. Machines **6**(4), 50 (2018)
35. Patil, S.S., Bewoor, A.K., Kumar, R., Ahmadi, M.H., Sharifpur, M., PraveenKumar, S.: Development of optimized maintenance program for a steam boiler system using reliability-centered maintenance approach. Sustainability **14**(16), 10073 (2022)
36. Ross, T.J.: Fuzzy Logic with Engineering Applications. Wiley (2009)
37. Samanta, B.M.S.K., Sarkar, B., Mukherjee, S.K.: Reliability centred maintenance (RCM) for heavy earth-moving machinery in an open cast coal mine. CIM Bull. 104–107 (2001)
38. Senapati, T., Yager, R.R.: Fermatean fuzzy sets. J. Ambient. Intell. Humaniz. Comput. **11**, 663–674 (2020)
39. Setiawan, D., Jusolihun, N., Cahyo, W.N.: Maintenance system design on air jet loom (AJL) machine using reliability centered maintenance (RCM) method. In: IOP Conference Series: Materials Science and Engineering, vol. 673, no. 1, p. 012102. IOP Publishing (2019)
40. Shaheen, B.W., Németh, I.: Integration of maintenance management system functions with industry 4.0 technologies and features—a review. Processes **10**(11), 2173 (2022)
41. Smarandache, F.: Neutrosophic set—a generalization of the intuitionistic fuzzy set. J. Defense Resour. Manag. **1**(1), 107–116 (2010)
42. Srinath, L.S.: Reliability Engineering, pp. 1641–1654. Affiliated East-West Press, New Delhi, Delhi (1991)
43. Suryono, M.A.E., Rosyidi, C.N.: Reliability centred maintenance (RCM) analysis of laser machine in filling Lithos at PT X. In: IOP Conference Series: Materials Science and Engineering, vol. 319, no. 1, p. 012020. IOP Publishing (2018)
44. Tabesh, M., Roozbahani, A., Hadigol, F., Ghaemi, E.: Risk assessment of water treatment plants using fuzzy fault tree analysis and Monte Carlo simulation. Iran. J. Sci. Technol. Trans. Civil Eng. 1–16 (2021)
45. Taha, H.A., Yacout, S., Shaban, Y.: Online failure analysis and autonomous risk control scheme for electric buses. Eng. Fail. Anal. **154**, 107629 (2023)
46. Tale, T.: Reliability centered maintenance (RCM): methodology and benefits. Int. J. Comm. Manag. **4**(3), 1–8 (2019)
47. Torra, V.: Hesitant fuzzy sets. Int. J. Intell. Syst. **25**(6), 529–539 (2010)
48. Tortorella, G.L., Fogliatto, F.S., Cauchick-Miguel, P.A., Kurnia, S., Jurburg, D.: Integration of industry 4.0 technologies into total productive maintenance practices. Int. J. Prod. Econ. **240**, 108224 (2021)
49. Wang, T., Hu, G., Cho, S.: A reliability-centered maintenance framework for distribution grids based on fault-tree analysis. In: PHM Society Asia-Pacific Conference, vol. 4, no. 1 (2023)
50. Werbińska-Wojciechowska, S., Werbińska-Wojciechowska, S.: Preventive maintenance models for technical systems. In: Technical System Maintenance: Delay-Time-Based Modelling, pp. 21–100 (2019)
51. Yager, R.R.: Pythagorean fuzzy subsets. In: 2013 Joint IFSA World Congress and NAFIPS Annual Meeting (IFSA/NAFIPS), pp. 57–61. IEEE (2013)
52. Yazdi, M., Nikfar, F., Nasrabadi, M.: Failure probability analysis by employing fuzzy fault tree analysis. Int. J. Syst. Assur. Eng. Manag. **8**, 1177–1193 (2017)
53. Zadeh, L.A.: Fuzzy sets. Inf. Control **8**(3), 338–353 (1965)

Chapter 16
On a Cost Analysis of Extended Warranty with Imperfect Repair Using Phase Type Quasi-Renewal Processes

R. Shenbagam

Abstract The present competitive market situation requires manufacturers to constantly provide superior service and support. As a result, warranty concerns become an important tool for improving product marketability. This research analysis delves into the warranty period for a system, which is divided into two parts: the basic warranty period, which utilizes a fixed free repair renewing warranty, and the extended warranty period, which employs the k-policy. System failures are rectified via imperfect repair through phase type quasi-renewal processes when the system fails during the basic warranty and extended warranty periods, respectively. When the system fails, the failed component is replaced with a new one and a new warranty period is associated with the renewing fixed warranty model. In the extended warranty model, the operating and repair times are modeled to use an alternative phase type quasi-renewal process. Failures occur at any point in time. The system's condition after the repair has not been as good as before the repair. By using k-policy, an explicit expression for the long-run average cost rate is obtained for the extended warranty model. To illustrate the usefulness of the proposed model, a numerical example and sensitivity analysis are provided.

Keywords Renewing warranty · Extended warranty · k-policy · Renewal reward theorem

16.1 Introduction

In any national economy, the analysis of optimal maintenance and replacement approaches, incorporating warranty cost analysis for repairable deteriorating systems, is essential. Age replacement policies, which appear to be similar to perfect repairs, were the subject of the initial maintenance analysis. A perfect repair, which

---

R. Shenbagam (✉)
Department of Mathematics and Actuarial Science, B.S. Abdur Rahman Crescent Institute of Science and Technology, Vandalur, Chennai, Tamil Nadu, India
e-mail: senbagam@ymail.com; shenbagam@crescent.education

J. Maiti et al. (eds.), *Recent Advances in Industrial and Systems Engineering*, Lecture Notes on Multidisciplinary Industrial Engineering,
https://doi.org/10.1007/978-981-96-8323-9_16

is modeled by a renewal process, restores a failed component to its original state. In real-world applications, repairs could produce an intermediate condition known as a general repair rather than one that conforms to either of the two stages. Some researchers define "general repair" as "imperfect maintenance" or "imperfect repair." (e.g., [1–4]).

Under such conditions, a model known as stochastically deteriorating systems which treat sequential operation durations as stochastically non-increasing and subsequent repair times as stochastically non-decreasing is suitable. One potential solution to deal with stochastically deteriorating systems is to employ monotone processes. This viewpoint is appropriate for such systems because, as the system undergoes consecutive repairs, the operating periods of the system become increasingly shorter, while the repair durations increase longer and longer, ultimately nearing infinity and the system becomes unrepairable. Lam [5] developed the geometric process repair model to examine the repair-replacement issue with a failing system. Additional research on the geometric process model in maintenance analysis has been conducted by [6–8], and [9].

Following that, the assumption of randomly distributed variables with identical distributions was substituted with the assumption of independent inter-arrival periods; [3, 4] named this process a quasi-renewal process. This approach, also known as the (α, β) model, increases the next repair time to a multiple of β ($\beta > 1$) of the previous working and repair durations, respectively, and decreases the working time after each maintenance to a fraction α ($0 < \alpha < 1$). This tedious method has inspired several scholars in the field of reliability [10–13].

Neuts [14] initially suggested the notion of substituting phase type distributions (PH-distributions) for generic distributions in a model, which forms the basis of the phase type approximation. Phase type distribution group includes exponential, Coxian, and gamma distributions, as well as convolutions and combinations of these distributions. The phase type distribution is an example of a matrix-analytic method (MAM), where the distribution of the random variable, together with its moments, density function, and cumulative distribution function, can all be expressed in terms of this matrix. Phase type distribution has been widely employed in many domains, including, but not limited to, queueing theory, finance, biostatistics, survival analysis, and telecommunications.

The previously indicated feasible methods encompass every category of failed systems, and an effort is made to perform a thorough analysis of these systems from the perspectives of reliability and availability. Additionally, it seems sense to believe that this kind of analysis would be insufficient in the absence of Asmussen and Rolski [15] state that while determining the matrix exponential for a general matrix may be difficult, determining the sub-intensity matrices seen in many algorithms may be done with ease. The phase type distribution was initially combined with a geometric process by [7]. This was then done by Perez-Ocon and Montetoro-Cazorla [16], Castro and Perez-Ocon [17], and Yu et al. [18]. A generalized monotone process named the phase type quasi-renewal process was proposed by [19] for a deteriorating system under warranty. Using a phase type quasi-renewal process in a deteriorating system with a repairman's multiple vacation policy, Sarada and Shenbagam [20] derived an optimal replacement policy N^* analytically. Using phase type quasi-renewal processes, [21] studied a preventive repair model with stochastic lead time.

An agreement between a manufacturer and a customer that commits the former to fix any flaws or malfunctions in the product for a predetermined amount of time is called a warranty policy. These guidelines protect against too-aggressive legal claims and marketing goals for consumers. Items that wear directly proportional to their use (e.g., car tyres, batteries) typically have pro-rata warranties available. Customers are shielded from shoddy craftsmanship and sporadic malfunctions by this type of guarantee.

Within the design of such systems, a renewal warranty is one of the many methods that have become popular for replacing failing systems with a product warranty. Renewing and non-renewing warranties are also separated. An initial warranty is usually granted under a renewing warranty if a product malfunctions within the warranty period. The warranty starts with every repair. For electrical and mechanical equipment that are reasonably priced and/or have a high initial failure rate, warranty renewals are commonplace. Under a non-renewing warranty, the manufacturer only offers competent service during the first warranty period; the original warranty is not extended. In addition, a lot of products, especially home appliances, come with an extended warranty policy, which gives the customer a lot of options at W (warranty), the warranty's expiration date. Practitioner interest in extended warranties has increased as a result.

Although stochastic warranty modeling is a relatively new idea [22, 23] used two warranty programs to examine the effects of extended policies, or warranty renewal plans, on a company's market share. [24] proposed an extended warranty model with options so that customers may select the optimum coverage. The temperature switch-cum-sensor 77 (TSS) one-dimensional renewable free replacement warranty policy was examined by Mondal et al. [25], who discovered that the cost of the warranty rises with the length of the warranty's validity. Jung and Park [26] were able to identify the most effective periodic preventive maintenance plans after the warranty expired by comparing the benefits of renewing and non-renewing the warranty.

A hybrid warranty model that included discounted costs to calculate the estimated maintenance cost was presented by [27] for pro-rata warranty renewal (RPRW). While there are many research articles on optimal maintenance schedules in different contexts, most of them are predicated on either a repair or a replacement warranty. For a repairable deteriorating system, [28] established an optimum periodic replacement approach under a free-repair warranty. Phase type quasi-renewal processes were used by Sarada and Shenbagam [19] to examine two warranty models: extended and fixed warranty models with inspections for a repairable failing system. Chien and Yu-Hung [29] determined the best preventive maintenance plan for a system covered by a free-repair warranty using a non-homogeneous pure birth process. Fatemeh et al. [30] proposed a policy plan for a repairable system's warranty period, which includes several minimal repairs and service duration. Madhu [31] analyzed the stochastic process generated by unit failures, as well as the resulting cash flows, for products with cumulative warranties, wherein the seller provides a warranty on the total useful life of a batch of components rather than the lifetime of individual components via phase type distribution.

Thus far, no reliability analyst has looked at the possibility of extending the warranty period under the k-policy through a phase type quasi-renewal process in addition to fixed free repair renewing the warranty period throughout the base time. As a result, the thrust of this article is geared toward providing the significance of the combination of fixed free repair renewing warranty in the base period and extended warranty for a repairable deteriorating system via phase type quasi-renewal process in an attempt to establish the importance of the simplified computational implementation of the phase type distribution and motivated by the above features. The standard renewal reward theorem is used to establish an explicit formula for the long-run average cost rate under the system's k-policy. Figure 16.1 illustrates an enhanced overview of the suggested model.

An outline of the paper's structure is provided below. In Sects. 16.2 and 16.3, the suggested model is explained and the underlying presumptions are mentioned. An explicit definition of the long-run average cost rate is provided in Sect. 16.4. A numerical illustration of Sect. 16.4 is also included in Sect. 16.5. Concluding observations are included in Sect. 16.6.

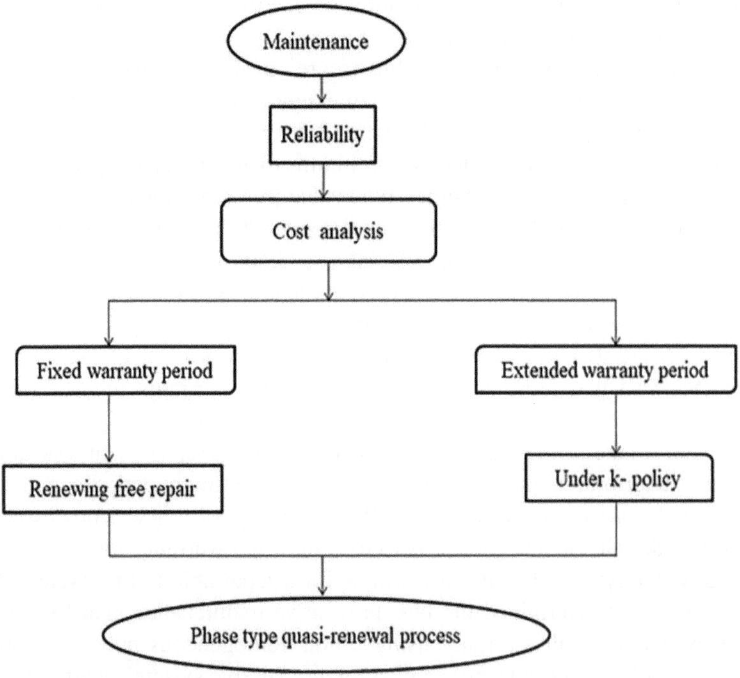

Fig. 16.1 Overview of the proposed model

16.2 Basic Definitions

16.2.1 Notations

The present work's notations are listed below.

W_n	The time between (n-1)th and nth repair, n = 1,2,3,…
P_n	Time of the nth repair, n = 1,2,3…
a	Phase type quasi-renewal process parameter for operating times
b	Phase type quasi-renewal process parameter for repair times
C_{oW}	Operating time cost under base warranty period
C_{rW}	Repair time cost under base warranty period
C_{oE}	Operating time cost under an extended warranty period
C_{rE}	Repair time cost under extended warranty period
C_R	Replacement time cost
$N(W)$	Phase type quasi-renewal counting process associated with the free repair in (0,W]
$E(C)$	Expected cost in a cycle
$E(L)$	Expected length of a cycle
$C(k, W)$	Long-run average cost rate for the extended warranty model
$C_1(k, W)$	Long-run average cost rate for the extended warranty model (Special case)
$E(C_0)$	Expected cost in a cycle (Special case)
$E(L_0)$	Expected length of a cycle (Special case)

Definition: Continuous Phase Type Distribution A probability distribution $F(.)$ on $[0, \infty)$ is a phase type distribution if and only if it is the distribution of the time until absorption in a finite Markov process with a single absorbent state. The infinitesimal generator can be expressed as,

$$Q = \begin{pmatrix} T & T^0 \\ 0 & 0 \end{pmatrix}$$

and a probability vector (α, α_{m+1}) exists such that

$$F(x) = 1 - \alpha \exp(Tx)e \text{ for } x \geq 0.$$

Here α is the initial probability vector and "e" is the column vector of ones.

Definition: Phase Type Quasi-Renewal Process A sequence of non-negative random variables $\{X_n, n = 1, 2, ...\}$ is said to be a phase type quasi-renewal process, if they are independent and the distribution function of $\{X_n\}$ is given by

$$F_{X_n}(x) = F_{X_1}\left(\frac{x}{a^{n-1}}\right) = 1 - \alpha \exp\left(\frac{Tx}{a^{n-1}}\right)e,$$

For $n = 1,2,\ldots,$ where $a > 0$ is called the ratio of the phase type quasi-renewal process.

16.3 Model Assumptions

Assume that the product is repairable and that it comes with a free-repair policy that renews the warranty for period W and an extended warranty period that lasts for policy K. All failures are either rectified through imperfect failure or replaced (catastrophic failure) by the manufacturer with a new product with a new warranty term W and at no cost to the client during the fixed free repair renewing warranty period (Base warranty period). Furthermore, it is assumed that when the warranty (W) expires, the manufacturer will provide the customer with an option to have the system sent for imperfect repairs whenever a failure arises, at an additional cost. When the warranty period W ends, the customer will also have the option to have the system replaced when the $(k + 1)$th repair occurs, which will occur k times before the first repair.

The following are some more assumptions:

i. Failures occur at random times.
ii. After a failure occurs the system is sent for repair which takes a non-negligible amount of time and is not as good as new.
iii. Let us assume that the operating time $\{W_n\}$ follows a phase type quasi-renewal processes with representations $\left(\alpha, \frac{T}{a^{n-1}}\right)$.
iv. Similarly, the repair time $\{P_n\}$ follows a phase type quasi-renewal process with representations $\left(\beta, \frac{Q}{b^{n-1}}\right)$.
v. If $0 < a < 1$ and $b > 1$ it models imperfect repairs.
vi. Let C_{oW} be the operating time cost under a base warranty period. Further, let C_{oE} and C_{rE} be the operating time cost and repair cost under an extended warranty period with C_{R} as the replacement time cost.
vii. The operating times, repair times, and replacement times are assumed to be independent.

16.4 Cost Analysis

Under the k-policy, an explicit formulation for the long-run average cost rate for the extended warranty model is obtained in this section.

The long-run average cost per unit time [32]

$$C(k, W) = \frac{\text{Expected cost in a cycle}}{\text{Expected length of a cycle}}$$

The length of a cycle under the k-policy is given by,

$$E(L) = E[N(W)] + E\left[\sum_{i=N(W)+1}^{N(W)+k} (W_i + P_i)\right] \tag{16.1}$$

where $\{N(t); t > 0\}$ represents the phase type quasi-renewal counting process associated with the free repair in (0,t]. The average phase type quasi-renewal process operating and repair times are given by,

$$E(W_i) = -\alpha\left(\frac{T}{a^{i-1}}\right)^{-1} e_1 \text{ and } E(P_i) = -\beta\left(\frac{Q}{b^{i-1}}\right)^{-1} e_2$$

$$E(L) = E[N(W)] - \left\{\sum_{n=1}^{\infty}\sum_{i=n+1}^{n+k}\left(\alpha\left(\frac{T}{a^{i-1}}\right)^{-1} e_1\right)P(N(W)=n)\right\}$$
$$-\left\{\sum_{n=1}^{\infty}\sum_{i=n+1}^{n+k}\left(\beta\left(\frac{Q}{b^{i-1}}\right)^{-1} e_2\right)P(N(W)=n)\right\} \tag{16.2}$$

where $P(N(W) = n) = K_n(W) - K_{n+1}(W)$, K_n is the convolution of $\{Z_n\}$ so that its Laplace transform is given by

$$K_n^*(s) = \frac{1}{s}\prod_{i=1}^{n}\left(\frac{\gamma\left(\frac{T^0}{a^{i-1}}\right)}{\mathrm{sI} - \left(\frac{T}{a^{i-1}}\right)}\right)$$

$$\Rightarrow K_n(W) = L^{-1}\left[\frac{1}{s}\prod_{i=1}^{n}\gamma\left(\frac{(-T/a^{i\ 1})}{\mathrm{sI}-(T/a^{i-1})}\right)e\right]$$

Next, the expected cost in a cycle can be expressed as,

$$E(C) = E(C_{\mathrm{W}}) + E(C_{\mathrm{E}}) + E(C_{\mathrm{R}}) \tag{16.3}$$

where $E(C_{\mathrm{W}})$ is the expected cost of the system under warranty period, $E(C_{\mathrm{E}})$ is the expected cost of the system under extended warranty period and $E(C_{\mathrm{R}})$ is the replacement cost of the system.

$$E(C_{\mathrm{W}}) = C_{\mathrm{o,W}}E[N(W)] = C_{\mathrm{o,W}}\sum_{n=1}^{\infty} nP(N(W)=n)$$

$$E(C_{\mathrm{E}}) = -C_{\mathrm{o,E}}\left\{\sum_{n=1}^{\infty}\sum_{i=n+1}^{n+k}\left(\alpha\left(\frac{T}{a^{i-1}}\right)^{-1} e_1\right)P(N(W)=n)\right\}$$

$$- C_{\mathrm{r,E}} \left\{ \sum_{n=1}^{\infty} \sum_{i=n+1}^{n+k} \left(\beta \left(\frac{Q}{b^{i-1}} \right)^{-1} e_2 \right) P(N(W) = n) \right\}$$

and $E(C_{\mathrm{R}}) = C_{\mathrm{r}}$.

Now (16.3) can be written as,

$$\begin{aligned} E(C) &= C_{\mathrm{o,W}} E[N(W)] \\ &\quad - C_{\mathrm{o,E}} \left\{ \sum_{n=1}^{\infty} \sum_{i=n+1}^{n+k} \left(\alpha \left(\frac{T}{a^{i-1}} \right)^{-1} e_1 \right) P(N(W) = n) \right\} \\ &\quad - C_{\mathrm{r,E}} \left\{ \sum_{n=1}^{\infty} \sum_{i=n+1}^{n+k} \left(\beta \left(\frac{Q}{b^{i-1}} \right)^{-1} e_2 \right) P(N(W) = n) \right\} + C_{\mathrm{r}} \end{aligned}$$

Using (16.2) and (16.3), the long-run average cost rate for the extended warranty model can be expressed as

$$C(k, W) = \frac{\left\{ C_{\mathrm{o,W}} E[N(W)] - C_{\mathrm{o,E}} \left\{ \sum_{n=1}^{\infty} \sum_{i=n+1}^{n+k} \left(\alpha \left(\frac{T}{a^{i-1}} \right)^{-1} e_1 \right) P(N(W) = n) \right\} - C_{r,E} \left\{ \sum_{n=1}^{\infty} \sum_{i=n+1}^{n+k} \left(\beta \left(\frac{Q}{b^{i-1}} \right)^{-1} e_2 \right) P(N(W) = n) \right\} + C_{\mathrm{r}} \right\}}{\left\{ E[N(W)] - \left\{ \sum_{n=1}^{\infty} \sum_{i=n+1}^{n+k} \left(\alpha \left(\frac{T}{a^{i-1}} \right)^{-1} e_1 \right) P(N(W) = n) \right\} - \left\{ \sum_{n=1}^{\infty} \sum_{i=n+1}^{n+k} \left(\beta \left(\frac{Q}{b^{i-1}} \right)^{-1} e_2 \right) P(N(W) = n) \right\} \right\}} \tag{16.4}$$

In what follows, (16.4) is further analyzed through a numerical example incorporating the sensitivity analysis.

16.4.1 Special Case

Assume that, for such a deteriorating system, a warranty cost model with non-negligible repair times and a non-renewing warranty duration under the fixed warranty model (W) is appropriate, retaining the fundamental assumptions of the model mentioned above. The expected length of a cycle and the expected costs incurred in the model are shown below.

$$E(C_0) = E(C_{\mathrm{W}}) + E(C_{\mathrm{E}}) + E(C_{\mathrm{R}}) \tag{16.5}$$

$$\begin{aligned} E(C_0) &= E\left[\sum_{i=1}^{N(W)} (W_i + P_i) \right] + E\left[\sum_{i=N(W)+1}^{N(W)+k} (W_i + P_i) \right] + E(C_R) \\ &= -C_{\mathrm{o,W}} \left\{ \sum_{n=1}^{\infty} \sum_{i=1}^{n} \left(\alpha \left(\frac{T}{a^{i-1}} \right)^{-1} e_1 \right) P(N(W) = n) \right\} \end{aligned}$$

$$
\begin{aligned}
&- C_{\mathrm{r,W}}\left\{\sum_{n=1}^{\infty}\sum_{i=1}^{n}\left(\beta\left(\frac{Q}{b^{i-1}}\right)^{-1}e_2\right)P(N(W)=n)\right\}\\
&- C_{\mathrm{o,E}}\left\{\sum_{n=1}^{\infty}\sum_{i=n+1}^{n+k}\left(\alpha\left(\frac{T}{a^{i-1}}\right)^{-1}e_1\right)P(N(W)=n)\right\}\\
&- C_{\mathrm{r,E}}\left\{\sum_{n=1}^{\infty}\sum_{i=n+1}^{n+k}\left(\beta\left(\frac{Q}{b^{i-1}}\right)^{-1}e_2\right)P(N(W)=n)\right\}\\
&+ C_{\mathrm{r}}E(L)
\end{aligned}
\tag{16.6}
$$

$$
\begin{aligned}
E(L_0) = &-\left\{\sum_{n=1}^{\infty}\sum_{i=1}^{n}\left(\alpha\left(\frac{T}{a^{i-1}}\right)^{-1}e_1\right)P(N(W)=n)\right\}\\
&-\left\{\sum_{n=1}^{\infty}\sum_{i=1}^{n}\left(\beta\left(\frac{Q}{b^{i-1}}\right)^{-1}e_2\right)P(N(W)=n)\right\}\\
&-\left\{\sum_{n=1}^{\infty}\sum_{i=n+1}^{n+k}\left(\alpha\left(\frac{T}{a^{i-1}}\right)^{-1}e_1\right)P(N(W)=n)\right\}\\
&-\left\{\sum_{n=1}^{\infty}\sum_{i=n+1}^{n+k}\left(\beta\left(\frac{Q}{b^{i-1}}\right)^{-1}e_2\right)P(N(W)=n)\right\}
\end{aligned}
$$

Thus, the long-run average cost rate is

$$
C_1(k, W) = \frac{E(C_0)}{E(L_0)}
$$

Equation (16.6) is same as $E(L)$ of Eqs. (16.2) and (16.3) of Sarada et al. [13].

16.5 Numerical Illustration

This section contains a numerical example to illustrate the model's applicability.

Let us assume that the parameter values are $a = 0.244$, $b = 1.333$,

$$
U = \begin{pmatrix} -0.9 & 0.5 \\ 0 & -0.4 \end{pmatrix}, T = \begin{pmatrix} -0.8 & 0.5 \\ 0 & -0.5 \end{pmatrix}, Q = \begin{pmatrix} -0.2 & 0.1 \\ 0 & -0.2 \end{pmatrix},
$$

$$
C_{oW} = 3, C_{oE} = 10, C_{rE} = 5, C_r = 2, \alpha = \beta = \begin{pmatrix} 1 & 0 \end{pmatrix}, e_1 = e_2 = \begin{pmatrix} 1 & 1 \end{pmatrix}'
$$

Table 16.1 and Figs. 16.2 and 16.3 show the increasing pattern of $C(k,W)$ for different values of (k,W) when the above-mentioned parameter values have been used in Eq. (16.4). Tables 16.2, 16.3 and Figs. 16.4, 16.5 also indicate that when

"a" increases, the long-run average cost rate ($C(k,W)$) decreases, but an increase in "b" increases the long-run average cost rate ($C(k,W)$). As a result of the foregoing observations, managers may be able to analyze system performance while developing a renewed and extended warranty model under the k-policy.

16.6 Conclusion

This research work attempted to establish a phase type quasi-renewal process for a repairable deteriorating system in a fixed free repair renewing warranty and extended warranty model. Product failures are rectified via imperfect repair when the product fails during the base warranty and extended warranty periods, respectively. When the system fails, the failed component is replaced with a new one and a new warranty period is connected under the renewing fixed warranty model. In the extended warranty model, the operating and repair times are modeled using an alternative phase type quasi-renewal process. Failures can happen at any time. The system is not in as good a condition as it was before the repair. Using the k-policy, an explicit expression for the long-run average cost rate is established for the extended warranty model. Theoretical observations are confirmed by numerical demonstration. The use of a phase type quasi-renewal process in the modeling of a fixed free repair renewing period and extended warranty for a repairable deteriorating system is a key element of this study. The developed model, it is emphasized, is not theoretically interesting, but it may appeal to practitioners due to its ease of application.

Table 16.1 $C(k,w)$ for a different combination of k and W

	$w = 1$	2	3	4	5	6	7	8	9	10
$k = 1$	8.4274	8.4491	8.5543	8.6697	8.7790	8.8787	8.9679	9.0471	9.1170	9.1788
2	9.2994	9.3539	9.4221	9.4851	9.5403	9.5880	9.6289	9.6638	9.6938	9.7196
3	9.7600	9.7847	9.8107	9.8337	9.8533	9.8698	9.8837	9.8955	9.9054	9.9139
4	9.9290	9.9370	9.9350	9.9520	9.9578	9.9627	9.9668	9.9702	9.9731	9.9756
5	9.9805	9.9828	9.9961	9.9869	9.9886	9.9899	9.9910	9.9920	9.9928	9.9934
6	9.9949	9.9955	9.9990	9.9966	9.9970	9.9974	9.9976	9.9979	9.9981	9.9983
7	9.9987	9.9988	9.9997	9.9991	9.9992	9.9993	9.9994	9.9995	9.9995	9.9996
8	9.9997	9.9997	9.9997	9.9998	9.9998	9.9998	9.9998	9.9998	9.9998	9.9998
9	9.9999	9.9999	9.9999	9.9999	9.9999	9.9999	9.9999	9.9999	9.9999	9.9999
10	10.0000	10.0000	10.0000	10.0000	10.0000	10.0000	10.0000	10.0000	10.0000	10.0000

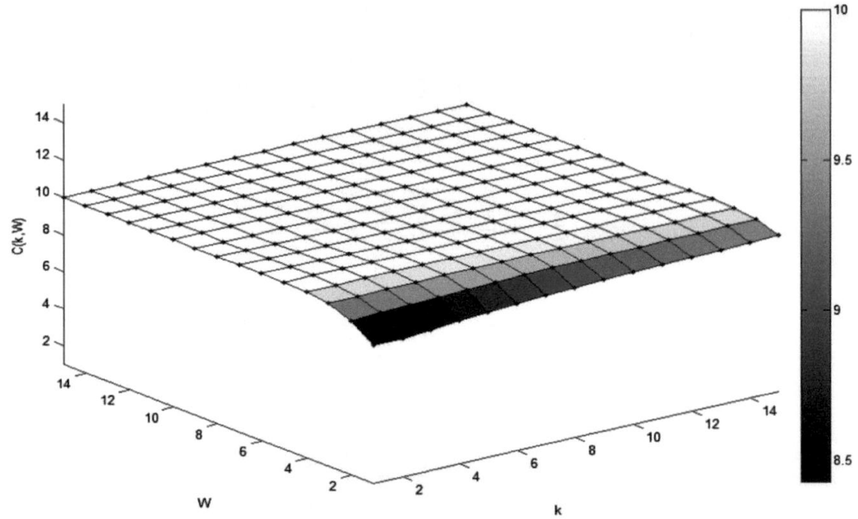

Fig. 16.2 Surface plot for $C(k,W)$

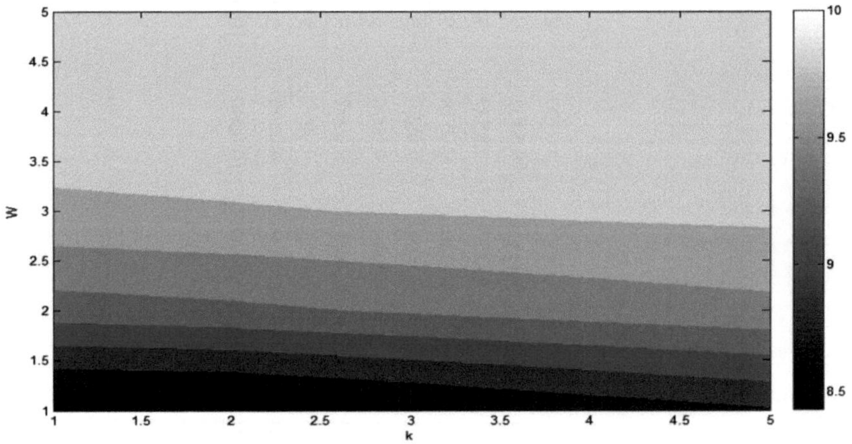

Fig. 16.3 Contour plot for $C(k,W)$

Table 16.2 $C(k,W)$ for different "a" values

$a = 0.3333$						$a = 0.2857$				
W k	1	2	3	4	5	$W = 1$	2	3	4	5
1	8.0837	8.0916	8.1690	8.2851	8.4012	8.2584	8.2654	8.3689	8.4861	8.6001
2	8.8938	8.9567	9.0482	9.1374	9.2187	9.1121	9.1725	9.2530	9.3292	9.3973
3	9.4782	9.5228	9.5737	9.6204	9.6614	9.6420	9.6760	9.7132	9.7465	9.7753
4	9.7854	9.8068	9.8292	9.8492	9.8665	9.8751	9.8885	9.9021	9.9141	9.9243
5	9.9187	9.9273	9.9361	9.9438	9.9504	9.9597	9.9643	9.9687	9.9727	9.9760

Table 16.3 $C(k,W)$ for different "b" values

$b = 2.00$						$b = 1.05$				
	$w = 1$	2	3	4	5	$w = 1$	2	3	4	5
$k = 1$	8.8199	8.8099	8.8947	8.9909	9.0826	8.1616	8.1957	8.3076	8.4279	8.5430
2	9.5714	9.5967	9.6403	9.6818	9.7185	9.0632	9.1364	9.2207	9.2983	9.3667
3	9.8751	9.8852	9.8990	9.9117	9.9225	9.6339	9.6720	9.7096	9.7425	9.7706
4	9.9671	9.9700	9.9737	9.9771	9.9800	9.8788	9.8929	9.9060	9.9171	9.9266
5	9.9917	9.9924	9.9934	9.9942	9.9950	9.9634	9.9678	9.9719	9.9753	9.9781

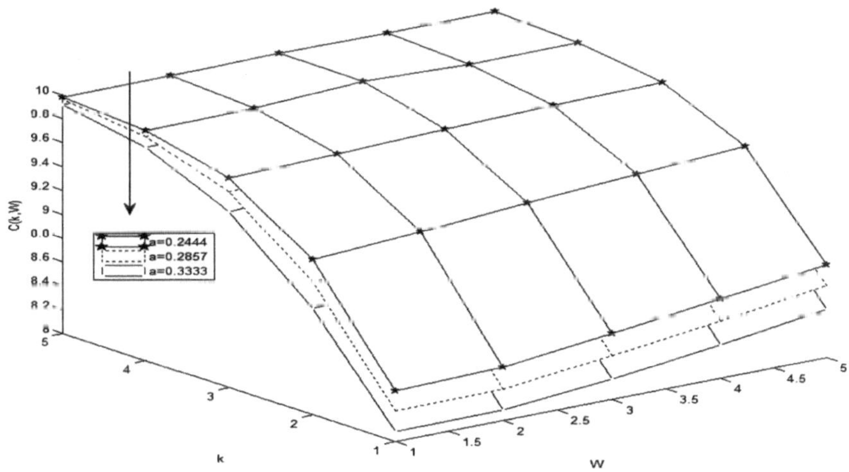

Fig. 16.4 Surface plot for $C(k,W)$ with different "a" values

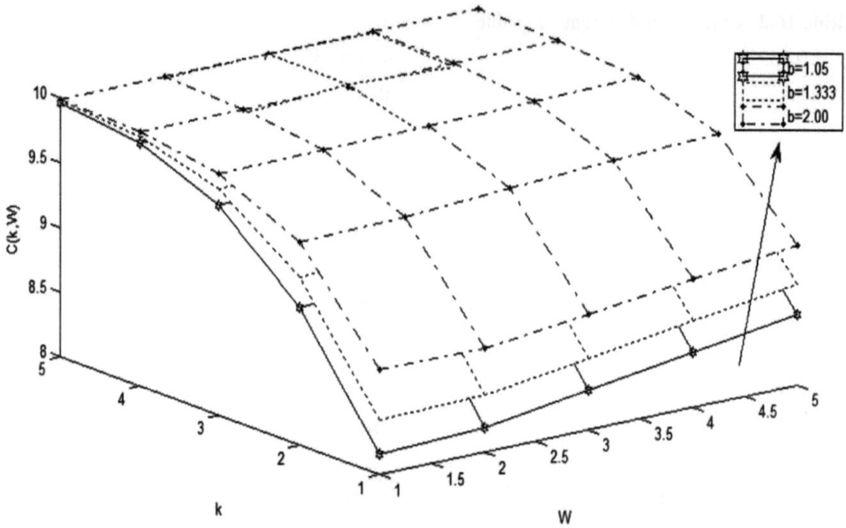

Fig. 16.5 Surface plot for $C(k,W)$ with different "b" values

Declaration of Interest Statement

Conflict of Interest The authors declare that they have no conflict of interest.

References

1. Brown, M., Proschan, F.: Imperfect repair. J. Appl. Prob. **20**(4), 851–859 (1983)
2. Wang, H.: A survey of maintenance policies of deteriorating systems. Eur. J. Oper. Res. **139**(3), 469–489 (2002)
3. Wang, H., Pham, H.: A quasi-renewal process and applications in imperfect maintenance. Int. J. Sys. Sci. **27**(3), 1055–1062 (1996)
4. Wang, H., Pham, H.: Correspondence changes to: a quasi-renewal process and its applications in imperfect maintenance. Int. J. Sys. Sci. **28**(12), 1329 (1997)
5. Lam Y.: Geometric processes and replacement problem. Acta Mathe. Appl. Sin. **4**, 366–377 (1988)
6. Lam, Y.: The geometric process and its applications. World Scientific Publishing cooperative, Singapore (2007)
7. Neuts, M.F., Perez-Ocon, R., Castro, I.T.: Repairable models with operating and repair times governed by Phase type distribution. Adv. Appl. Proba. **32**(2), 468–479 (2000)
8. Stanley, A.D.J.: On Geometric processes and repair replacement problems. Micro. Reli. **33**(4), 489–491 (1993)
9. Wang, G.J., Zhang, Y.L.: A geometric process repair model for a two-component system with shock damage interaction. Int. J. Sys. Sci. **40**(11), 1207–1215 (2009)
10. Chukova, S., Hayakawa, Y.: Warranty cost analysis: quasi—renewal inter-repair times. Int. J. Quali. Reli. Manag. **22**(7), 687–698 (2005)
11. Samatli-Pac, G., Taner M.R.: The role of repair strategy in warranty cost minimization: an investigation via quasi-renewal processes. Eur. J. Oper. Res. **197**(2), 632–641 (2009)

12. Park, M., Pham, H.: Warranty cost analyses using quasi-renewal processes for multicomponent systems. IEEE Trans. Sys. man. cybe. Part A: Sys. Hum. **40**(6), 1329–1340 (2010)
13. Sarada, Y., Mubashir Unnissa, M., Ranjani, R.: Warranty cost analysis using alternating quasi-renewal processes with a warranty option. Int. J. Sys. Sci. **43**(3), 507–517 (2012)
14. Neuts, M.F.: Matrix geometric solutions in stochastic models: an algorithmic approach. The John Hopkins University Press, Baltimore (1981)
15. Asmussen, S., Rolski, T.: Computational methods in risk theory: a matrix-algorithmic approach. Insur.: Math. Econ. **10**(4), 259–274 (1992)
16. Perez-Ocon, R., Montoro-Cazorla, D.: Transient analysis of a repairable system using phase type distributions and geometric process. IEEE Trans. Reliab. **53**(2), 185–192 (2004)
17. Castro, I.T., Perez-Ocon, R.: Reward optimization of a repairable system. Reli. Eng. Sys. Saf. **91**(3), 311–319 (2006)
18. Yu, M., Tang, Y., Liu, L., Cheng, J.: A phase type geometric process repair model with spare device procurement and repairman's multiple vacations. Eur. J. Oper. Res. **225**(2), 310–323 (2013)
19. Sarada, Y., Shenbagam, R.: On warranty cost analysis using alternating phase type quasi-renewal processes. Int. J. Reli. Qua. Saf. Eng. **22**, 15500266 (2015)
20. SaradaY, S.R.: Optimization of repairable systems with Repairman's multiple vacations. Int. J. Quali. Reli. Manag. **35**, 266–290 (2018)
21. Sarada, Y., Shenbagam, R.: Optimization of a repairable deteriorating system subject to random threshold failure using preventive repair and stochastic lead time. Reli. Eng. Sys. Saf. **205**, 107229 (2021)
22. Blischke, W.R., Murthy, D.N.P.: Product Warranty Handbook, New York (1995)
23. Mitra, A., Patankar, J.G.: Market share and warranty costs for renewable warranty programs. Int. J. Pro. Econo. **50**, 155–168 (1997)
24. Lam, Y., Lam, P.: An extended warranty policy with options open to consumers. Eur. J. Oper. Res. **131**(3), 514–529 (2001)
25. Mondal, S., Pal, S., Manna, D.K.: Cost estimation under renewing warranty policy—an application. Qual. Eng. **16**, 93–98 (2003)
26. Jung, G.M., Park, D.H.: Optimal maintenance policies during the post-warranty period. Reli. Eng. Sys. Saf. **82**, 173–185 (2003)
27. Madhu, J., Sandhya, M.: Discounted costs for repairable units under hybrid warranty. Appl. Math. Comput. **173**, 887–901 (2006)
28. Yeh, R.H., Chen, M.Y., Lin, C.Y.: Optimal periodic replacement policy for repairable products under free-repair warranty. Eur. J. Oper. Res. **176**(3), 1678–1686 (2007)
29. Chien, Y.H.: The optimal preventive maintenance policy for an NHPBP repairable system is under a free-repair warranty. Reli. Eng. Sys. Saf. **188**, 444–453 (2019)
30. Hooti, F., Ahmadi, J., Longobardi, M.: Optimal extended warranty length with a limited number of repairs in the warranty period. Reli. Eng. Sys. Saf. **203**, 107111 (2020)
31. Rao, M.: Cumulative free replacement warranty with phase type lifetime distributions. Com. Indu. Eng. **162**, 107771 (2021)
32. Ross, S.M.: Stochastic processes. Wiley, New York (1996)

Chapter 17
Advancing Safety in Petrochemical Enterprises: Intelligent HAZOP Analysis and Risk Mitigation Using Data Mining and Automation

Gourab Kumar Bagchi, Jhareswar Maiti, and Harshad Shrigondekar

Abstract Amidst a surge in production equipment accidents, posing grave societal and industrial ramifications, the imperative to fortify preventive measures and safety protocols grows pivotal. Learning from accident cases, their lessons, and preventive strategies assumes paramount significance to enhance the inherent safety of petrochemical enterprises. The well-established Hazard and Operability (HAZOP) analysis method has been instrumental in identifying risks and averting accidents within the petrochemical sector. However, the reservoir of insights embedded in completed HAZOP reports often remains untapped and underutilized. Moreover, the assessment of novel processes leans heavily on expert's intuition, a subjective and time-intensive endeavor. This has prompted a push for automation using Natural Language Processing (NLP) and Machine Learning (ML). The present work introduces an intelligent HAZOP analysis methodology, underpinned by data mining principles. This innovative approach offers adaptability to modest sized units and companies with limited size of data, and provides the foundational automated framework for identifying risks, preventing accidents, and facilitating emergency response in petrochemical facilities.

Keywords HAZOP study · Natural language processing · Machine learning · Chemical process safety

G. K. Bagchi · J. Maiti (✉) · H. Shrigondekar
Centre of Excellence in Safety Engineering and Analytics, Indian Institute of Technology Kharagpur, Kharagpur, India
e-mail: jmaiti@iem.iitkgp.ac.in

G. K. Bagchi
e-mail: bagchigourab@kgpian.iitkgp.ac.in

J. Maiti et al. (eds.), *Recent Advances in Industrial and Systems Engineering*, Lecture Notes on Multidisciplinary Industrial Engineering,
https://doi.org/10.1007/978-981-96-8323-9_17

17.1 Introduction

A Hazard and Operability (HAZOP) report is a comprehensive documentation of potential deviations, causes, and consequences identified during a HAZOP study. It outlines risks, safeguards, and recommendations for process safety enhancement. Thorough analysis of the HAZOP report provides valuable insights into operational vulnerabilities, enabling effective risk management and process optimization.

In the realm of the chemical process industry, data falls into two distinct categories: one being well-organized data encompassing process parameters, and the other being the more loosely-structured data consisting of accident records and Hazard and Operability (HAZOP) assessment narratives. Leveraging the capabilities of Machine Learning (ML) and Natural Language Processing (NLP) to scrutinize this data holds the potential to extract valuable insights and enhance the overall safety of processes [1]. While recent advancements in safety analysis predominantly revolve around structured data, it's intriguing to note that accident reports have garnered more prominence than HAZOP reports when it comes to delving into the realm of unstructured data studies [2]. The structured data studies like fault detection have been already analyzed by many [3, 4]. HAZOP studies unveil potential deviations from intended design objectives, delve into the underlying triggers of these deviations, and assess the potential outcomes and impacts they might yield [5]. Performing a Hazard and Operability (HAZOP) analysis requires a significant investment of time and resources [6, 7]. Automation of HAZOP studies involves the integration of knowledge representation and logical reasoning processes [7]. Previously reported graph, ontology-based methods are time consuming, and due to their susceptibility to errors, automated HAZOP studies tend to be costlier than traditional manual HAZOP analyses [7].

The cornerstone of a conventional HAZOP study is the brainstorming process, which proves challenging to replicate in automated HAZOP methods. Brainstorming facilitates a collaborative environment where team members with diverse expertise contribute a multitude of ideas stemming from their knowledge and insights [1]. Natural language processing can replicate the whole brainstorming process by training past HAZOP data and predict the desirables automatically. An NLP model using advanced BERT-like techniques are used for extracting meaningful insights from HAZOP reports [8]. Also, intelligent HAZOP report analysis using data mining techniques have been reported [9]. All these techniques are also in the research stage and requires huge numbers of data and complex models to implement.

This paper reported an NLP and ML-based risk prediction method for replicating the HAZOP expert system in intelligent automation based. The article represents the following.

- A data mining-driven intelligent HAZOP analysis method is introduced with the aim of uncovering the interconnections among the entire dataset.
- The described methodology integrates natural language processing and machine learning classifiers for prediction of text to score or numeric values.

- This strategy will predict the severity and likelihood scores from text data.

The remaining sections are arranged as follows. Sect. 17.2 gives a brief review of the literature on HAZOP-based analysis. The implementation steps for the proposed methodology are described in Sect. 17.3. Section 17.4 describes the Results and discussion along with comparing the other works. Finally, Sect. 17.5 concluded with the advantages and future works.

17.2 Literature Review

A systematic procedure integrating automatic HAZOP analysis with structural modeling demonstrated the identification of hazards and mal-operations, leading to a method that simplifies hazards and mal-operations identification shown by Boonthum et al. [10]. For overall consequence assessment of chemical industry, a comprehensive framework of major losses was proposed by Maiti et al. [11], aiming to improve risk assessment quality. An ontology-based knowledge modeling framework combined with probabilistic assessment was introduced by Aziz et al. [12] for dynamic hazard identification and rapid risk estimation, using a semantic web-based approach. The introduction of a functional HAZOP methodology emphasized its potential for efficient hazard identification, aiding in root cause analysis and consequence assessment through computer-aided reasoning tools, was proposed by Rossing et al. [13]. The techniques all are needed expert opinions for each evaluation resulted higher time and cost expenditure. The automation of these industries safety strategy basically started with the advent of machine learning for text data [14]. The topic modeling technique like using advance transformer-based techniques BERT and GPT improves the accuracy of the process but also increases the complexity and precision of the models [15, 16]. Despite of being the development of these techniques the automation of HAZOP in a fully functional way is still under research and implementation of simple to complex automated methods are requisite for industrial safety improvement.

17.3 Implementation Method

The study employed machine learning algorithms that couldn't directly handle text data. To address this, textual information was transformed into a numerical format. In the realm of Natural Language Processing (NLP), there exist two primary approaches for modeling text: traditional and advanced methods. Conventional techniques encompass straightforward structures like Bag of Words (BOW) or Term Frequency-inverse Document Frequency (TF-IDF). On the other hand, more sophisticated methodologies such as Word2Vec or BERT leverage Machine Learning and Artificial Intelligence to effectively represent text datasets. Advanced techniques come into play when dealing with extensive data corpora. Word2Vec, for instance, employs neural networks to attribute vectorized forms to words, while the BOW model suits the conversion of text to numerical forms, especially in cases of smaller datasets. Machine learning models are utilized to make predictions about severity and likelihood scores. These models are developed during the learning phase of the predictive process and serve the purpose of predicting class labels. In the context of this research, the factors responsible for outcomes were employed as features to anticipate consequences. These factors, extracted from sentences in the HAZOP study worksheet, were transformed into feature vectors using the sentence embedding technique in Natural Language Processing (NLP), specifically the TF-IDF approach. Various machine learning models were applied, including Random Forest, Support Vector Machine (SVM), Gradient Boosting, Neural networks, Logistic Regression, Gaussian Naïve Bayes, and Ensembled model, to achieve this objective. The model efficiency (e.g., accuracy, precision, F1 scores etc.) are measured and confusion matrix are used for assessment.

$$\text{Accuracy} = \frac{(\text{TP} + \text{TN})}{(\text{P} + \text{N})}$$

$$\text{Precision} = \frac{\text{TP}}{(\text{TP} + \text{FP})}$$

$$\text{Recall} = \frac{\text{TP}}{\text{P}}$$

$$F1 - \text{Score} = \frac{(2 \times \text{Precision} \times \text{Recall})}{(\text{Precision} + \text{Recall})}$$

True positive (TP): These are the instances classified as positive by the classifier in an accurate manner.

True negative (TN): These are the instances that the classifier correctly identified as negative.

False positive (FP): These instances, which should have been classified as negative, were inaccurately labelled as positive.

False negative (FN): These instances, which should have been categorized as positive, were erroneously labelled as negative.

The visualization of the method introduced in the present work is presented in Fig. 17.1. This diagram outlines the automation flow of the HAZOP study and aids in comprehending the complete procedural sequence.

Python is progressively establishing itself as the predominant programming language in the realm of data science. Its widespread adoption spans both academic and industrial domains, owing to its user-friendly nature and robust scientific libraries. Throughout this study, the Python library scikit-learn was extensively used for machine learning purposes, and NLTK library is used for natural language processing. The proposed methodology for severity prediction for text consequence data is shown in Fig. 17.2.

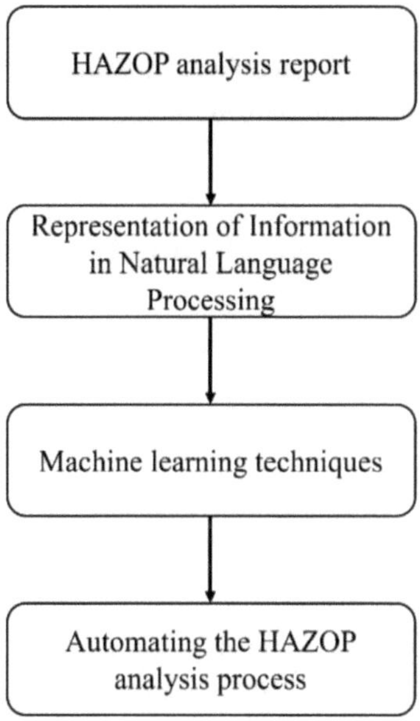

Fig. 17.1 Automating the HAZOP process using NLP and ML

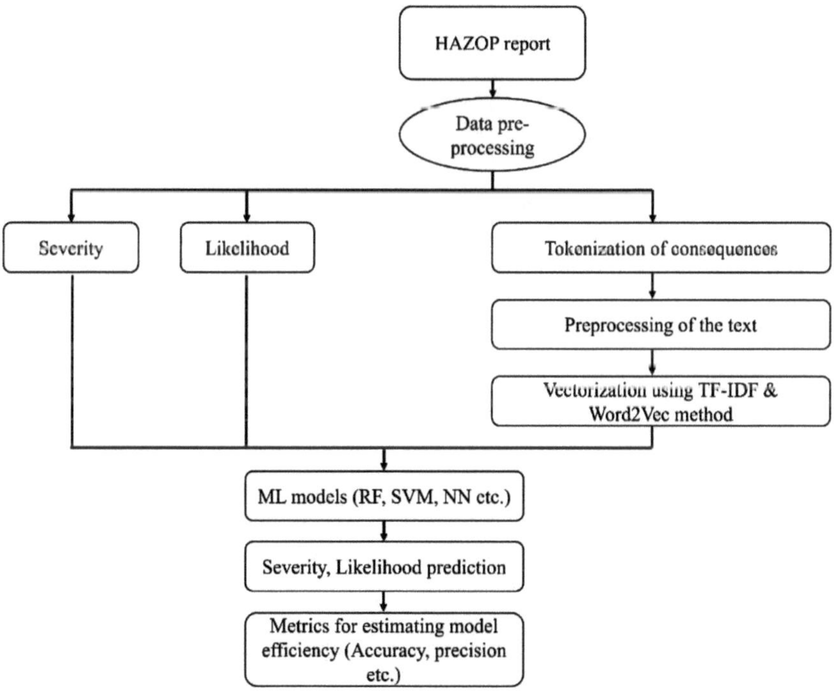

Fig. 17.2 Overall proposed methodology

In the present work, the effectiveness of ML models was assessed using the train-test split technique. A training dataset of 80% and test dataset of 20% is considered in the present work.

17.4 Results and Discussion

This section describes the results and outcomes of each of the steps and methods described above. The NLTK library is used for the text data preprocessing. The TF-IDF vectorization method is implemented which basically takes as bag of words techniques. For the bag of words technique, each consequence text words have been encoded as integers. A sample of encoded form is shown below:

('accumul': 34), ('air': 35), ('ap': 36), ('asset': 37), ('atmospher': 38), ('bad': 39), ('benzen': 40), ('beu': 41), ('beyond': 42), ('blow': 43), ('boot': 44), ('bottom': 45), ('bp': 46), ('c6': 47), ('carryov': 48), ('case': 49), ('caus': 50), ('cavit': 51), ('choke': 52), ('clean': 53), ('column': 54), ('condens': 55), ('consequ': 56), ('consumpt': 57), ('contain': 58), ('cool': 59), ('could': 60), ('damag': 61), ('decreas': 62), ('degrad': 63), ('delay': 64), ('design': 65), ('dispers': 66), ('distil': 67), ('disturb': 68), ('downstream': 69), ('drain': 70), ('dri': 71), ('drum': 72), ('due': 73), ('ed': 74), ('effect': 75), ('enter': 76), ('environment': 77), ('equip': 78), ('exchang': 79), ('experi': 80), ('exposur': 81), ('failur': 82), ('fatal': 83), ('feed': 84), ('financi': 85), ('fire': 86), ('flameout': 87), ('flare': 88), ('flow': 89), ('ga': 90), ('gasket': 91), ('get': 92), ('greater': 93), ('hc': 94), ('header': 95), ('heat': 96), ('heavi': 97), ('high': 98), ('hydrocarbon': 99), ('identifi': 100), ('immedi': 101), ('impact': 102), ('improp': 103), ('inadequ': 104), ('increas': 105), ('injuri': 106), ('instrument': 107), ('lead': 108), ('leak': 109), ('leakag': 110), ('less': 111).

The sample of vectorization using TF-IDF method which has been obtainable by scikit-learn library is shown below.

[[0 0 0 ...1...0.....1..... 0 0 0]
[0 0 0 ..1......0....... 0 0 0]
...
[0 0 0 ....1....0........ 0 0 0]
[0 0 0...1...0...... ... 0 0 0]]

This is a common approach in Natural Language Processing (NLP) for transforming sentences into numerical vectors. In this specific analysis, every consequence is depicted as a vector comprising numerous components. These vectors can now be harnessed to train machine learning models and make predictions for severity and likelihood levels.

Various ML models have been applied to gauge the reliability of the approach suggested in the present work. Additionally, robust metrics were utilized to assess the classifier's efficacy in forecasting deviations. Dependable metrics were utilized to assess the performance of the models in accurately predicting severity levels. Table 17.1 describes the performance of the models.

Table 17.1 ML classifier performances

SL No.	ML models and scores	
	ML models	Accuracy
1	Random forest accuracy Random forest precision Random forest recall Random forest F1-score	0.65 0.73 0.65 0.62
2	SVM accuracy Precision Recall F1-score	0.71 0.64 0.70 0.65
3	Gradient boosting accuracy Gradient boosting precision Gradient boosting recall Gradient boosting F1-score	0.63 0.68 0.60 0.59
4	Bagging neural network accuracy Bagging neural network precision Bagging neural network recall Bagging neural network F1-score	0.59 0.54 0.52 0.51
5	Bagging logistic regression accuracy Bagging logistic regression precision Bagging logistic regression recall Bagging logistic regression F1-score	0.52 0.57 0.52 0.42
6	Voting naive bayes accuracy Voting naive bayes precision Voting naive bayes recall Voting naive bayes F1-score	0.60 0.65 0.60 0.54
7	Ensemble model accuracy (RF + NN + GB)	0.65

It has been observed that SVM has outperformed others. The model predicts text data to numeric scores and only single feature (consequence) from the HAZOP data has been considered for the present work. The proposed method is a unique on its way and shows similar result to other comparable studies [9]. The primary diagonal of the confusion matrix comprises elements aligned from the top-left to the bottom-right. Within this diagonal lie the correctly predicted labels, representing accurate classifications. Figure 17.3 shows the confusion matrix for all the proposed models.

Most of the HAZOP report analysis on data mining are based on large dataset. The present work focused on a single text variable and its severity and likelihood scores to train the models. This method could be beneficial for any industry where single unit text data and its numeric scores are given. The accuracy for SVM model using one single level text (consequence) obtained (71%) can be comparable to other work reported in the literature such as Wang et al. [9] (accuracy severity: 73%) where two levels have been selected (cause, consequence) with more complicated techniques. The NLP modeling technique shown here used simple technique which

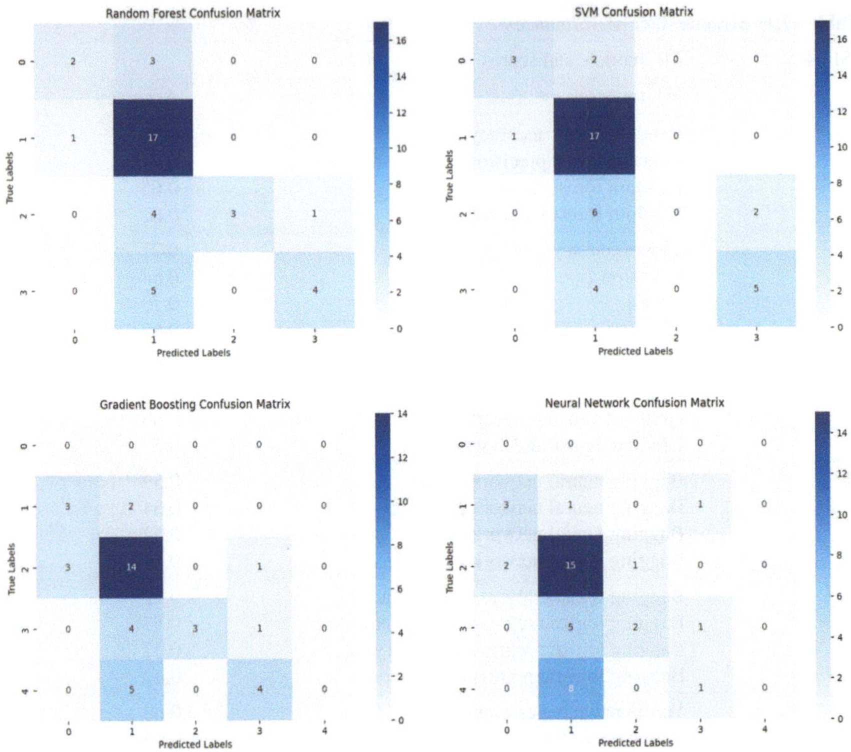

Fig. 17.3 Confusion matrix for each model

is TF-IDF method where other advanced methods like word2vec or transformer-based models like BERT can be used to train the expert system. For ML models to perform effectively in HAZOP study automation, it's imperative to train them using HAZOP study reports. The success of this automation process is intricately tied to the caliber and trustworthiness of the input dataset. The variables in the input dataset plays a significance role in the model's performance. Here only one variable or level is used whereas more levels could increase the accuracy of the models. The embedding of the sentences also impacts on the accuracy of the model's performance, where bigger sentences result poor performance. Here each consequence has large sentences resulted sacrificing the accuracy of the models.

17.5 Conclusion

The present work reports an automated system for predicting risk which is a combination of severity and likelihood scores from HAZOP report, by integrating natural language processing (NLP) and machine learning (ML) techniques. This method can predict the severity, likelihood scores from a new consequence data in text format like replicating an expert system. This proposed approach represents a unique automated HAZOP expert system that has the capability of replicating the entire HAZOP process and can result enormous cost effectiveness, and instant outcome. The NLP technique used in the present work is well-established and simple in implementation. The present work involves less variables and large sentences, thus the comparison with other model performances reported in literature is acceptable depending on the structure of the data. The SVM model provides the best results, and also ensemble technique is used combining random forest, neural network and gradient boosting. The effectiveness of the suggested approach is contingent upon the excellence and dependability of the input dataset. The future model also considered using ensembled transformer-based NLP techniques.

Acknowledgements We acknowledge the Centre of Excellence in Safety Engineering and Analytics (CoE-SEA) (www.iitkgp.ac.in/department/SE), IIT Kharagpur, and Safety Analytics and Virtual Reality (SAVR) Laboratory (www.savr.iitkgp.ac.in) of Department of Industrial and Systems Engineering, IIT Kharagpur for experimental/computational and research facilities for this work. We are also thankful to Haldia Petrochemicals Limited for providing the data for this work.

References

1. Ekramipooya, A., Boroushaki, M., Rashtchian, D.: Application of natural language processing and machine learning in prediction of deviations in the HAZOP study worksheet: a comparison of classifiers. Process Saf. Environ. Prot. **176**, 65–73 (2023). https://doi.org/10.1016/j.psep.2023.06.004
2. Ahadh, A., Binish, G.V., Srinivasan, R.: Text mining of accident reports using semi-supervised keyword extraction and topic modeling. Process Saf. Environ. Prot. **155**, 455–465 (2021). https://doi.org/10.1016/j.psep.2021.09.022
3. Amin, M.T., Khan, F., Ahmed, S., Imtiaz, S.: Risk-based fault detection and diagnosis for nonlinear and non-Gaussian process systems using R-vine copula. Process Saf. Environ. Prot. **150**, 123–136 (2021). https://doi.org/10.1016/j.psep.2021.04.010
4. Arunthavanathan, R., Khan, F., Ahmed, S., Imtiaz, S.: A deep learning model for process fault prognosis. Process Saf. Environ. Prot. **154**, 467–479 (2021). https://doi.org/10.1016/j.psep.2021.08.022
5. Batres R (n.d.) An ontology approach to support HAZOP studies
6. Baybutt, P.: A critique of the Hazard and Operability (HAZOP) study. J. Loss Prev. Process Ind. **33**, 52–58 (2015). https://doi.org/10.1016/j.jlp.2014.11.010
7. Single JI, Schmidt J, Denecke J (2019) State of research on the automation of HAZOP studies. J Loss Prev Process Ind **62**. https://doi.org/10.1016/j.jlp.2019.103952
8. Feng, X., Dai, Y., Ji, X., Zhou, L., Dang, Y.: Application of natural language processing in HAZOP reports. Process Saf. Environ. Prot. **155**, 41–48 (2021). https://doi.org/10.1016/j.psep.2021.09.001
9. Wang, F., Gu, W.: Intelligent HAZOP analysis method based on data mining. J. Loss Prev. Process Ind. **80**, (2022). https://doi.org/10.1016/j.jlp.2022.104911
10. Boonthum, N., Mulalee, U., Srinophakun, T.: A systematic formulation for HAZOP analysis based on structural model. Reliab. Eng. Syst. Saf. **121**, 152–163 (2014). https://doi.org/10.1016/j.ress.2013.08.008

11. Arunraj, N.S., Maiti, J.: A methodology for overall consequence modeling in chemical industry. J. Hazard. Mater. **169**(1–3), 556–574 (2009). https://doi.org/10.1016/j.jhazmat.2009.03.133
12. Aziz, A., Ahmed, S., Khan, F.I.: An ontology-based methodology for hazard identification and causation analysis. Process Saf. Environ. Prot. **123**, 87–98 (2019). https://doi.org/10.1016/j.psep.2018.12.008
13. Rossing, N.L., Lind, M., Jensen, N., Jørgensen, S.B.: A functional HAZOP methodology. Comput. Chem. Eng. **34**(2), 244–253 (2010). https://doi.org/10.1016/j.compchemeng.2009.06.028
14. Aggarwal CC (n.d.) Machine learning for text
15. Devlin J, Chang M-W, Lee K, Google KT, Language AI (n.d.) BERT: pre-training of deep bidirectional transformers for language understanding. https://github.com/tensorflow/tensor2tensor
16. Grootendorst M (2022) BERTopic: neural topic modeling with a class-based TF-IDF procedure. http://arxiv.org/abs/2203.05794

Chapter 18
Uncertainty Quantification of the External Ballistic 6-DOF Model of Artillery Projectile by Using Regression-Based Sensitivity Analysis Methods

Ranit Senapati and Biswajit Mahanty

Abstract While designing artillery external ballistics, it is essential to effectively quantify the influence of multi-source uncertain parameters on the external ballistics, which can improve the overall attack efficiency of the projectile. An artillery firing system is complex owing to the dependencies among multiple inputs and outputs, which can be explained through a set of equations known as the 6-DOF model that are to be solved numerically. In the absence of a properly defined input–output relationship, sensitivity analysis using the standardized regression coefficient (SRC) provides satisfactory results even for non-linear models. The sensitivity indices obtained can help reduce design costs and improve computational efficiency by improving structural design, parameter identification, and reliability-based design.

Keywords Uncertainty quantification · Product design · Sensitivity analysis · External ballistics · Artillery system

18.1 Introduction

Artillery serves as the primary weapon for firepower suppression in contemporary battlefields. Its prominence can be attributed to various advantages: affordability, dependable construction, well-established technology, straightforward operation, ample ammunition supply, and formidable killing potential. Considering its strong

R. Senapati (✉) · B. Mahanty
Department of Industrial and Systems Engineering, Indian Institute of Technology Kharagpur, Kharagpur, India
e-mail: ranit.rt@iitkgp.ac.in

B. Mahanty
e-mail: bm@hijli.iitkgp.ac.in

J. Maiti et al. (eds.), *Recent Advances in Industrial and Systems Engineering*, Lecture Notes on Multidisciplinary Industrial Engineering,
https://doi.org/10.1007/978-981-96-8323-9_18

performance in terms of cost-effectiveness, technological maturity, and combat effectiveness, artillery holds a pivotal position in modern warfare. Enhancing the precision of artillery fire can substantially elevate its lethality, decrease ammunition consumption, and profoundly influence the overall course of the battle [7].

The examination of uncertainty and sensitivity constitutes integral components in the analysis of complex systems. More precisely, uncertainty analysis involves identifying the uncertainties in analysis outcomes arising from uncertainties in analysis inputs. On the other hand, sensitivity analysis entails discerning the individual contributions of uncertainty analysis inputs to the overall uncertainty in analysis results. Epistemic uncertainty emerges due to a lack of knowledge regarding the appropriate value to assign to a quantity assumed to have a fixed value in a particular analysis context. In contrast, aleatory uncertainty arises from inherent randomness in the behavior of the studied system [1].

Within a complex artillery external ballistic system, numerous uncertain parameters emanate from multiple sources, encompassing aspects like projectile mass, muzzle velocity, elevation angle, azimuth angle, axial wind speed, and crosswind speed. These variables contribute significantly to the variations observed in projectile performance. The precise analysis of the impact of uncertainty on projectile performance parameters and the subsequent quantification of the importance ranking of these uncertain parameters are pivotal steps in enhancing the accuracy of the artillery system's attack capabilities. The main objective of this study is to understand the influence of uncertainties in some of the projectile properties such as mass, moment of inertia; firing conditions such as muzzle velocity, elevation angle, and azimuth angle; and uncontrollable factors such as wind speed, which depends on the present weather pattern. The total mass of the projectile (M) is the sum of the mass of the cylindrical metal shell, the mass of the explosive charge, and the mass of the fuze—which may vary from shell to shell due to manufacturing tolerances. The muzzle velocity (V) is dependent on the type of propellant, percentage of propellant burned, etc., and can also be subjected to uncertainty. The elevation angle (E) and azimuth angle (A) are set by the operating personnel and are subjected to change due to the calculations made by the commander after determining the enemy position. Axial wind speed (W_x) and crosswind speed (W_z) are uncontrollable factors and depend on the weather conditions present at the time of firing.

Four external ballistic characteristics are considered in this study—time of flight, range, maximum height, and deviation, which are calculated by solving the 6-DOF model numerically. The sensitivity indices obtained by performing sensitivity analysis provide an understanding of how the uncertainties in the input are propagated to the output and contribute to the changes in the precision of an artillery projectile strike.

18.2 Design/Methodology/Approach

18.2.1 6-DOF Equations of Motion

The 6-DOF model gives the most accurate position and attitude of a projectile in flight with the assumption that the projectile is a finite length rigid body with its mass distributed throughout its geometry. The direction cosines are assumed to translate with the center of gravity (CG) without rotation. The complete derivation of the equations stated below are taken from [2, 4].

$$\begin{aligned}\frac{\mathrm{d}V_1}{\mathrm{d}t} &= -\tilde{C}_D v_1 + \tilde{C}_{L_\alpha}(v^2 i_1 - v v_1 \cos\alpha_t) \\ &\quad - \tilde{C}_{N_{p\alpha}}(v_3 i_2 - v_2 i_3) \\ &\quad + \tilde{C}_{N_q}(h_2 i_3 - h_3 i_2) + g_1 + \Lambda_1 \end{aligned} \tag{18.1}$$

$$\begin{aligned}\frac{\mathrm{d}V_2}{\mathrm{d}t} &= -\tilde{C}_D v_2 + \tilde{C}_{L_\alpha}(v^2 i_2 - v v_2 \cos\alpha_t) \\ &\quad - \tilde{C}_{N_{p\alpha}}(v_1 i_3 - v_3 i_1) \\ &\quad + \tilde{C}_{N_q}(h_3 i_1 - h_1 i_3) + g_2 + \Lambda_2 \end{aligned} \tag{18.2}$$

$$\begin{aligned}\frac{\mathrm{d}V_3}{\mathrm{d}t} &= -\tilde{C}_D v_3 + \tilde{C}_{L_\alpha}(v^2 i_3 - v v_3 \cos\alpha_t) \\ &\quad - \tilde{C}_{N_{p\alpha}}(v_2 i_1 - v_1 i_2) \\ &\quad + \tilde{C}_{N_q}(h_1 i_2 - h_2 i_1) + g_3 + \Lambda_3 \end{aligned} \tag{18.3}$$

$$\begin{aligned}\frac{\mathrm{d}h_1}{\mathrm{d}t} &= (\tilde{C}_{lp} + \tilde{C}_{1\delta}) i_1 + \tilde{C}_{M_\alpha}(v_2 i_3 - v_3 i_2) \\ &\quad + \tilde{C}_{M_{p\alpha}}(v_1 - v i_1 \cos\alpha_t) \\ &\quad + \tilde{C}_{M_q}\left[h_1 - \left(\frac{I_p p}{I_T}\right) i_1\right] \end{aligned} \tag{18.4}$$

$$\begin{aligned}\frac{\mathrm{d}h_2}{\mathrm{d}t} &= (\tilde{C}_{lp} + \tilde{C}_{1\delta}) i_2 + \tilde{C}_{M_\alpha}(v_3 i_1 - v_1 i_3) \\ &\quad + \tilde{C}_{M_{p\alpha}}(v_2 - v i_2 \cos\alpha_t) \\ &\quad + \tilde{C}_{M_q}\left[h_2 - \left(\frac{I_p p}{I_T}\right) i_2\right] \end{aligned} \tag{18.5}$$

$$\begin{aligned}\frac{\mathrm{d}h_3}{\mathrm{d}t} &= (\tilde{C}_{lp} + \tilde{C}_{1\delta}) i_3 + \tilde{C}_{M_\alpha}(v_1 i_2 - v_2 i_1) \\ &\quad + \tilde{C}_{M_{p\alpha}}(v_3 - v i_3 \cos\alpha_t) \end{aligned}$$

$$+ \tilde{C}_{M_q}\left[h_3 - \left(\frac{I_p p}{I_T}\right)i_3\right] \tag{18.6}$$

where the coefficients are given by
$\tilde{C}_D = \frac{\rho v S C_D}{2m}, \tilde{C}_{L_\alpha} = \frac{\rho S C_{L\alpha}}{2m}, \tilde{C}_{N_{p\alpha}} = \frac{\rho S d C_{N_{p\alpha}} p}{2m}, \tilde{C}_{N_q} = \frac{\rho v S d (C_{N_q} + C_{N_{\dot\alpha}})}{2m}, \tilde{C}_{lp} = \frac{\rho v S d^2 C_{lp} p}{2I_T},$
$\tilde{C}_{1\delta} = \frac{\rho v^2 S d \delta_F C_{1\delta}}{2I_T}, \tilde{C}_{M_\alpha} = \frac{\rho v S d C_{M\alpha}}{2I_T}, \tilde{C}_{M_{p\alpha}} = \frac{\rho S d^2 C_{M_{p\alpha}} p}{2I_T}, \tilde{C}_{M_q} = \frac{\rho v S d^2 (C_{M_q} + C_{M_{\dot\alpha}})}{2I_T}$. $[V_1, V_2, V_3]$ are the components of velocity vector V in x, y, and z directions, respectively, with respect to the fixed earth component system. Similarly, $[h_1, h_2, h_3]$ are the components of the angular momentum vector in x, y, and z directions. If the wind velocity is represented by $\boldsymbol{w} = [w_1, w_2, w_3]$, the relative velocity of the projectile can be given by $\boldsymbol{v} = \boldsymbol{V} - \boldsymbol{w} = [v_1, v_2, v_3]$ and $v = \sqrt{v_1^2 + v_2^2 + v_3^2}$ represents the magnitude of the relative velocity vector.

The six coupled equations can be solved numerically with the help of MATLAB to calculate the position and attitude of the projectile in space and time. Since the direction cosines also change along the trajectory, three additional equations are obtained for each along with three more equations for the three coordinates of the CG of the projectile.

$$\frac{di_1}{dt} = h_2 i_3 - h_3 i_2 \tag{18.7}$$

$$\frac{di_2}{dt} = h_3 i_1 - h_1 i_3 \tag{18.8}$$

$$\frac{di_3}{dt} = h_1 i_2 - h_2 i_1 \tag{18.9}$$

$$\frac{dx}{dt} = V_1 \tag{18.10}$$

$$\frac{dy}{dt} = V_2 \tag{18.11}$$

$$\frac{dz}{dt} = V_3 \tag{18.12}$$

with $[i_1, i_2, i_3]$ representing the unit vectors of the coordinate system attached to the projectile and is also assumed to be rotating with it. $[x, y, z]$ represents the position of the projectile at any point of time t with reference to the earth-fixed coordinate system.

The initial conditions at time $t = 0$ are given in Eqs. (18.13)–(18.20) to begin the numerical integration procedure. The values of the various parameters used are given in Table 18.1.

$$\overline{V}_0 = V_0 \begin{bmatrix} \cos\varphi_0 \cos\theta_0 \\ \sin\varphi_0 \cos\theta_0 \\ \sin\theta_0 \end{bmatrix} \begin{bmatrix} \hat{e}_1 \ \hat{e}_2 \ \hat{e}_3 \end{bmatrix} \tag{18.13}$$

$$\hat{i}_0 = \begin{bmatrix} i_{1_0} \\ i_{2_0} \\ i_{3_0} \end{bmatrix} = \begin{bmatrix} \cos(\varphi_0 + \alpha_0)\cos(\theta_0 + \beta_0) \\ \sin(\varphi_0 + \alpha_0)\cos(\theta_0 + \beta_0) \\ \sin(\theta_0 + \beta_0) \end{bmatrix} \begin{bmatrix} \hat{e}_1 \ \hat{e}_2 \ \hat{e}_3 \end{bmatrix} \tag{18.14}$$

$$\begin{aligned} \hat{j}_0 &= \begin{bmatrix} j_{1_0} \\ j_{2_0} \\ j_{3_0} \end{bmatrix} \\ &= \frac{1}{\sqrt{Q}} \begin{bmatrix} -\cos^2(\theta_0 + \beta_0)\sin(\varphi_0 + \alpha_0)\cos(\varphi_0 + \alpha_0) \\ \cos^2(\theta_0 + \beta_0)\cos^2(\varphi_0 + \alpha_0) + \sin^2(\theta_0 + \beta_0) \\ -\sin(\theta_0 + \beta_0)\cos(\theta_0 + \beta_0)\sin(\varphi_0 + \alpha_0) \end{bmatrix} \begin{bmatrix} \hat{e}_1 \ \hat{e}_2 \ \hat{e}_3 \end{bmatrix} \end{aligned} \tag{18.15}$$

$$\hat{k}_0 = \begin{bmatrix} k_{1_0} \\ k_{2_0} \\ k_{3_0} \end{bmatrix} = \frac{1}{\sqrt{Q}} \begin{bmatrix} -\sin(\theta_0 + \beta_0) \\ 0 \\ \cos(\theta_0 + \beta_0)\cos(\varphi_0 + \alpha_0) \end{bmatrix} \begin{bmatrix} \hat{e}_1 \ \hat{e}_2 \ \hat{e}_3 \end{bmatrix} \tag{18.16}$$

where the quantity Q is defined as:

$$Q = \sin^2(\theta_0 + \beta_0) + \cos^2(\theta_0 + \beta_0)\cos^2(\varphi_0 + \alpha_0) \tag{18.17}$$

$$\frac{d\hat{i}_0}{dt} = \begin{bmatrix} \omega_{k_0} j_{1_0} - \omega_{j_0} k_{1_0} \\ \omega_{k_0} j_{2_0} - \omega_{j_0} k_{2_0} \\ \omega_{k_0} j_{3_0} - \omega_{j_0} k_{3_0} \end{bmatrix} \begin{bmatrix} \hat{e}_1 \ \hat{e}_2 \ \hat{e}_3 \end{bmatrix} \tag{18.18}$$

$$\overline{h}_0 = \begin{bmatrix} h_{1_0} \\ h_{2_0} \\ h_{3_0} \end{bmatrix} \begin{bmatrix} \hat{e}_1 \ \hat{e}_2 \ \hat{e}_3 \end{bmatrix} = \begin{bmatrix} \frac{I_p p_0}{I_T} i_{1_0} + i_{2_0} i_{3_0}{}' - i_{3_0} i'_{2_0} \\ \frac{I_p p_0}{I_T} i_{2_0} + i_{3_0} i_{1_0}{}' - i_{1_0} i'_{3_0} \\ \frac{I_p p_0}{I_T} i_{3_0} + i_{1_0} i_{2_0}{}' - i_{2_0} i'_{1_0} \end{bmatrix} \begin{bmatrix} \hat{e}_1 \ \hat{e}_2 \ \hat{e}_3 \end{bmatrix} \tag{18.19}$$

$$\begin{aligned} \overline{\Lambda} &= \begin{bmatrix} \Lambda_1 \\ \Lambda_2 \\ \Lambda_3 \end{bmatrix} \begin{bmatrix} \hat{e}_1 \ \hat{e}_2 \ \hat{e}_3 \end{bmatrix} \\ &= 2\Omega \begin{bmatrix} -V_2 \cos L \sin AZ - V_3 \sin L \\ V_1 \cos L \sin AZ + V_3 \cos L \cos AZ \\ V_1 \sin L - V_2 \cos L \cos AZ \end{bmatrix} \begin{bmatrix} \hat{e}_1 \ \hat{e}_2 \ \hat{e}_3 \end{bmatrix} \end{aligned} \tag{18.20}$$

The twelve equations can be solved using the Runge Kutta (RK4) numerical method with twelve initial conditions one for each variable. The density function used is according to the standard atmospheric model. Based on the data table generated by numerical integration, time of flight is taken as the time when y changes sign from positive to negative, range is the corresponding x value at that time, max. height is

Table 18.1 List of data used for numerical simulation

Parameter name	Symbol	Type	Value
Mass of projectile (kg)	m	Stochastic	Unif[14,16]
Diameter of projectile (m)	D	Deterministic	0.105
Axial moment of inertia (kg–m^2)	I_p	Deterministic	0.02326
Transverse moment of inertia (kg–m^2)	I_T	Deterministic	0.23118
Length of projectile (m)	L	Deterministic	0.4947
Muzzle velocity (m/s)	V_0	Stochastic	Unif[473,513]
Rifling rate of gun (caliber/rev)	n	Deterministic	25
Initial yaw rate (rad/s)	q_0	Deterministic	0
Initial pitch rate (rad/s)	r_0	Deterministic	1.98
Drag coefficient	C_D	Deterministic	0.243
Lift coefficient	C_L	Deterministic	1.76
Magnus force coefficient	$C_{N_{p\alpha}}$	Deterministic	0.215
Pitch damping force coefficient	C_{nq}	Deterministic	0
Roll damping coefficient	C_{lp}	Deterministic	– 0.0108
Overturning moment coefficient	$C_{m\alpha}$	Deterministic	3.76
Magnus moment coefficient	$C_{Y_{p\alpha}}$	Deterministic	0.381
Pitch damping moment coefficient	C_{mq}	Deterministic	– 9.3
Elevation angle	φ_0	Stochastic	Unif[45,70]
Azimuth angle	θ_0	Stochastic	Unif[– 30,30]
Initial pitch angle	α_0	Deterministic	0
Initial yaw angle	β_0	Deterministic	3
Latitude of firing site (Positive in Northern Hem) = L	L	Deterministic	0
Azimuth angle of fire (from North)	AZ	Deterministic	0
Wind speed (m/s)	W	Stochastic	Unif[3.5,7.5]

the maximum value of y attained, and deviation is the z value at the time of flight. The range and elevation are validated for the 105 mm M1 HE projectile for which all the aerodynamic coefficients are known (Fig. 18.1).

18.2.2 Regression-Based Method

Non-parametric methods using linear regression have been used most widely in the development of global sensitivity analysis methods receiving extensive research attention. The success of non-parametric methods can be in part attributed to the fact that the sensitivity measures can be directly estimated from a given Monte Carlo

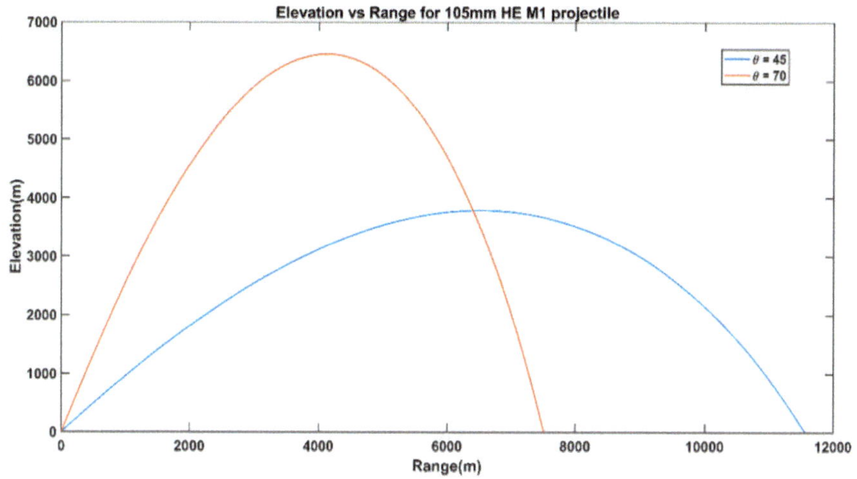

Fig. 18.1 Elevation versus range plot for 105 mm HE M1 projectile

sample. Consequently, analysts can perform Monte Carlo simulation and subsequently compute the sensitivity measures through post processing of the resulting input–output sample [5]. Various sampling strategies can be employed, such as random sampling, importance sampling, and Latin hypercube sampling. Latin hypercube sampling is particularly favored when dealing with computationally intensive models due to its efficient stratification properties, enabling the extraction of considerable uncertainty and sensitivity information with a relatively modest sample size. Conversely, when the model is not computationally demanding and multiple evaluations can be conducted, random sampling performs comparably to Latin hypercube sampling (LHS). In the current study, LHS is chosen as the preferred sampling method to generate the input matrix of uncertain variables (mass, muzzle velocity, elevation angle, azimuth angle, axial wind speed, cross wind speed), which are then passed sequentially to the generated 6 DOF model code to generate the four output parameters (time of flight, range, maximum height, and lateral deviation).

The method assumes that the input–output black box model can be well approximated by the linear model $g(X) \approx b_0 + \sum_{i=1}^{n} b_i X_i$; then the sensitivity of $g(X)$ to X_i is naturally represented by the standardized regression coefficient (SRC)

$$SRC_i = b_i {}^{\sigma_i}/_{\sigma_Y} \tag{18.21}$$

where σ_i is the standard deviation in model input X_i and σ_Y is the standard deviation of output Y. An alternative widely used non-parametric sensitivity measure is represented by the Pearson correlation coefficient

$$\mathrm{PEAR}_i = \frac{\mathrm{Cov}(Y, X_i)}{\sigma_i \sigma_Y} = \frac{\sum_{j=1}^{N} \left(X_{ji} - \overline{X}_i\right)\left(Y_j - \overline{Y}\right)}{\sqrt{\sum_{j=1}^{N} \left(X_{ji} - \overline{X}_i\right)^2} \sqrt{\sum_{j=1}^{N} \left(Y_j - \overline{Y}\right)^2}} \tag{18.22}$$

Typically, the capacity of non-parametric methods to encompass variability is quantified using R_Y^2, which represents the model coefficient of determination. A low R_Y^2 value indicates a suboptimal regression fit, as suggested by Campolongo and Saltelli [3]. In such instances, assessing the influence of input variables based on sensitivity indices such as SRCs might be impractical. Generally, in scenarios where the input–output relationship involves non-linearities and interactions, as noted by Saltelli and Marivoet [6], the regression fit tends to be less robust.

Employing rank transformation can enhance the dependability of non-parametric methods. The Spearman correlation coefficient is obtained by substituting the model output with its associated ranks and calculating the Pearson correlation. This sensitivity measure is adept at identifying non-linear monotonic relationships. The sign of the Spearman correlation coefficient indicates whether Y exhibits a statistically increasing trend with X_i. Additionally, if $Y = g(\boldsymbol{X})$ represents a univariate monotonically increasing function of X, the Spearman correlation coefficient equals unity. A rank transformation involves substitution of the raw data with the corresponding ranks applied both to the independent variable X_i and the dependent variable Y. The rank transformed input column is X_i^R and the corresponding rank transformed output column is Y^R from which the non-parametric ranked sensitivity measures can be computed by applying the linear regression model

$$g^R = b_0^R + \sum_{i=1}^{N} b_i^R X_i^R \tag{18.23}$$

with the standardized rank regression coefficients (SRRC) and Spearman correlation coefficients given as

$$\text{SRRC}_i = b_i^R \frac{\sigma_i^R}{\sigma_Y^R} \tag{18.24}$$

$$\text{SPEAR}_i = \frac{\text{Cov}\left(Y^R, X_i^R\right)}{\sigma_Y^R \sigma_{X_i}^R} \tag{18.25}$$

18.3 Results and Discussion

The numerical 6-DOF model is run for a total of $N = 1000$ samples with the input parameters generated by LHS to create the Monte Carlo sample. The equations obtained by applying linear regression fit to each of the output variables with respect to the input variables illustrates how uncertainty is propagated from the input to the output variables. The coefficients of linear regression fit are given in Table 18.2.

Applying linear regression to time of flight with respect to the six input parameters, the values of $R^2 = 0.7958$ and $R_{\text{adj}}^2 = 0.7946$ are obtained thus showing a good

Table 18.2 Coefficients of linear regression fit

	Time of flight	Range	Maximum height	Deviation
Mass	0.6041	275.5290	92.1200	– 17.6620
Muzzle velocity	0.0852	21.9690	13.1502	1.2470
Elevation angle	0.6770	– 147.1140	100.7211	8.6410
Azimuth angle	0.0041	– 9.4540	0.3584	300.2570
Axial wind speed	0.0185	– 62.8690	2.2863	– 1.1570
Cross wind speed	0.0834	26.3630	12.9996	– 70.3810

regression fit. Further the values of SRC, SRRC, Pearson and Spearman coefficients are computed, which show that the effects of uncertainty on elevation angle and muzzle velocity are the most influential in predicting the uncertainty in the time of flight. It is interesting to note that the uncertainty levels of uncontrollable factors axial wind speed and cross wind speed do not appear to have any effect on the time of flight.

On the other hand, the linear regression fit of range with the six uncertain parameters shows an average fit with the values of $R^2 = 0.6664$ and $R^2_{\text{adj}} = 0.6643$. The not-so-good fit may be attributed to the presence of non-linearity in the function. On computation of the sensitivity coefficients, it can be seen that the elevation angle followed by the muzzle velocity are the dominant factors controlling the uncertainty in range. The coefficient values with respect to elevation angle are all highly negative thus suggesting a strong linear monotonically decreasing relationship between range and elevation angle. Axial wind speed seems to have a negative effect on range in the given interval, while cross wind speed has no significant effect on range. Since, the coefficient of determination is low in this case, the linear regression method applied here should be supplemented by additional methods such as variance decomposition in order to handle non-linearity.

In case of maximum height, the linear regression fit yields the values of $R^2 = 0.7948$ and $R^2_{\text{adj}} = 0.7935$ showing a good linear fit. Once again, the elevation angle and the muzzle velocity are the dominant factors controlling the uncertainty in maximum height. Increase in elevation angle and muzzle velocity increases the height of the projectile.

In case of deviation, applying the linear regression model leads to the values of $R^2 = 0.9905$ and $R^2_{\text{adj}} = 0.9904$, which shows that the model is highly linear. Furthermore, the uncertainty in deviation is almost due to the uncertainty in azimuth angle followed by small contribution from cross wind speed as is evident from the SRC, SRRC, Pearson and Spearman coefficients (see Figs. 18.2, 18.3, 18.4 and 18.5). It is clear that the deviation increases with increase in cross wind speed. It can be minimized by setting the value of azimuth angle properly.

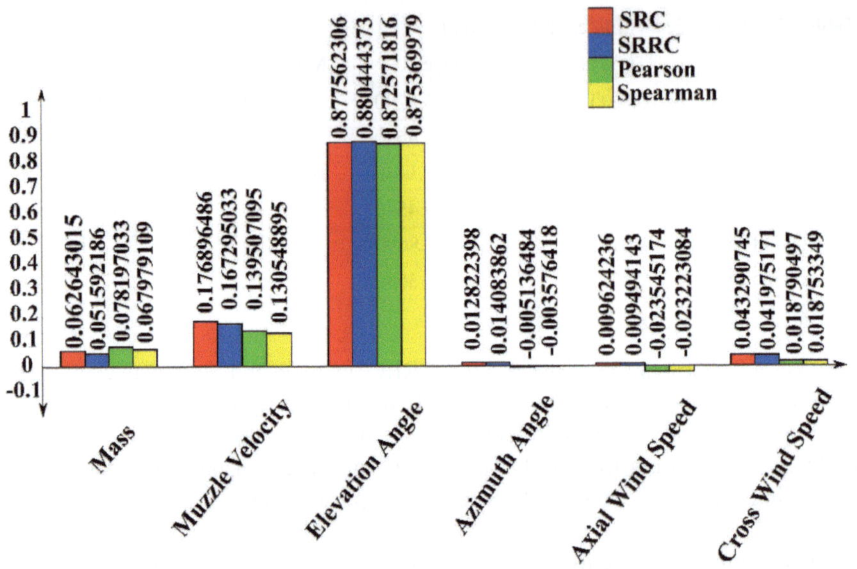

Fig. 18.2 Comparison of SRC, SRRC, Pearson and Spearman coefficients for time of flight

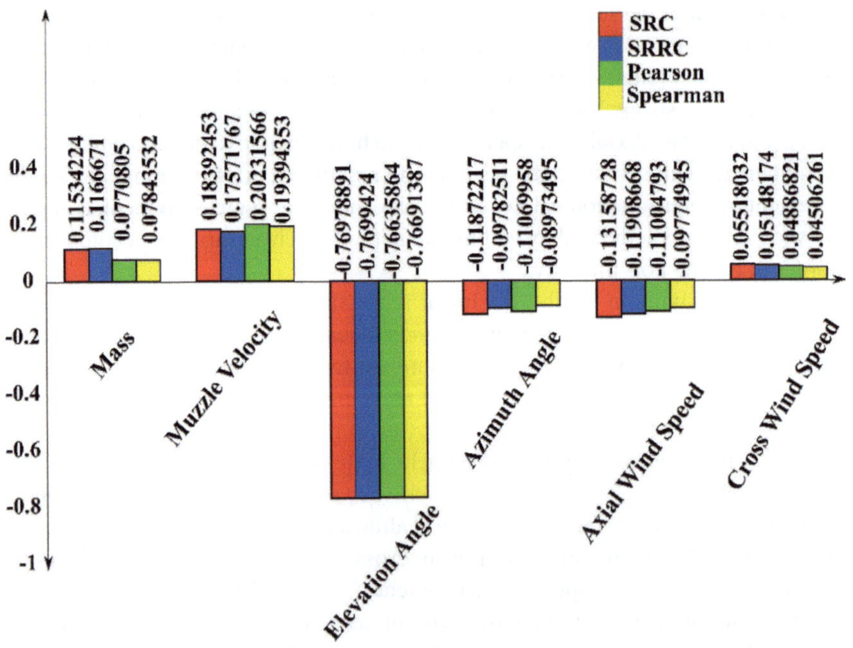

Fig. 18.3 Comparison of SRC, SRRC, Pearson and Spearman coefficients for range

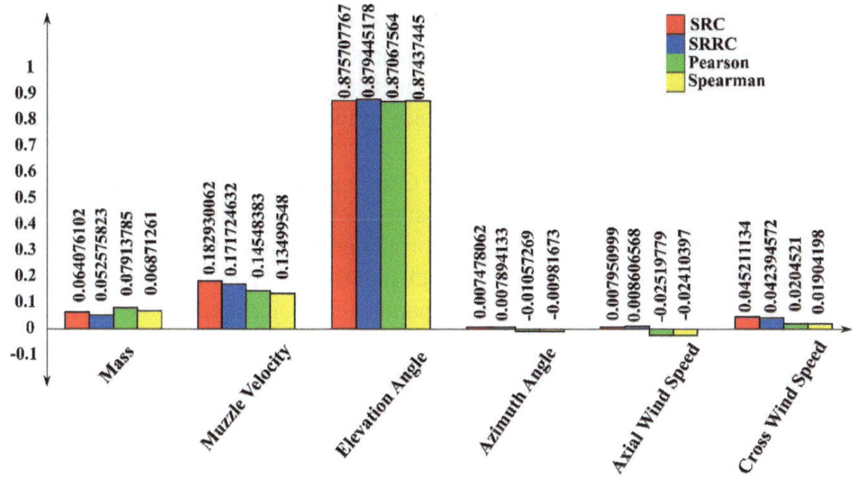

Fig. 18.4 Comparison of SRC, SRRC, Pearson and Spearman coefficients for maximum height

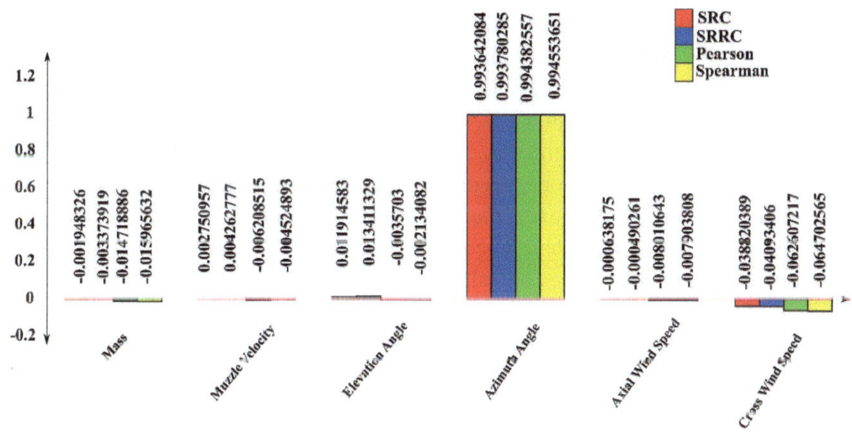

Fig. 18.5 Comparison of SRC, SRRC, Pearson and Spearman coefficients for deviation

18.4 Conclusion

The study uses the method of non-parametric linear regression-based sensitivity indices to study the effect of uncertainties in six parameters namely mass, muzzle velocity, elevation angle, azimuth angle, axial wind speed, and cross wind speed on the performance of an artillery projectile measured by the four parameters of time of flight, range, maximum height, and lateral deviation. The motion of the projectile is simulated here using the 6-DOF model, which is the most accurate model of projectile trajectory discovered till date. Literature review has revealed that there is a lack of studies on the propagation of uncertainty from input to output parameters and sensitivity analysis in the case of an artillery projectile theoretically represented by the 6-DOF model. The current study attempts to fill this gap. The novelty of the work lies in the application of the linear regression method to evaluate the effect of varying input parameters on the performance of the projectile in term of the output parameters.

Using regression analysis on a Monte Carlo sample of 1000 simulation run of the 6-DOF model, the standardized regression coefficient (SRC), standardized rank regression coefficient (SRRC), Pearson correlation coefficient, and Spearman rank correlation coefficients are obtained for each input–output pair. The rationale behind the choice of these particular six input parameters is justified as the uncertainty in mass and muzzle velocity is brought about by manufacturing defects/tolerances, uncertainty in elevation angle and azimuth angle are due to user controlled operating conditions, and uncertainty in axial as well as cross wind speed is environmental and uncontrollable. The study thus encompasses all types of uncertain variables that are possible in a given system. Projectiles designed to attack infantry forces use a timed fuze to control the explosion when the projectile reaches near the enemy position; hence determining the uncertainty in time of flight due to input uncertainties is of paramount importance in priming the fuze to maximize damage potential. An increase in the elevation angle increases the time of flight but decreases the range of the projectile making the two objectives contradictory and hence the operating conditions must be set accordingly. On the other hand, deviation is found to be almost entirely dependent on azimuth angle, hence we can conclude from our analysis that we can reduce deviations entirely by setting the proper value of azimuth angle.

This study has considered only six out of many possible parameters influencing the performance of a projectile; hence, the study can be expanded to cater to an increasing list of uncertain variables. Some uncertainties are due to manufacturing errors and operation errors, which can be controlled; but some like the wind speed, temperature and humidity are uncontrollable and thus provide an opportunity for further study involving robust optimization. Also, the uncertainties are obtained in this study by assuming a uniform distribution around the commonly used mean values. The effect of different probability distributions around different mean values may also provide another avenue for further research.

References

1. Borgonovo, E., Plischke, E.: Sensitivity analysis: a review of recent advances. Eur. J. Oper. Res. **248**(3), 869–887 (2016)
2. Carlucci, D.E., Jacobson, S.S.: Ballistics: Theory and Design of Guns and Ammunition. CRC Press (2018)
3. Campolongo, F., Saltelli, A.: Sensitivity analysis of an environmental model: an application of different analysis methods. Reliab. Eng. Syst. Saf. **57**(1), 49–69 (1997)
4. McCoy, R.: Modern Exterior Ballistics: The Launch and Flight Dynamics of Symmetric Projectiles. Schiffer Publishing (1999)
5. Saltelli, A., Ratto, M., Andres, T., Campolongo, F., Cariboni, J., Gatelli, D., Saisana, M., Tarantola, S.: Global Sensitivity Analysis: The Primer. Wiley (2008)

6. Saltelli, A., Marivoet, J.: Non-parametric statistics in sensitivity analysis for model output: a comparison of selected techniques. Reliab. Eng. Syst. Saf.Saf. **28**(2), 229–253 (1990)
7. Tong, N., Liu, Q., Han, X., Wu, X., Zhang, Z.: Uncertainty analysis and sensitivity estimation on an artillery external ballistic system. J. Mech. Des. **144**(10), 101702 (2022)

Chapter 19
Quality Inspection Experimentation on Image Processing with CPUs and GPUs: A Case Study

Yifei Zhang, Ramiro Rodriguez Buno, Chih-Yin Lin, Yuan-Ke Tsai, and Theodore T. Allen

Abstract Camera and Graphical Processing Unit technology are enabling a new generation of image inspection and quality control charting methods in manufacturing. A critical aspect of meeting processing time requirements is exploiting the specific computer architecture relating to Graphical Processing Units (GPUs) and Central Processing Units (CPUs). Using a real work automotive body-in-white inspection problem, we compare alternative processing methods. The results show dramatic improvements from exploiting combinations of GPUs and CPUs.

Keywords Automotive assembly line · Quality inspection · Computer vision · Image processing · OpenCV · Performance · CUDA · NVIDIA GPU · RTX 3070TI · Raspberry Pi

19.1 Introduction

Computer vision applications will play a critical role in a wide range of technologies and applications for quality monitoring of production in manufacturing [1]. Image processing speed in computer systems is a fundamental enabler of these new technologies for different purposes: motion detection, pattern recognition, measurements, 3D reconstruction, artificial neural networks. Recent developments in Graphic Processing Units (GPUs) and new uses of it for math applications have attracted considerable interest from the manufacturing industry and potential applications to optimize operations and reduce costs [2].

Y. Zhang · R. Rodriguez Buno (✉) · T. T. Allen
The Ohio State University, Columbus, OH, USA
e-mail: rodriguezbuno.1@osu.edu

C.-Y. Lin · Y.-K. Tsai
National Taiwan University of Science and Technology, Taipei City, Taiwan

J. Maiti et al. (eds.), *Recent Advances in Industrial and Systems Engineering*, Lecture Notes on Multidisciplinary Industrial Engineering,
https://doi.org/10.1007/978-981-96-8323-9_19

A new stream of data in the form of video or images will help to solve monitoring activities in a very unexpensive way for 100% of production, without the necessity of costly statistical sampling for human manpower in time and effort. At the same time, it requires new computer resources to handle that stream of data and to be processed in real-time.

Continued efforts are needed to make real-time quality monitoring of production more accessible to all manufacturing companies. There are several important improvements which need to be made on the overall production system: new hardware equipment in workstations, strong network infrastructure, cybersecurity measures and monitoring, servers for new data streams, internal development of software at the same time as new insights from quality analytics are obtained, and a new culture for manufacturing 4.0 in the entire organization.

The specific objective of this study was to benchmark the computational time to process a batch of images to measure gap and flush of body in whites in an assembly line. This study set out to find a possible solution to **process those images under a very demanding time constraint for real-time monitoring of production**. This study was exploratory, to survey different hardware solutions available on the current market and being processed with OpenCV with GPU (CUDA support) and without it (only CPU).

19.2 Literature Review

The range of industrial activities using image processing is growing rapidly, involving real-time quality control and automatic visual inspection [3–5]. Key quality parameters predominantly revolve around aspects such as color, size, shape, and the detection of defects. Image processing and computer vision techniques provide online quality monitoring in situations that pose challenges for human inspectors. Certain studies try to propose method to guide or enhance image processing tasks in quality inspection. Zhang [6] proposes a framework for exploratory image data analysis to guide image processing tasks, identify deviations, and generate hypotheses regarding quality aspects. Chen [7] introduces a data-driven method to enhance image-based automatic inspection for detecting defects in IC wire bonding.

More recent attention has focused on the provision of deep learning agents for the automotive assembly line quality monitoring. The existing literature on deep learning for the manufacturing industry pays particular attention to object detection, semantic segmentation, and anomaly detection [8]. Most deep learning methods rely on convolutional neural networks (CNNs) as their foundation. CNNs employ a multilayered architecture that progressively condenses the images into the most pertinent representations. Various architectures have emerged in recent years, such as AlexNet [9], GoogleNet and VGGNet [10], ResNet [11], DenseNets [12], and Xception [13].

These advanced deep learning models are advantageous because they handle both feature extraction and the subsequent classification or prediction tasks, reducing the need for manual feature definition and extraction. However, the complexity of these models presents challenges in terms of interpretability and demands significant annotation effort [1, 6]. Moreover, to alleviate the computational demands associated with utilizing complex deep learning models, it is common to resize the image to a lower resolution [14]. However, this approach may not be appropriate for applications that require high-detail and high-resolution, such as precise dimensional measurements. Therefore, classical image processing methods and the utilization of GPU acceleration merit additional exploration and research for industrial automotive inspection.

Over the past decade, the explosion in the use of inexpensive (compared to previous technologies) computer equipment, camera, and network connections, along with major advances in embedded systems, CPU, and GPUs have allowed new applications for computer vision algorithms in different industries. To date, there are two trends in research on this topic that are of interest to us: applications of computer vision to manufacturing industry with GPU acceleration and performance studies on the different solutions available.

GPU acceleration comes from the benefits of powerful parallel computing and high memory bandwidth. GPUs have been widely used to accelerate deep learning with advances in frameworks to support deep learning models on GPU, such as Tensorflow, PyTorch, etc., and GPU-accelerated libraries, such as NVIDIA released cuDNN in 2014 [15, 16]. Recent studies propose solution of visual inspection with the GPU-accelerated deep neural network, such as detection of dirties, scratches, and wears on surface parts [17], identification of product defects of bottles [18] and workpieces defect segmentation [19]. More research extends to image processing and machine learning including use GPUs to accelerate contrast enhancement and Random Forest for the detection file defects [20].

For the performance studies, Park et al. [21] deduce key factors in design and evaluation of diverse image processing algorithms (including stereo matching, linear feature extraction, image encoding, and rendering) on massively parallel GPUs using CUDA. Hangün, in 2017, [22] highlights that there are huge benefits for processing high-resolution images with GPU, being between 10 × to 12 × times faster than with CPU for histogram equalization of image and threshold problems. Patidar et. al, in 2020, [23] studied solutions of using CPU, Multi-threaded CPU, OpenCV, and GPU with CUDA support for face detection algorithms, concluding that the use of GPUs with OpenCV is almost 100 × times faster than CPU. Sehairi, in 2023, [24] compared different versions of ARM multicore processors and embedded GPUs, finding that for high-resolution videos, the GPU is 4 × times faster than the CPU. These studies were conducted exclusively on embedded systems (Raspberry Pi 3, Jetson TK1, and Jetson Nano), while the earlier research typically involved PC desktop stations. There is no preview literature comparing both solutions for the same study.

Overall, these studies suggest limitations on computing power for real-time applications (with images up to 720 × 720 pixels and video analysis up to 1280 × 720 pixels). As Sehairi [24] writes: *"For video analysis for 1920 × 1080 pixels, it was not in real-time due to great mass of data that carries HD and Full-HD videos"*. In this paper, we explore the potential solution for real-time dimension inspection with high-resolution image sequence processing.

19.3 Materials and Methods

Data for this study were collected using a 16 MP camera, installed in a car manufacturing plant, to take a batch of 30 images, then after image filtering, they are reduced to 12–15 images for processing per Body-In-White car, to measure the gap and flush of the left side front gap between the fender and the driver's door. To guarantee the accuracy of the gap and flush measurements derived from images, we have collected high-resolution images with dimensions of 4656 × 3496 pixels.

This present study utilizes different hardware:

1. Raspberry Pi 4 8 GB RAM (4 CPU cores) with Raspberry Pi OS
2. Apple Mac Studio with M1 chip (20 CPU cores)
3. PC with AMD × 64 (8 CPU cores) and NVIDIA GTX 1060 GPU with Windows 10
4. TopGro Intel i7 (8 CPU cores) and NVIDIA GTX 1650 GPU with Windows 11 Pro
5. Asus Intel i9 (16 CPU cores) and NVIDIA RTX 3070Ti GPU with Windows 11 Pro

The criteria for selection of this hardware were as follows: the initial hardware solution was the Raspberry Pi 4; some computers were already available and the last two PCs were purchased specifically for the relatively advanced GPUs.

For the analysis, we used a sample comprising two batches of 15 images each, representing two different cars. These two batches were processed using various hardware options, and the average processing time was then calculated.

19.3.1 Initial Hardware Solution: Raspberry Pi 4 8 GB

The initial hardware solution could not process the high-resolution (16 MP) batch of images per car. The batch is composed of 15 images, and it requires more storage and computing power to process. The disadvantage of this option is the total time required to process each batch. It takes 9 min per car to run image processing, while the goal is to complete the entire cycle of taking and processing the images in less than 56 s per car. With this very inexpensive hardware, it is possible to take the batch of images in around 10 s, but the time required to process these images makes it impossible to do Real-Time monitoring of the assembly line production.

19.3.2 Experiment for Camera Speed, Time Between Pictures

The camera speed is important to capture the targeted area to measure in the Body-In-White from different angles. The camera speed was tested with a python script for the Raspberry Pi 4 using Picamera2 library and in the Asus ROG i9 using OpenCV.

Table 19.1 Experiment 1, configuration of options

Options	Image filtering (HOG SAD)	Temple matching (Pixel-level NCC)
All CPU	CPU	CPU
GPU filter + CPU TM	GPU	CPU
All GPU	GPU	GPU

19.3.3 Experiment Configurations for Image Processing

There are multiple image processing procedures in the quality inspection with vision measurement task. Two of these procedures that can benefit from GPU acceleration include image filtering and template matching (TM). In image filtering, each image in the batch is compared with a reference image based on similarity score. This is achieved by calculating the Sum of Absolute Difference (SAD) of Histogram of Oriented Gradients (HOG) features. For template matching, the objective is to detect the Region of Interest (ROI) by computing similarity between each sliding window with a predefined template. This process utilizes two distinct methods: pixel-level Normalized Cross Correlation (NCC) and HOG gradient image NCC.

The image size in our experiments is 4656 × 3496 pixels (16 MP). Processing a sequence of images of this size poses a computational challenge, without compromising resolution, which is crucial for maintaining high accuracy of image filtering and ROI detection.

We conducted two sets of experiments to investigate the computational performance of various configurations involving CPU and GPU. In the first set of experiments (E1), image processing involved HOG SAD for image filtering and pixel-level NCC for template matching. The HOG features and NCC score were computed using opencv.HOGdescriptor() and opencv.matchTemplate() with and without CUDA support. In the second set of experiments (E2), HOG SAD was employed for image filtering, while HOG gradient image NCC was used for template matching. The NCC score on the HOG gradient image was calculated using the hog function from scikit-image; however, the CUDA support version is unavailable. Consequently, we lack GPU results for both filtering and matching in E2.

19.3.3.1 Experiment 1 (E1)

See Table 19.1.

19.3.3.2 Experiment 2 (E2)

See Table 19.2.

Table 19.2 Experiment 2, configuration of options

Options	Image filtering (HOG SAD)	Temple matching (HOG NCC)
All CPU	CPU	CPU
GPU filter + CPU TM	GPU	CPU
All GPU	No test (Function of computing HOG gradient image is not available in OpenCV CUDA for template matching)	

19.4 Results

19.4.1 Experiment: Camera Speed, Time Between Pictures

Overall, these results from Table 19.3 for the time between pictures in average of 100 pictures, for the Asus ROG i9 is 0.17 s; indicate that it is 4 × faster than previous solution with Raspberry Pi of 0.67 s (Fig. 19.1).

Table 19.3 Camera speed, average time between pictures for 100 images

Options	Mean time (s)	Std dev (s)
Raspberry Pi 4 with Picamera2	0.67	0.06
Asus ROG with OpenCV	0.17	0.03

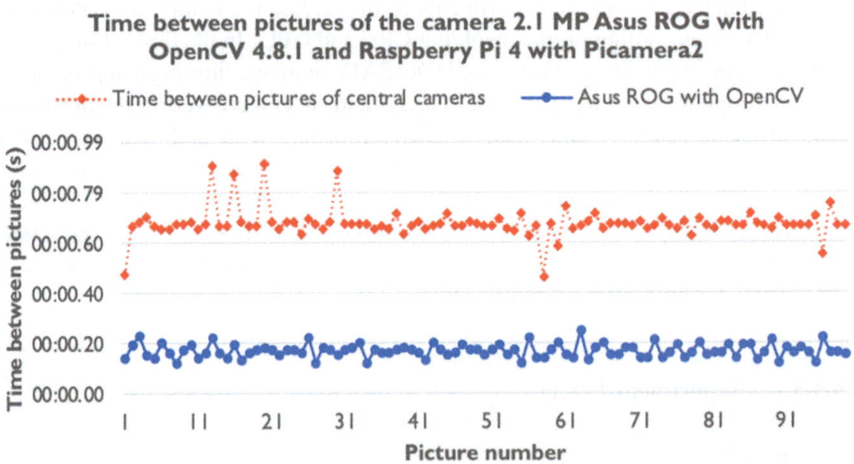

Fig. 19.1 Time between pictures for Asus ROG with OpenCV and Raspberry Pi 4 with Picamera2

19.4.2 Experiment 1: Benchmarking for HOG Filtering + Pixel-Level TM

In the first set of Experiment E1, Tables 19.4, 19.5, and 19.6 present the computational time in seconds for all-CPU, combination of GPU filter and CPU TM, and all-GPU setups across four computers with various parallel lines configurations. In the "Parallel Lines (Filter, TM)" section, the first number represents the count of parallel lines with multiprocessing for image filtering, and the second number indicates the count of parallel lines for template matching. Three levels of image filtering are conducted by thresholding HOG SAD. Our primary focus is on Column "HOG SAD < T3 (12–15 images)", which involves filtering out 12–15 images from a sequence of 30. This quantity is sufficient to maintain the accuracy of our dimension estimation model. Although Columns "HOG SAD < T3(4-5images)" and "HOG SAD < T2 (8images)", with their reduced number of images, can sometimes decrease computational time, this reduction often comes at the cost of diminished accuracy in dimension estimation.

We have detailed the CPU cores and GPU models for each computer in each table. The MAC Studio without GPU support but with a 20-core CPU, serves as our reference point for comparing the performance of each setup.

Table 19.4 All CPU, total running time (only CPU for HOG filtering and Pixel-level TM)

Computer	CPU cores	GPU model	Parallel lines (Filter, TM)	HOG SAD < T1 (4–5 images)	HOG SAD < T2 (8 images)	HOG SAD < T3 (12–15 images)
Mac Studio M1	20	Not support	(30, #filtered images)	15.40	16.10	17.10
AMD × 64	8	GTX 1060	(30, #filtered images)	38.00	39.50	Freeze issue, memory allocation
AMD × 64	8	GTX 1060	(30, min (8, #filtered images))	36.80	40.10	44.00
AMD × 64	8	GTX 1060	(30, 4)	37.50	39.50	46.50
TopGro i7 PC	8	GTX 1650	(4,4)	37.00	40.50	47.10
TopGro i7 PC	8	GTX 1650	(8,4)	37.70	39.90	46.00
Asus ROG i9 workstation	16	RTX 3070Ti	(30, #filtered images)	28.30	30.80	25.60
Asus ROG i9 workstation	16	RTX 3070Ti	(8,6)	23.60	30.80	25.03

Table 19.5 Combination of GPU and CPU, total running time (GPU for HOG filtering and CPU for Pixel-level TM)

Computer	CPU cores	GPU model	Parallel lines (Filter, TM)	HOG SAD < T1 (4–5 images)	HOG SAD < T2 (8 images)	HOG SAD < T3 (12–15 images)
Mac Studio M1	20	Not support	(30, #filtered images)	15.37	16.12	17.10
AMD × 64	8	GTX1060	(30, #filtered images)	21.50	25.00	Freeze issue, memory allocation
AMD × 64	8	GTX1060	(30, min (8, #filtered images))	24.00	27.50	28.00
AMD × 64	8	GTX1060	(30,4)	30.00	26.50	30.57
TopGro i7 PC	8	GTX1650	(8,4)	16.75	23.75	23.30
Asus ROG i9 workstation	16	RTX 3070Ti	(8,6)	8.51	10.40	11.84

Table 19.6 All GPU, total running time (Both GPU for HOG filtering and Pixel-level TM)

Computer	CPU cores	GPU model	Parallel lines (Filter, TM)	HOG < T1 (4–5 images)	HOG < T2 (8 images)	HOG < T3 (12–15 images)
Mac Studio M1	20	Not support	(30, #filtered images)	15.37	16.12	17.10
AMD × 64	8	GTX 1060	(30, #filtered images)	20.50	Freeze issue, memory allocation	Freeze issue, memory allocation
AMD × 64	8	GTX 1060	(30, min (8, #filtered images))	20.00	22.50	23.00
AMD × 64	8	GTX 1060	(30,4)	20.90	21.90	23.20
TopGro i7 PC	8	GTX 1650	(4,4)	14.64	16.99	18.90
TopGro i7 PC	8	GTX 1650	(8,4)	15.00	16.35	19.14
Asus ROG i9 workstation	16	RTX 3070Ti	(6,6)	7.61	8.56	9.15
Asus ROG i9 workstation	16	RTX 3070Ti	(8,6)	8.04	8.95	9.15

As shown in Table 19.4, there is a clear trend where more CPU cores result in reduced computational time. Additionally, the configuration of multiprocessing setups does have an impact, though the variance is not significant when the count is set to 4, 6, and 8 for template matching on computers with 8-cores CPU (AMD × 64 and TopGro i7 PC). However, increasing this count to 12–15 leads to freeze issues. For instance, setting the parallel lines of template matching as the number of images 12–15 (i.e., HOG SAD < T3) on AMD × 64 may cause memory allocation problems. It appears that having a very large number of parallel lines, exceeding the number of physical CPU cores, may not be an optimal choice for multiprocessing in image processing tasks.

If we only replace image filtering with GPU, as demonstrated in Table 19.5, there is a noticeable decrease in computational time across all setups with GPU support. Further, by replacing both image filtering and template matching with GPU processing, we achieve a substantial reduction in computational time, show in Table 19.6. This is particularly evident in the case of the Asus ROG i9 workstation (16 cores CPU + RTX 3070Ti GPU), where we managed to bring the time down to 9.15 s. This is 2.5 times faster compared to using CPUs alone. A similar trend is also observable in Fig. 19.2a. When compared to the slowest time recorded for the i7 PC, this represents a 5 × improvement.

In Fig. 19.2a, the x-axis shows each hardware option, of interest here is the reduction in processing time across all options when transitioning from utilizing only CPU to employing a combination of GPU and CPU, specifically GPU filter + CPU TM, and eventually to an all-GPU approach. The Mac Studio M1, lacking Nvidia GPU support, only provides results for the implementation with all CPUs.

In Fig. 19.2b, it is observed that the TopGro i7 PC (8-cores CPU + GTX 1650 GPU) show computational speeds are comparable to those of with Mac studio M1 with 20-cores CPU. The TopGro i7 PC achieves a processing time of 18.9 s using all GPU, while the Mac Studio M1 records 17.1 s with CPU alone.

In summary, these results show that it is possible to implement an online real-time quality monitoring of the production through a batch of images for each BIW for 100% of the units.

19.4.3 New Hardware Solutions: PC with I9 (16 Cores CPU) with GPU RTX 3070Ti

The PC with i9 processor (16 cores CPU) with GPU (RTX 3070Ti) acceleration takes less than 10 s to process the batch of images. Now, with the same time to take the batch of images and this processing time, it is possible to make it REAL-TIME.

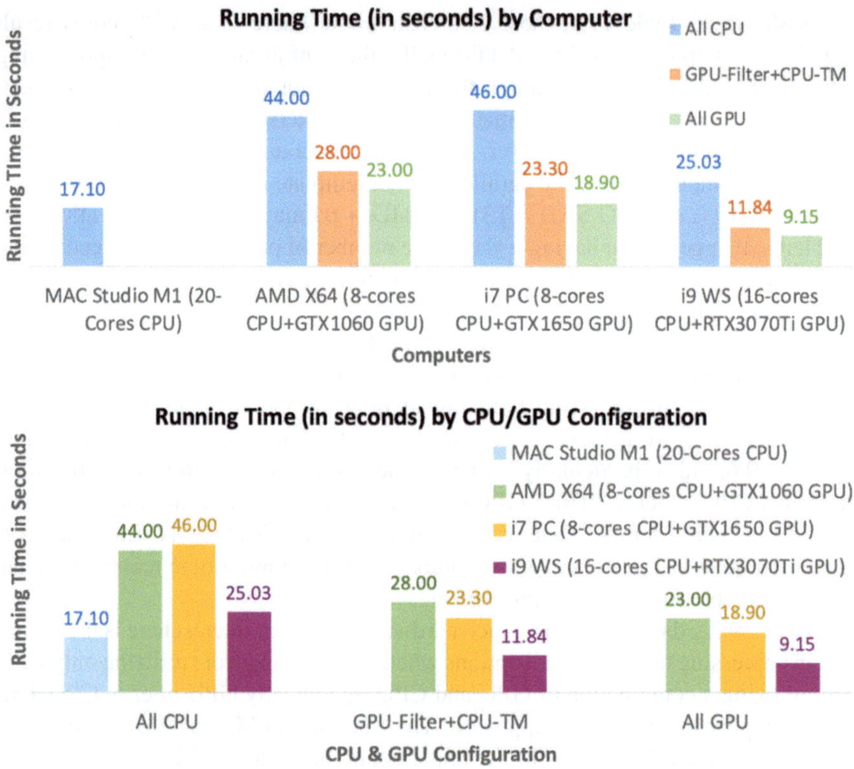

Fig. 19.2 Benchmarking for HOG filtering (30 images) + Pixel-Level TM (12 ~ 15 images)

19.4.4 Experiment 2: Benchmarking for HOG Filtering + HOG TM

The best we achieved is 48.8 s/car with the PC i9 workstation (16 cores + RTX 3070TI GPU).

There are no results for all GPU as the function of computing HOG gradient image required for template matching is not available in OpenCV CUDA. Nevertheless, Fig. 19.3 displays the reduction in computational time achieved by switching from CPU to GPU for image filtering in the three setups that support GPU.

19.5 Cost–Benefit Analysis

This section evaluates the financial and potential benefits associated with aforementioned different configurations for developing and implementing Vision System in this vehicle quality control project. We structure the Cost–Benefit Analysis (CBA)

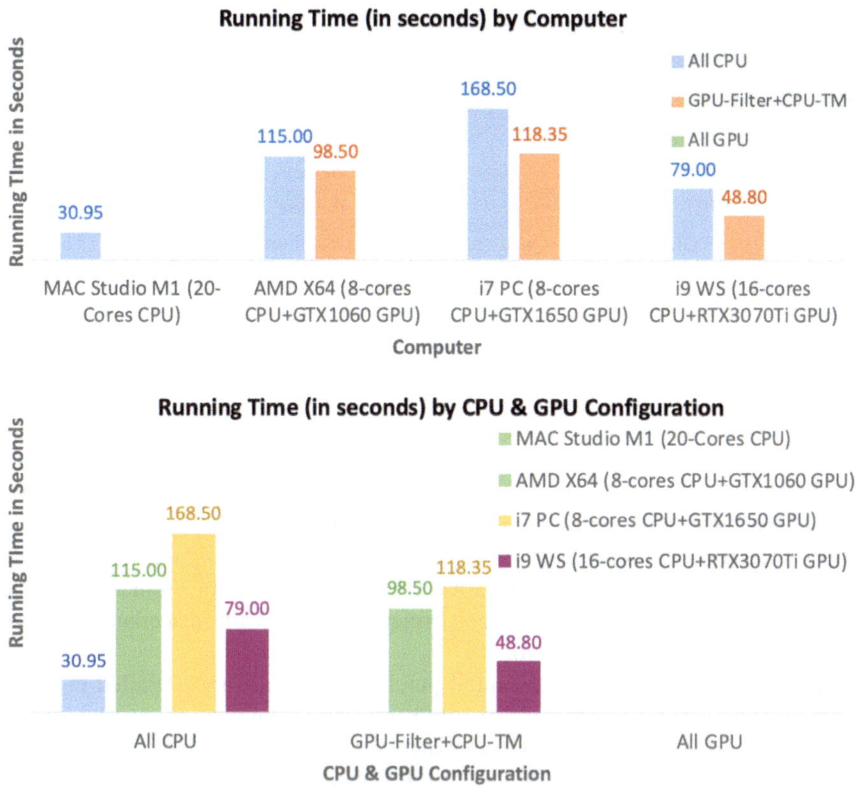

Fig. 19.3 Experiment 2, Benchmarking for HOG filtering + HOG TM

into two main parts: costs and benefits. By detailing the initial development, operational costs, and hardware specifications, alongside the expected benefits, this section aims to provide decision-makers with a comprehensive basis for making informed choices.

19.5.1 Cost Analysis

The selection of the i9 with RTX3070Ti as the final choice was based on a thorough evaluation of the development and operational costs associated with each configuration. This analysis ensures that the decision supports both current needs and future scalability.

Table 19.7 Hardware costs

Hardware	Purpose	Cost (USD)
Mac Studio M1	Initial develop	3000
AMD × 64 GTX1060	Initial develop	1300
i7 with GTX1650	Deploy	1300
i9 with RTX3070Ti	Release	2859
Raspberry Pi 4 (8 GB RAM) *2 + Quad-camera*2	Initial set up	950

19.5.1.1 Development Cost

The development cost encompasses personnel cost, hardware expenses, and miscellaneous costs, such as travel to the factory. We will focus the comparison on hardware expenses to highlight the distinctions between configurations. This emphasis is justified because personnel and miscellaneous costs are relatively consistent across different hardware configurations, as we develop and migrate code for each setup. The primary variation arises from the additional code development required for hardware with GPU acceleration. The project was carried out by a team of four individuals, consisting of three developers and one project manager with miscellaneous costs totaling approximately USD 1200.

Hardware expenses cover the initial purchases of hardware. Detailed comparisons of costs and evaluations of different hardware configurations are provided in Tables 19.7 and 19.8. These comparisons suggest that the i9 with RTX3070Ti, despite its higher cost, is a promising option. The Raspberry Pi 4 (8 GB RAM) *2, while affordable, is limited for large-scale computation. The AMD × 64 GTX1060 and i7 with GTX1650 are moderate choices, offering medium cost and suitable performance for current needs but may not support more complex tasks if image processing requirements expand.

19.5.1.2 Operational Costs

The operational costs encompass energy consumption, maintenance, and software updates, show in Table 19.9. Energy consumption is assessed based on the Thermal Design Power specified in product documentation. Maintenance costs are evaluated by considering the complexity of components, heat production, and the potential of needing component replacements.

The Mac Studio M1 needs lower maintenance due to its integrated design and minimal heat generation. The AMD × 64 GTX1060 and i7 with GTX1650, both older GPUs, generate more heat, potentially need occasional replacements of thermal interfaces or GPU drivers. The i9 with RTX3070Ti, while offering high performance suitable for intensive tasks, also leads to high maintenance costs due to its substantial heat output and the likely frequent need for thermal paste replacement. Raspberry Pi devices, known for their low power and minimal heat generation, are easy to maintain

Table 19.8 Hardware detail comparison

Hardware	Lifespan (Years)	Compatibility	Evaluation
Mac Studio M1	5–7	MacOS, limited × 86 software compatibility via Rosetta2	High cost; Low system compatibility; not ideal for high-performance tasks
AMD × 64 GTX1060	4–6	Windows, Linux, broad × 86/ × 64 software support	Requires additional code development specifically for GPU acceleration; Adequate for basic to moderate complexity tasks
i7 with GTX1650	4–6	Windows, Linux, broad × 86/ × 64 software support	Requires additional code development specifically for GPU acceleration; not ideal for high-performance tasks
i9 with RTX3070Ti	4–7	Windows, Linux, suited for high-performance computing	Requires additional code development specifically for GPU acceleration; Highly recommended for computationally demanding tasks
Raspberry Pi 4 (8 GB RAM) *2 + Quad-camera*2	3–5	Raspbian, Ubuntu	Low cost; Insufficient for large-scale calculation, particularly when processing extensive images under time constraints; Not suitable as a primary computing device

Table 19.9 Operational costs for each hardware

Hardware	Energy consumption	Maintenance cost	System updates
Mac Studio M1	Low (40–100 W)	Low	Medium cost (Regular updates with moderate complexity considering potential compatibility issue)
AMD × 64 GTX1060	Medium (200–300 W)	Medium	Medium cost (Regular updates with moderate complexity for driver updates and OS)
i7 with GTX1650	Low-Medium (150–500 W)	Medium	Medium cost (Regular updates with moderate complexity for driver updates and OS)
i9 with RTX3070Ti	High (400–500 W)	High	High cost (Regular updates with high complexity to maintain top performance)
Raspberry Pi 4 (8 GB RAM) *2 + Quad-camera*2	Very Low (5–10 W)	Low	Low cost (Less complexity in updates due to its lower performance and nature of the platform)

and replace but fall short in handling intensive image processing tasks due to their limited capabilities.

All the tools developed for our project utilize open-source libraries, including OpenCV and scikit-learn, and are built using Python and the free version of MySQL. There are no license fees currently. The variation in operational costs from software updates primarily stems from the labor cost to manage these updates and to handle specific configurations during their implementation.

Initially, our strategic investment focused on exploring various hardware options to identify the optimal setup for scaling up vision system deployment. After a thorough evaluation detailed in Tables 19.7, 19.8 and 19.9, we chose the Intel Core i7 with NVIDIA GeForce GTX 1650 for development phases and Intel Core i9 with NVIDIA GeForce RTX for the release phases. This decision effectively balances cost with performance needs, ensuring an efficient deployment of the Vision System.

19.5.2 Benefits Analysis

The total benefits of vision inspection for quality control are manifold, including reduced dependence on manual labor. Traditional methods require significant manpower and often involve only sampling a few vehicles rather than inspecting each one, making a direct cost–benefit comparison between manual and vision inspection challenging. However, assuming that a routine inspection is necessary for all vehicles and considering two inspection lines handling over 100 vehicles daily, the implementation of a vision system can replace the labor of at least four workers involved in measuring dimensions, data acquisition, and analysis. This equates to a reduction in labor costs amounting to USD 51,000 per year, based on the local hourly wage of USD 35.

The system excels in quickly detecting non-conforming dimensions in vehicle gaps and flush areas, ensuring deviations from design specifications are identified early in the production process for immediate corrective action. Additionally, automation enables continuous, uninterrupted inspections along the production line.

For comparing the benefit of different configuration of vision inspection system, we compared the processing speeds in Table 19.9, where GPU acceleration is utilized if available. The processing capability are ranked from fastest to slowest: i9 with RTX 3070Ti, Mac Studio M1, i7 with GTX 1650, followed by two Raspberry Pi 4 units (each with 8 GB RAM). The i9 equipped with an RTX 3070Ti not only leads in speed but also offers the potential for expansion with more advanced image processing techniques. Investing in the i9 with RTX3070Ti may result in higher initial costs but ensures compatibility with future technologies and increasing demands in image processing.

19.6 Discussion

An initial objective was to identify the necessary hardware configuration to satisfy the strict requirements of the project in the given time constraint for image processing.

The first question in this study sought to determine if it was possible to process the images in under 30 s using current hardware in the market. The finding confirmed this possibility with fastest processing time achieved at 9.15 s per car, resulting in an overall cycle time of 19 s per car.

The results for the experiment on the camera speed (time between pictures) show that the time interval between capturing pictures is not a limitation for the monitoring system (0.17 s per picture saved), when comparing to the total time taken for a car to pass in front of the camera (56 s). The batch of images must be synchronized with the position of the desired location within the Body-In-White targeted.

This research uses three strategies to maintain data privacy. First, we only collect and save necessary data to minimize the risk of handling sensitive information. Second, we established access control to restrict the data usage and access only to research team members and stakeholders. Third, we only use trusted networks, such as plant network to minimize the risk of unauthorized access, data breaches, or other security threats that could compromise the integrity of the data being handled or transmitted. Considering that data threats may keep changing and being created, prioritizing cybersecurity is essential. For example, monitoring data continuously, and conducting security assessments testing regularly are crucial to prevent potential risks.

19.7 Limitations

This study considered computers already owned and two new ones purchased specifically for this research application, based on the hardware already available on the market. New studies can include top-of-the-line hardware to test the limits of computing speed, as well as set a maximum time and then find the most inexpensive solution to it.

The major limitation of this study is the hardware and software requirements for image processing of high-resolution photos of a moving Body-In-White in the assembly line. With the evolution of new technologies and hardware, different specialized solutions can help in the future.

This paper does not engage in any particular environmental setup for a manufacturing plant nor its variations (lighting conditions, workers passing in front of the camera, anomalies in the assembly line or sudden stops). All these factors must be considered when applying the technological solution to a particular case.

All processing and imaging were conducted locally in the singular hardware tested. These studies may not be applicable to cases of multiple cameras in different locations doing an inspection on the same product (in our case, the BIW) and being processes in the manufacturing facility servers (or even with cloud computing). For example, other types of hardware solutions could include different architecture of GPUs from different providers (e.g., AMD).

These findings may be somewhat limited by the sample size. A larger sample can be utilized to improve the results when comparing different alternative methods and to better reflect the particular instances of the different situations that can happen at the assembly line. This can give a better generalization of the results across different applications of computer vision in manufacturing environments.

Notwithstanding these limitations, the study suggests that it is possible to achieve a real-time solution for monitoring quality features of the production without interruptions or the need to take a sample out of the assembly line. These new methodologies, quality control and hardware make possible the monitoring of 100% of the production.

19.8 Conclusions

The aim of the present research was to determine a method to process the images in a timely manner, within a short time frame set by the current conditions in the automotive manufacturing plant. The results suggest that with current market-available hardware, it is possible to conduct real-time quality monitoring of 100% of the production. The study is limited to a small sample of body-in-white images and the hardware solutions in hand. Future studies should analyze complete production of an entire shift to discover the implications of the real-world instances of a manufacturing environment (not only in some experiments), to address possible conflicts with irregular stops in the assembly line, associate worker passing in front of the system, changes in light conditions and more unknown situations.

A natural progression of this work is to analyze further growth in devices taking pictures simultaneously and connecting to a central server and later complexity, as new implementations grow with new equipment and monitoring goals.

Acknowledgements This work was supported by Dr. Ted Allen and the Honda group for the vision system development.

References

1. De Ketelaere, B., Wouters, N., Kalfas, I., Van Belleghem, R., Saeys, W.: A fresh look at computer vision for industrial quality control. Qual. Eng. **34**(1) (2022)
2. NVIDIA: Reinvent manufacturing—from design to delivery, 26 January 2024. [Online]. Available: https://www.nvidia.com/en-us/industries/manufacturing/
3. Sriratana, W., Narknam, N., Apichitanon, R., Tammarugwattana, N.: Application of webcam for inspection of rice grain quality by using image processing technique. In: 2020 20th International Conference on Control, Automation and Systems (ICCAS) (2020)

4. Yüksel, C.T.P.V., Tetik, Y.E., Yılmaz, M., Özdemir, Ö.C.: Automated pipe inspection based on image processing. In: ECON 2021—47th Annual Conference of the IEEE Industrial Electronics Society, pp 1–5 (2021)
5. Kaewmorakot, C.T.P.U., Poolperm, L., Prongphimai, P., Tansriwong, S.: The development of prototype of automated bottle caps visual inspection system. In: 2022 International Electrical Engineering Congress (iEECON) (2022)
6. Zhang, Y., Allen, T.T., Rodriguez Buno, R.: Exploratory image data analysis for quality improvement hypothesis generation. Qual Eng (2024)
7. Chen, J.: A data-driven method for enhancing the image-based automatic inspection of IC wire bonding defects. Int. J. Prod. Res. 4779–4793 (2021)
8. Mazzetto, M.E.A.: Deep learning models for visual inspection on automotive assembling line. Int. J. Adv. Eng. Res. Sci. (IJAERS) **7**(4), 473–494 (2020)
9. Alex Krizhevsky, I.S.G.E.H.: ImageNet classification with deep convolutional neural networks. Adv. Neural Inf. Process. Syst. **25** (2012)
10. Karen Simonyan, A.Z.: Very deep convolutional networks for large-scale image recognition. In: 2015 3rd international conference on learning representations (ICLR) (2015)
11. Kaiming He, X.Z.S.R.J.S.: Deep residual learning for image recognition. In: IEEE Conference on Computer Vision and Pattern Recognition (CVPR) (2016)
12. Huang, G., Liu, Z., Van Der Maaten, L. and Weinberger, K.Q.: Densely connected convolutional networks. In: Conference on Computer Vision and Pattern Recognition (CVPR) (2017)
13. Chollet, F.: Xception: deep learning with depthwise separable convolutions. In: IEEE Conference on Computer Vision and Pattern Recognition CVPR, (2017)
14. Talebi, H., Milanfar, P.: Learning to resize images for computer vision tasks. In: Proceedings of the IEEE/CVF International Conference on Computer Vision (ICCV), pp. 497–506 (2021)
15. Jeon, W., Ko, G., Lee, J., Lee, H., Ha, D., Ro, W.W.: Chapter six—deep learning with GPUs. Adv. Comput. **122**(0065–2458), 167–215 (2021)
16. Pandey, M., Fernandez, M., Gentile, F., Isayev, O., Tropsha, A., Stern, A.C., Cherkasov, A.: The transformational role of GPU computing and deep learning in drug discovery. Nat. Mach. Intell. **4**, 211–221 (2022)
17. Park, J.K., Kwon, B.K., Park, J.H., Kang, D.J.: Machine learning-based imaging system for surface defect inspection. Int. J. Precis. Eng. Manuf. Green Technol. **3**, 303–310 (2016)
18. Wang, J., Fu, P., Gao, R.X.: Machine vision intelligence for product defect inspection based on deep learning and Hough transform. J. Manuf. Syst. **51**, 52–60 (2019)
19. Li, Z., Lin, H.W., Liu, Y.Y., Chen, C., Xia, Y.F.: An industrial defect detection algorithm based on CPU-GPU parallel call. Multimed Tools Appl. **82**, 44191–44207 (2023)
20. Ando, H., Niitsu, Y., Hirasawa, M., Teduka, H., Yajima, M.: Improvements of classification accuracy of film defects by using GPU-accelerated image processing and machine learning frameworks. In: 2016 Nicograph International (NicoInt), pp. 83–87 (2016)
21. Park, I.K., Singhal, N., Lee, M.H., Cho, S., Kim, C.: Design and performance evaluation of image processing algorithms on GPUs. IEEE Trans. Parallel Distrib. Syst. **22**(1), 91–104 (2011)
22. Hangün, B., Eyecioğlu, Ö.: Performance comparison between OpenCV built in CPU and GPU functions on image processing operations. Int. J. Eng. Sci. Appl. **1**(2), 34–41 (2017)
23. Patidar, S., Singh, U., Patidar, A., Munsoori, R.A., Patidar, J.: Comparative study on face detection by GPU, CPU and OpenCV. In: Second International Conference on Computer Networks and Communication Technologies: ICCNCT 2019 (2019)
24. Sehairi, K., Chouireb, F.: Implementation of motion detection methods on embedded systems: a performance comparison. Int. J. Technol. **14**(3), 510–521 (2023)

MIX
Papier aus verantwortungsvollen Quellen
Paper from responsible sources
FSC® C105338

If you have any concerns about our products,
you can contact us on
ProductSafety@springernature.com

In case Publisher is established outside the EU,
the EU authorized representative is:
Springer Nature Customer Service Center GmbH
Europaplatz 3, 69115 Heidelberg, Germany

Printed by Libri Plureos GmbH
in Hamburg, Germany